Centrifuge Modelling for Civil Engineers

Centrifuge Modelling for Civil Engineers

Gopal Madabhushi

CRC Press is an imprint of the
Taylor & Francis Group, an **informa** business
A SPON BOOK

CRC Press
Taylor & Francis Group
6000 Broken Sound Parkway NW, Suite 300
Boca Raton, FL 33487-2742

Printed on acid-free paper
Version Date: 20140620

International Standard Book Number-13: 978-0-415-66824-8 (Paperback)

Library of Congress Cataloging-in-Publication Data

Madabhushi, Gopal.
Centrifuge modelling for civil engineers / author, Gopal Madabhushi.
pages cm
Includes bibliographical references and index.
ISBN 978-0-415-66824-8 (paperback)
1. Geotechnical engineering--Methodology. 2. Structural analysis (Engineering)--Equipment and supplies. 3. Soil-structure interaction--Simulation methods. 4. Centrifuges. I. Title.

TA705.M245 2014
624.028'4--dc23 2014002075

Visit the Taylor & Francis Web site at
http://www.taylorandfrancis.com

and the CRC Press Web site at
http://www.crcpress.com

To my parents for allowing me the freedom to pursue my dreams.
To my beloved Raji for supporting me every step of the way.
To Spandana for becoming an independent young doctor and for not being too demanding on my time and Srikanth for coming of age and discussing intelligently the contents of this book.

Contents

Foreword

Every textbook reflects the character and experience of its author. Gopal Madabhushi arrived in Cambridge in 1988 to study soil dynamics for a PhD under Andrew Schofield. He used both finite element simulations and centrifuge models to investigate the dynamic response of towers on spread foundations resting on saturated sands. In his thesis he was at pains to discuss the required nature and remoteness of the system boundary, whether numerical or physical, if the interaction of interest was to be displayed without untoward distortion. Having achieved that, he went on to introduce some rather simple single-degree-of-freedom characterizations that could capture the essence of the behaviour that he had observed, without disregarding the complex and varying response of the soil. This mixture of interests, abilities, and concerns provided a firm foundation for his future academic career, all but one year of which has been spent in Cambridge. It also forms the basis of this book, which is much enhanced by Professor Madabhushi's generous use of examples drawn from the shelves of the Roscoe Library at the Schofield Centre, which record the successful exploits of scores of centrifuge modellers who have worked there since the 1970s.

The reader is invited in Chapter 1 to consider whether the conventional routines of civil engineering, with its limit analyses and safety factors, will provide an adequate level of assurance regarding the safety and serviceability of the planned facility. If performance is to be guaranteed, the designer is advised to consider both numerical and physical modelling to clarify the issues. The complementarity of these techniques is considered further in Chapter 2, where the best attributes of each are drawn out. This provides a purposeful context for a book on physical modelling, whether it is to be read by an advanced student or by a project engineer who is wondering what benefits there may be in additionally commissioning centrifuge tests.

Chapters 3 to 6 cover the basic principles of centrifuge modelling, starting from a revision of circular motion in Chapter 3 in case the reader has forgotten the difference between centripetal accelerations and centrifugal forces, or has previously skated too quickly over Coriolis. Scaling factors and the meaning of "prototype" in relation to "model" are carefully discussed in Chapter 4, including how to harmonize the timescales arising from different behavioural mechanisms—so central to the modelling of seismic liquefaction phenomena which Professor Madabhushi has done much to clarify. The various alternative configurations of geotechnical centrifuges are introduced in Chapter 5, together with the classes of problems for which they are best suited. And Chapter 6 covers the various respects in which a centrifuge differs from a planet, with its inside-out body force that converges to a point "above" the soil surface, for example, and the Coriolis "wind" that seems to blow fast-moving particles sideways as they "fall." Of course, all this is done properly, from first principles.

Chapters 7 to 9 cover the essential technology that enables centrifuge modelling, and do so bearing in mind the needs of a nonexpert. So while soil containers are introduced in Chapter 7, so are the means of getting the soil into the containers in the desired condition. The author takes special care to explain how to create saturated sand layers of known relative density, and saturated clay layers of known stress history, prior to illustrating how to verify that the model soils have the intended characteristics through the use of in-flight investigation tools such as CPTs and seismic waves. Chapter 8 deals with instrumentation and image analysis, and Chapter 9 with data acquisition, again from the perspective of an interested and intelligent novice.

Chapters 10 to 14 cover some of the geotechnical applications that have benefited from insights that arose through centrifuge modelling shallow foundations in Chapter 10, retaining walls in Chapter 11, piles in Chapter 12, and in Chapter 13 sequence-dependent construction interactions such as excavations with propping, or tunnel construction that affects existing piled foundations. The final Chapter 14 covers dynamic centrifuge modelling, with application to earthquake-induced soil–structure responses. This has been the central research mission of Professor Madabhushi, so it is no surprise that the issues are described in detail, and dispatched authoritatively.

Indeed, the whole book reflects the hands-on style of its author. So it should serve as an excellent primer and a thorough introduction to the topic of centrifuge modelling, as well as to its variety of uses in geotechnical and foundation engineering. As I retire from the position of director of the Schofield Centre, and as Gopal Madabhushi takes over from me, I wish him well. And I recommend this thoughtfully crafted book to its future readers.

Malcolm Bolton, FREng

Preface

The first time I ever heard the word centrifuge modelling was during the late afternoon lectures given by Professor Chandrasekaran at the Indian Institute of Technology, Bombay. The course was about finite element methods in geomechanics but Professor Chandrasekaran would philosophise that if we wanted a "real" understanding of failure mechanisms we should really be doing centrifuge modelling. On my arrival in Cambridge Professor Andrew Schofield drove me to the centrifuge centre to show me the beam and drum centrifuges. In those days I was still an ardent numerical modeller so I was looking for the big mainframe computers. Only a year later did I slowly come to realise the unique opportunity centrifuge modelling provided, and started my first centrifuge tests on tower structures subjected to earthquake loading. Over the last 25 years I have been involved in a wide variety of centrifuge tests and often complement them with numerical analysis. This combination of centrifuge modelling to clarify mechanisms and use of centrifuge data to verify numerical procedures is a powerful one and is reflected in this book.

The origins of this book took shape from my lecture course that introduces new research students to centrifuge modelling. It also benefits from the various short courses I gave at IIT Bombay with support from the Royal Society, United Kingdom, and the Department of Science and Technology, India, and the training course offered at the IFSTTAR centrifuge facility in Nantes, France, under the European Union–funded SERIES project. I felt that there was a need for a book that makes the principles of centrifuge modelling, the scaling laws and the techniques used accessible to practicing civil engineers. They can refer to this book and see how they can use it effectively in solving complex geotechnical problems. The applications chapters of this book should provide an overview of the kinds of problems that can be studied effectively using centrifuge modelling. For graduate research students it offers detailed descriptions of the equipment used in centrifuge modelling, the instrumentation that can be deployed in centrifuge models and the signal processing techniques required to make the best use of the centrifuge test data. Early in their research they can use the applications chapters to see what types of tests are possible.

The book begins with an outline of a generalized design process employed for civil engineering projects. With the modern shift towards performance-based design, it is inevitable that geotechnical engineers will be called upon to estimate deformations and not just the ultimate load-carrying capacities. This requires a good understanding of the fundamental failure mechanisms at play and the mobilised strains in the soil under a given loading. The second chapter emphasizes the need for both physical and numerical modelling and aims to establish the complementarity of the two techniques. Chapter 3 is a brief revision of basic mechanics that is useful for an understanding of centrifuge modelling. The use of polar coordinate systems for expressing acceleration and velocity of centrifuge models is quite useful and these are derived from first principles in this chapter.

Chapter 4 introduces the basic principles of centrifuge modelling and explains the concepts of a prototype and a centrifuge model. It builds up to the derivation of scaling laws that relate the behaviour of a prototype and the centrifuge model. Scaling laws for seepage velocity, consolidation time, force and so on, are derived. In addition, the ability of centrifuge testing itself to validate the scaling laws is presented using the so-called modelling of models. Chapter 5 introduces the various types of centrifuges, such as beam and drum centrifuges, along with examples of each type. The suitability of different centrifuges for various problems is also explained. Chapter 6 describes some of the errors and limitations of centrifuge modelling. Each of these is considered in detail and an effort is made to quantify them.

Centrifuge modelling requires some specialist equipment and these are presented in Chapter 7. This chapter describes the model containers and actuators that are used routinely in centrifuge testing. The techniques used for sample preparation are elaborated. In-flight characterization of soil samples is described using various methods such as miniature CPT and shear wave velocity measurement. Chapter 8 describes the instrumentation that is commonly used in centrifuge models. Of course this area is constantly growing and this chapter aims to provide a comprehensive overview of instrumentation currently being used. Chapter 9 deals with centrifuge data acquisition systems and signal processing techniques. It introduces the concepts of electrical noise removal using digital filtering with increasing levels of sophistication.

Chapters 10 to 14 cover different applications of centrifuge modelling. Each chapter reviews the physical testing pertaining to that problem and builds to show the types of experiments that were carried out and the main outcome of the centrifuge test data. Shallow foundations are considered in Chapter 10, and the main emphasis of this chapter is the visualization of the failure mechanism generated below the foundation. Retaining walls are considered in Chapter 11 and in this chapter the main focus is on the measurement of bending moments in the wall, lateral displacement suffered by the walls and soil deformations in front and behind the wall. Pile foundations are considered in Chapter 12 and the main emphasis here is the modelling of axial and laterally loaded piles. Piles wished into place and piles driven in-flight are also discussed. Chapter 13 deals with modelling of construction sequences in a centrifuge test. This chapter deals with constructing new geotechnical structures in the vicinity of existing infrastructure and effects the new build may have on the older structures.

Chapter 14 deals with modelling of dynamic events. One of the advantages of centrifuge modelling is that we can model devastating or extreme load scenarios well before they occur in real life. This allows us to predict accurately how either existing or new structures would behave under a strong earthquake event or under extreme wind loading caused by a hurricane. This chapter deals with additional scaling laws required for dynamic events and the specialist actuators and model containers required. Emphasis is placed on earthquake loading, although the same principles could be applied for other types of dynamic events. Examples of dynamic soil–structure interaction problems and soil liquefaction problems are considered.

Graduate students may wish to go through Chapters 1 to 9 and then pick the applications chapter that is closest to the problem they wish to investigate. Of course they can read other applications chapters later to provide a more comprehensive view of centrifuge modelling. Practicing engineers who want to learn about centrifuge modelling may wish to concentrate on Chapters 3 to 8 to obtain a good overview of the modelling techniques and instrumentation used and then move on to the applications chapters to see what types of problems can benefit from centrifuge modelling.

Finally I would like to say that much of the book relies on developments that took place in centrifuge modelling at Cambridge. This is mainly due to my own experiences, and readers must realize that there are other excellent centers where centrifuge modelling is flourishing. The contents of this book bring together the research carried out by many researchers. However, any omissions or errors in this book are entirely mine.

Acknowledgements

The list of acknowledgements is vast but I wish to start by thanking all the researchers past and present at the Schofield Centre, for the stimulating atmosphere they create. I wish to thank Prof. Andrew Schofield for introducing me to centrifuge modelling and for providing a historical perspective to this book. I must also thank Prof. Malcolm Bolton for supporting me over the years and writing the foreword to this book on very short notice. My other geotechnical colleagues in the Cambridge Soil Mechanics group, particularly Prof. Robert Mair must be acknowledged for their support. I must also thank Prof. V. S. Chandrasekaran for the interactions we had during the development of the Indian centrifuge as these gave me a good insight into the design aspects. Similarly I must thank Prof. Bruce Kutter for the interactions and use of their centrifuge facilities during my sabbatical at the University of California, Davis. I must thank Prof. Sarah Springman of ETH, Zurich, for her support over the years.

Special mention must be made of Dr. Ulas Cilingir for the finite element analyses used in Chapter 2. I must also thank Dr. Stuart Haigh for the discussions we had from the early stages of this book. I must thank Drs. Raji and Spandana Madabhushi and Mr. Srikanth Madabhushi for proofreading the draft chapters of this book. Dr. Giovanna Biscontin must also be thanked for the final proofreading of the book and her feedback, soon after her arrival from Texas A&M.

The success of centrifuge modelling depends to a large extent on excellent technical support, both mechanical and electronics/instrumentation. I have learned a lot over the years from the excellent technical team at Cambridge, from Chris Collison, Neil Baker and the late Steve Chandler to the current-day technicians at the Schofield Centre, John Chandler, Kristian Pather, Chris McGinnie, Mark Smith and Richard Adams. Special mention must also be made of the excellent administrative support provided by Anama Lowday.

Author

Dr Gopal Madabhushi is a professor of civil engineering at the University of Cambridge, UK and the director of the Schofield Centre. He has over 25 years of experience in the areas of soil dynamics and earthquake engineering. His expertise extends from dynamic centrifuge modelling to the time domain finite element analyses of earthquake engineering problems. He has an active interest in the areas of soil liquefaction, soil–structure interaction, and liquefaction-resistant measures and their performance. He has an active interest in the biomechanics of hip replacement surgeries.

He has acted as an expert consultant to the industry on many geotechnical and earthquake engineering problems, for example, Mott MacDonald, Royal Haskonig, and Ramboll-Whitby, UK. He has an active interest in post-earthquake reconnaissance work and has led engineering teams from the UK to the 921 Ji-Ji earthquake of 1999 in Taiwan, the Bhuj earthquake of 2001 in India, and many other missions. Dr Madabhushi served as the chairman of Earthquake Engineering Field Investigation Team (EEFIT) that runs under the auspicious of the Institute of Structural Engineers, London.

He was awarded the TK Hsieh award in 2005, 2010, and 2013 by the Institution of Civil Engineers, UK, the BGA medal in 2010 given by the British Geotechnical Association, the Shamsher Prakash Research Award in 2006, Medical Innovations Award in 2007, the IGS-AIMIL Biennial award in 2008, and the Bill Curtin Medal in October 2009 by the Institution of Civil Engineers, UK, for his contributions in the area of soil dynamics, tsunamis, and earthquake engineering. He has written 103 journal publications and 240+ papers in international conferences and workshops to date. He has authored a very successful book on the *Design of Pile Foundations in Liquefiable Soils* (Imperial College Press) and geotechnical chapters in the book *Designing to Eurocode 8* (Taylor & Francis).

A historical perspective

Geotechnical centrifuge model testing has a long history in the Department of Engineering at the University of Cambridge. For over 50 years several faculty members have formed a Soil Mechanics Group that has had 232 graduates writing PhD theses between 1951 and 2012. The first on a centrifuge modelling test was thesis number 25 in 1970 and the number of such tests has steadily increased. The author of this book, formerly my research student, graduated with a PhD in 1991, with thesis number 112, titled "Response of Tower Structures to Earthquake Perturbations"; it was one of the earliest of many contributions that followed in earthquake engineering. He later became a colleague, giving lectures to final-year students on centrifuge model testing that are the basis for this book. In this book the reader will learn about the basic principles of centrifuge modelling, along with a great variety of tests by him and others. This book expertly fills a gap on current centrifuge model testing in the literature of geotechnical engineering.

It is now 40 years since our early model tests. The early Japanese and British geotechnical centrifuge workers met our Soviet counterparts during the 1973 ISSMFE conference in Moscow. We arranged to meet at their Hydro project, met G. I. Pokrovsky, saw one of his very powerful centrifuges, and heard that tests it made had determined the choice of the location and the material used for their Nurek Dam in Tajikistan, at 300 m the world's tallest man-made dam. I was surprised when a railway engineer began to describe dynamic model tests, expecting Coriolis effects to have created difficulties; I asked if this was not so for any dynamic tests. At once Pokrovsky said, "No, it is all in my books." When I was given copies of the two books later I was surprised to see how well he had advanced Soviet dynamic testing; he compared model-cratering test data with U.S. atomic bomb crater data published by Vesic. We were later to learn that as an important defense scientist, Pokrovsky held the rank of Major General in the Red Army. At that time of the Cold War it was regrettable but inevitable that our groups could not make and maintain contact over the next 15 years. Independently the West developed earthquake and dynamic model testing with solid-state transducers giving new data for our computers to acquire and analyse. Research workers trained in the concepts behind and the performance of tests in centrifuges avoided the secrecy that hid Soviet model test experience. We got the ISSMFE to establish a technical committee and organized meetings that published many papers openly.

The modelling method is now well established. Past earthquake engineers had few observations at full scale. In $1/N$ small-scale model-tests an increased acceleration Ng acts on every grain of material in centrifuge models. A model structure is made of selected material. An observer, in safety, chooses what earthquake will perturb it. A series of models can be made of identical material, giving us very many possible research choices, and very many publication opportunities. Tests inconceivable at full scale can be proposed on this modelling principle. Once I tested a 1/300 scale model of a hazardous pollutant plume at 300 g for 100 hours in our drum centrifuge, hoping to win a research contract that modeled movement of a plume for 1000 years at full scale. All such proposals need a careful study to establish the physics of modelling. Soviet publications emphasize the importance of tests at different scales that "model a model test" in making such a study. The concept of modelling of models is introduced in Chapter 4 of this book.

Ladd and de Groot (2003) discussed a concept of soil strength that lies behind SHANSEP computation of the behaviour of soft ground; they base it on data of triaxial tests of samples representative of the site. In contrast in physical modelling I rely on the critical state concept that the strength of every aggregate of grains is the sum of two components (Schofield 2006); one due to dissipation of work in internal friction and one due to work done but not dissipated in interlocking. We use reconstituted soil such as Speswhite or E-grade kaolin to make models. I view apparent cohesion on Coulomb's slip surface as due to interlocking and dependent on the effective pressure in, and the relative density of, the aggregate of soil grains, rather than on soil chemistry. In this view an element in a model made of reconstituted soil that has followed an authentic stress path and is subject to a correctly scaled perturbation has properties that cause authentic behaviour. Our models exhibit mechanisms to be expected at full scale in the field. When engineers plan to use observational methods at full scale in the field, they should also make relevant tests at small scale in a centrifuge before the project and be prepared to make more tests if any emergency arises that needs study as the work progresses.

This book aims to cover all aspects of centrifuge modelling and make the principles of centrifuge modelling clear to practicing civil engineers and researchers new to centrifuge modelling. I wish the author every success with this book.

Andrew Schofield, FRS

Chapter 1

Modern geotechnical engineering design in civil engineering

1.1 INTRODUCTION

For most civil engineering problems, geotechnical engineering design starts by collecting detailed information on the soil strata at the site. A wide variety of site investigation techniques is used to collect information on the types of soil strata, their strength and stiffness properties, water content, topographical features of the site, and so on. In addition to this, the geotechnical engineers liaise with their structural engineering counterparts to obtain information on loading that will come on to the foundations and the allowable deformations that the structure can tolerate.

Using these two sets of information, a geotechnical engineer starts the foundation design. Of course, at the beginning of this process a choice of foundation types may be available. For example, for a normal building that has three or four floors, the superstructure loads can be carried by columns that are supported on individual pad foundations. Alternatively, a raft foundation may be considered. The ultimate choice of the type of foundation may depend on the in situ soil characteristics as well as allowable foundation settlements and/or rotations. This type of normal geotechnical design process may culminate by checking the solutions obtained against the relevant code requirements in existence such as Eurocode 7 (EC7) or Eurocode 8 (EC8), for example. A simplified flowchart for geotechnical design is presented in Figure 1.1.

Although the general procedure described above may be applicable to a vast majority of routine geotechnical designs, a significant number may require further analysis using some of the advanced methods. This may be required under a number of circumstances where the site investigations reveal poor soil conditions or when the expected loadings from the superstructure are abnormal. When the soil conditions are poor, if the soil behaviour is well researched and good constitutive models for the soil exist, finite element (FE) analyses can be carried out using well-established numerical codes. However, it is important that the constitutive models chosen for the soil are relevant for the type of loading. For example, certain soil models may work well for monotonic loading but the actual loading from the

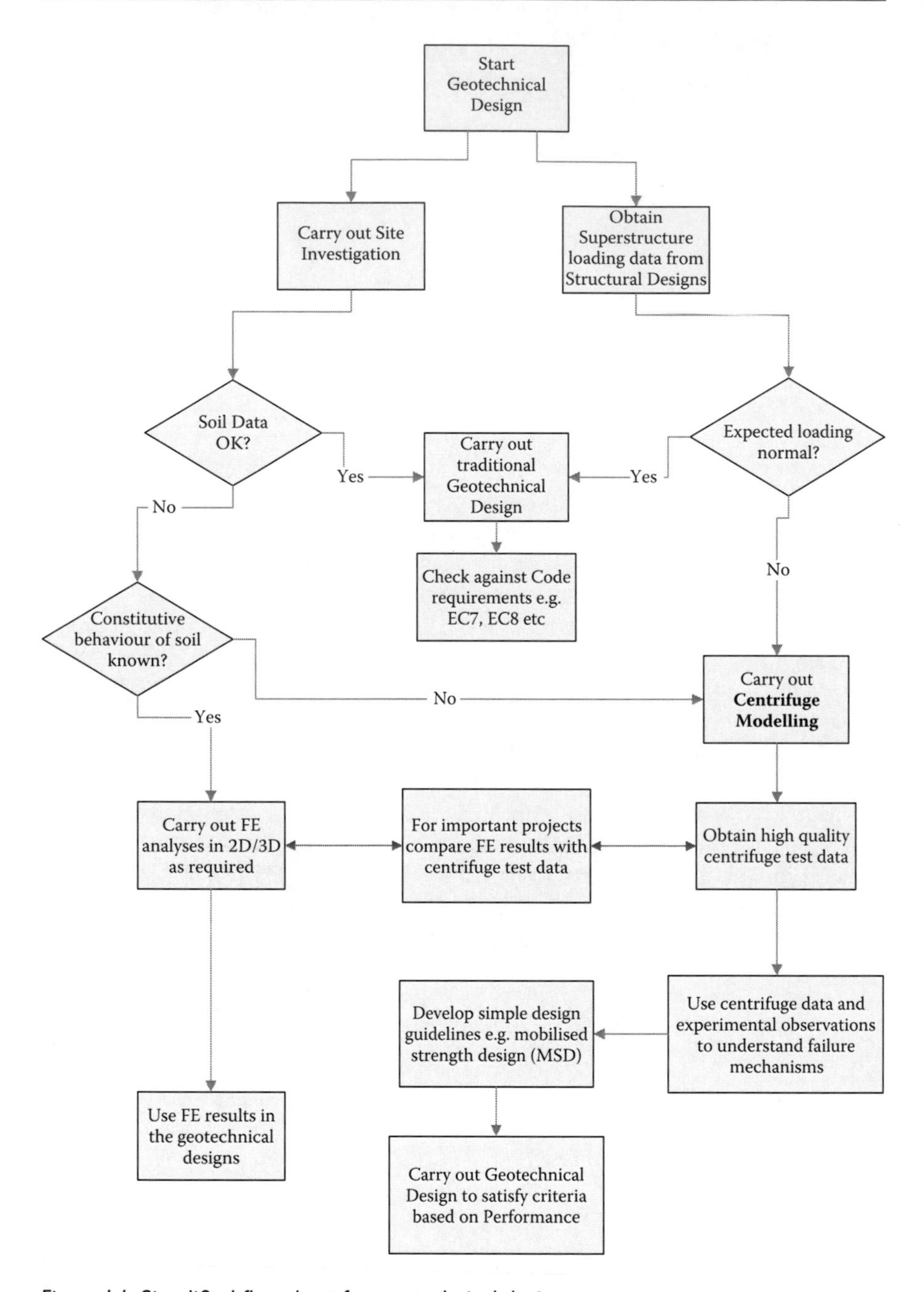

Figure 1.1 Simplified flowchart for geotechnical design.

superstructure may be cyclic in nature and the soil model's performance under cyclic loading may not be known. Further, the parameters required by the constitutive models have to be determined from good quality soil testing carried out on undisturbed samples recovered from the site or by conducting in situ testing at the site, such as SPT or CPT testing.

Now the question arises with regard to sites where the soil conditions are poor or unusual and there are no well-established constitutive models for those soils. Similarly the expected loading from the superstructure could be excessive or unusual. Under such circumstances centrifuge modelling can offer valuable insights into the foundation behaviour. The simplified flowchart for geotechnical design outlined in Figure 1.1 identifies these routes that may require centrifuge modelling.

It is true that carrying out physical model testing in a geotechnical centrifuge normally involves some amount of simplification of the actual field situations. The main challenge in creating centrifuge models is in simplifying the field situation in such a way that the centrifuge model is able to capture the essential behaviour of the foundations. This aspect will be described in more detail in Section 4.3 in Chapter 4. In most cases, centrifuge modelling will give rise to data that reveals the essential behaviour and the failure mechanisms of the foundations that might occur. This information can be used to develop appropriate geotechnical designs. Similarly, competing geotechnical designs can be tested to compare their performances in a given soil type and for a given set of loading conditions. Also for important or particularly difficult projects it may be worthwhile to carry out simplified centrifuge model testing and use these experimental data to validate the predictions from a FE code. This process can be used to fine tune the FE analyses until the FE code is able to produce matching results to the centrifuge test data. Such a validated FE code can then be used to carry out analyses of complex geometries that are more common in the field. The finite element method is particularly well suited to handle complex geometries.

Centrifuge modelling has another important role to play. The data from centrifuge tests together with the observation of failure mechanisms made during centrifuge testing can be used to develop novel design guidelines for particular classes of problems. In fact, the contribution of centrifuge modelling to geotechnical design guidelines in this regard has been substantial over the years. This path is also identified in the flowchart in Figure 1.1.

1.2 COMPLEX ROLE OF GEOTECHNICAL ENGINEERS

Adequate performance of foundations is of paramount importance if the superstructure is to perform as it is designed. The role of geotechnical engineers in any project is intertwined with that of structural engineers in almost all of the civil engineering projects.

1.2.1 Traditional safety factor–based design

It is common to base the geotechnical designs to withstand a given loading combination expected from the superstructure. This design philosophy relies on knowing the loads that are expected to be carried by the foundation and applying a suitable load factor for different load combinations.

Similarly, soil properties such as the peak and critical state friction angle, drained and undrained shear strength, and so on are known for the soil strata encountered at the site. Modern codes such as EC7 recommend partial factors to be applied to different soil parameters. The ultimate aim of the geotechnical designs carried out under this philosophy is to carry the loading imposed by the superstructure with an adequate safety factor. However, the actual displacements and/or rotations that occur under different load combinations are not considered.

The geotechnical engineer's role in this type of safety factor-based design methodology is well established. They have a linear interaction with structural colleagues in procuring the loading combinations expected from the superstructure and then carry out the geotechnical design using suitably factored soil strength parameters. This is quite straightforward in a majority of cases. The final design of the foundation is expected to carry the loads with an adequate factor of safety which is handed over to the structural engineers.

1.2.2 Performance-based design

In recent years the design philosophies employed in geotechnical engineering have changed rapidly and the concepts of performance-based design are being increasingly adopted. The role of geotechnical engineers will now involve estimating the deformations in the soil under the action of the applied load combination. This of course requires a good understanding of the soil stiffness as well as the soil strength. The settlements and/or rotations of the foundations caused by soil deformations govern the designs. There will be much closer collaboration with structural engineers in determining the allowable limits of the settlements and/or rotations by determining their effects on the superstructure. This type of interaction between the geotechnical and structural engineers is more iterative by its very nature.

The benefits of performance-based design are being realized in both geotechnical and structural fields. The final designs are well integrated, with both the behaviour of the foundation and its superstructure under a given loading being more predictable.

1.3 ROLE OF CENTRIFUGE MODELLING

Centrifuge modelling has been identified as one of the possible methods that can help with geotechnical design when the soil conditions are difficult, the constitutive models for the soil are not well defined or when the loading anticipated is unusual or extreme as shown in Figure 1.1. For these difficult or challenging cases, centrifuge modelling can have a role to play in helping the geotechnical engineer in both the design philosophies that are outlined above and are currently in use. In addition the centrifuge modelling of a class of problems such as retaining walls may be used to develop a new set of design guidelines by using the centrifuge test data and the physical observation of the failure mechanisms. This aspect has been utilised in a wide variety of boundary value problems at different centrifuge centers around the world.

1.3.1 Use of centrifuge modelling in safety factor–based designs

Let us consider a problem such as an offshore pile foundation being driven into stratified soils. The pile is anticipated to carry both axial and horizontal loads that vary with time depending on wave and wind loading on the superstructure. Let us assume that the constitutive models for the particular types of soils are not known under cyclic loading. It is therefore decided to conduct centrifuge modelling to determine the size of piles required to support the loading.

This would be a relatively straightforward problem to be studied in a geotechnical centrifuge and, in fact, one that was investigated widely at many research centers around the world. In the centrifuge model, soil strata to represent the layered nature of the site in the field can be recreated, perhaps using the soil samples from the coring tubes used in the site investigation. Model piles of varying lengths and diameters can be inserted into the centrifuge model during flight, and can be tested to obtain their ultimate axial and lateral capacities. From this experimental data, a suitable pile section can be chosen in the geotechnical design to provide an adequate safety factor against the anticipated loads in the field.

1.3.2 Use of centrifuge modelling in performance-based designs

Performance-based designs in geotechnical engineering rely to a large extent on our ability to estimate deformations in the ground under the applied loading. It is usually acknowledged that such an estimation of ground deformations is in general more challenging than performing a safety factor-based design using the concepts of ultimate limit state. The geotechnical engineering design becomes even more challenging when dealing with difficult soil conditions with unknown constitutive models and extreme or unusual loading from superstructure. However, these are the precise conditions under which centrifuge modelling becomes very useful and when performance-based designs are considered. The performance-based design concept allows for economic designs compared to safety factor-based designs by allowing some amount of ground deformation to be allowed. Centrifuge modelling can help geotechnical engineers determine the amount of ground deformation that will occur under a given loading, and also experiment to determine the limits of ground deformation beyond which the foundation may become unstable by developing a failure mechanism.

In addition to the direct use of centrifuge modelling to provide experimental data that can be used for geotechnical designs or in validating a particular design concept, the data from the centrifuge modelling can be used to develop a more fundamental understanding of the soil-structure interaction in a given problem. For example, Osman and Bolton (2004, 2006) have developed mobilised strength design (MSD) concepts with the primary aim of estimating soil strains and ground deformations that can be used in performance-based designs. Lam and Bolton (2009) applied the concepts of MSD to propped retaining walls based on their observations and experimental data obtained in a series of centrifuge tests. This is an example of the contribution and use of centrifuge modelling in helping develop design guidelines or procedures that can help geotechnical engineering practice. Again this pathway is identified in the simplified geotechnical design flowchart shown in Figure 1.1.

Chapter 2

Need for numerical and physical modelling

2.1 INTRODUCTION

In the previous chapter the major techniques available to a civil engineer for geotechnical designs and their interaction with structural designs were described. These interactions change depending upon the design philosophies being used; for example, whether the designs are to be carried out using a stress-based factor of safety approach or a deformation-linked performance-based approach. It was identified that when the soil conditions at a site are complex or if the structural loading anticipated is unusual, then prior investigation using numerical analyses or using centrifuge modelling may be required. In this chapter the role of physical modelling using geotechnical centrifuges and its complementarity to numerical modelling are explored.

The use of numerical analyses in modern-day civil engineering designs is well established and has in fact become a common practice. Most of these numerical analyses use well-established finite element (FE) or finite difference codes. These may be either commercially available codes such as ABAQUS, PLAXIS, FLAC, CRISP, etc., which are generalized programs for problems in geomechanics or those adopted for specific types of problems, such Geo-Slope for slope stability analysis or Wall-AP for retaining wall analysis. The complexity of the generalized FE codes can vary depending on their ability to handle the fully coupled two-phase formulation required for handling the solid and fluid phases of soils and the availability of advanced constitutive models that can simulate the nonlinear stress-strain behaviour of soil accurately. For dynamic problems in geomechanics more advanced FE codes such as Swandyne, Dynaflow, etc., are available that are able to capture the cyclic behaviour of soils and excess pore pressure generation during earthquake and other cyclic loading events.

2.2 USE OF NUMERICAL MODELLING IN FOUNDATION DESIGN

As explained above, currently numerical modelling in the form of FE analyses is used extensively in geotechnical engineering designs. This has become particularly useful with the advances in computing and the cheap and easy availability of both computing power and memory. In this section the ease of use of numerical methods and their critical role in the design of complex problems are presented.

Let us consider the simple case of a shallow foundation to be designed on a uniform, horizontal layer of soil with undrained shear strength c_u. For such a problem the plasticity–based, closed-form analytical solution is well known and the ultimate load capacity Q_u of the square foundation with size B can be determined as

$$Q_u = (\pi + 2)\, c_u\, B^2 \tag{2.1}$$

Depending on the column load applied to the foundation, the size of the foundation can be determined using Equation 2.1 with a suitable safety factor. Now let us consider a similar foundation to be supported on a layered soil as shown in Figure 2.1. The layered soil at this site consists of a surface layer with undrained shear strength c_u of 30 kPa and with a thickness

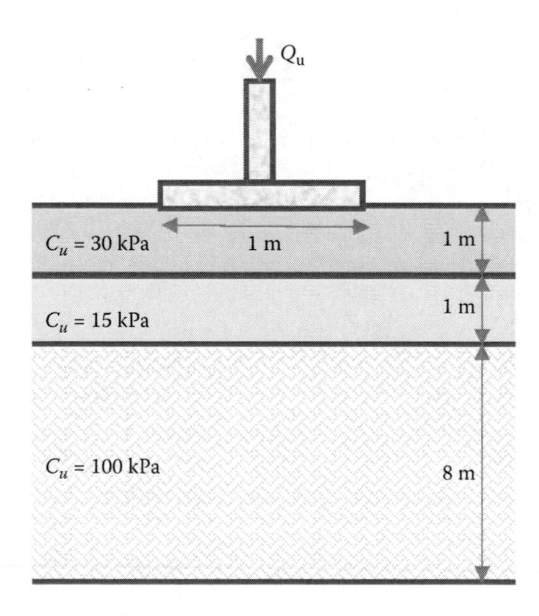

Figure 2.1 Simple footing on layered soil strata.

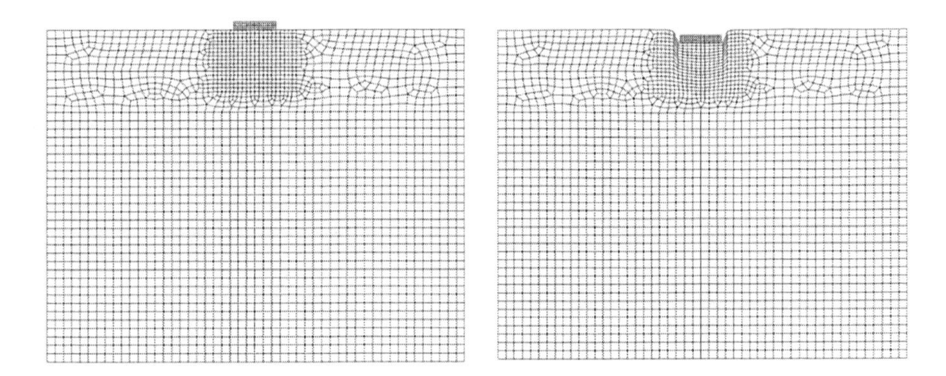

Figure 2.2 (a) Finite element discretization of the problem. (b) Deformed mesh (magnified by 2).

of 1 m. Underlying this layer, there is a very soft clay layer with undrained shear strength c_u of 15 kPa of 1 m thickness. A competent clay layer exists below this soft clay layer with an undrained shear strength of 100 kPa that extends to a depth of 8 m as shown in Figure 2.1. Let us assume that we are interested in ultimate capacity Q_u of a 1-m-wide footing placed on such a soil strata. We are only considering the undrained behaviour of the soil strata in this problem.

For such a problem, it is quite difficult to obtain a simple, closed-form solution. Using the upper and lower bound theorems of plasticity, it is possible to obtain a well-bounded solution but this could be quite tedious. An expedient way to find the ultimate capacity of the foundation in such a problem is to use the FE method.

As an example of the use of numerical methods, this problem is solved using the FE code ABAQUS. The boundary value problem shown in Figure 2.1 can be discretized as shown in Figure 2.2 using eight-noded quadrilateral elements. The stress-strain behaviour of all the soil layers was modeled using a simple plasticity model with friction angle set to zero and the yield stress set to the undrained shear strength for each layer as shown in Figure 2.1. Alternative constitutive models are also readily available in the FE code and are easy to use. The friction between the footing and the soil is also assumed to be small, that is, the footing is smooth. The load Q_u on the footing is increased in the analysis. In Figure 2.3 the increasing value of Q_u is plotted against the normalised settlement as a percentage identified by the legend "Footing only." The normalised settlement can be taken as a representative normal strain in the soil. In this plot it can be seen that the bearing capacity does not reach a peak value but increases with increasing strain. This can be expected as the clay below the foundation yields and the softer clay layer below that squeezes out with increasing values of the bearing pressure generated by the load Q_u. The geotechnical engineer may have

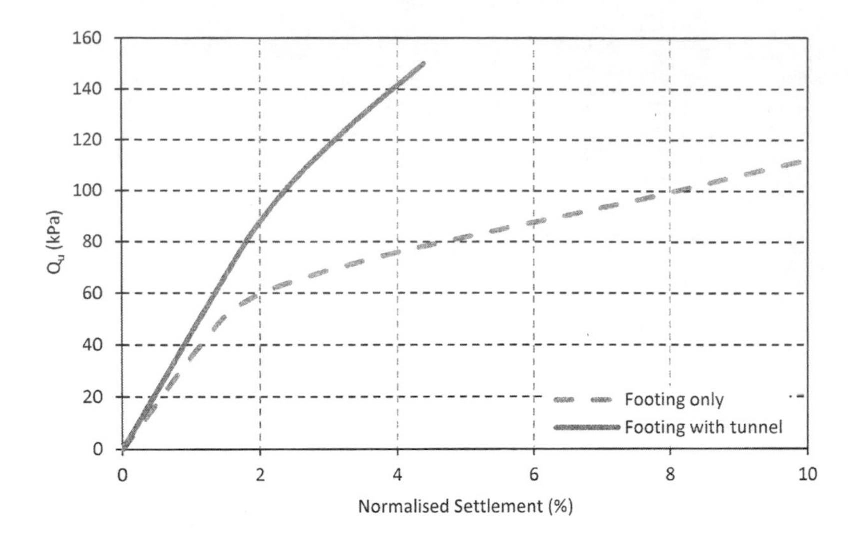

Figure 2.3 Load versus normalied settlement plot.

to make a judgment on the allowable strain and choose the corresponding value of Q_u as the allowable bearing pressure on the foundation. For example, we can assume that the settlement may be limited to 20 mm which corresponds to a normal strain of 2 percent and choose the value of Q_u to be 60 kPa from Figure 2.3.

In addition to obtaining a solution to the boundary value problem as described above, the FE analysis offers some additional advantages. For example, the deformed mesh can be plotted to give an overview of the failure mechanism. In addition to this we can also obtain the stress contours. In Figure 2.2(b) the deformed mesh is plotted. In Figure 2.4 the Von Mises stress contours are plotted for 2 percent strain. In Figure 2.2(b), it can be seen that soil elements just below the foundation deform significantly. These deformations also extend into the middle soft clay layer that gets squeezed out. The deformations in the stiffer, bottom clay layer are quite small but the stress contours in this layer are higher and more concentrated. This mechanism can be better visualised by plotting all the soil elements that have reached yield. In Figure 2.5 the elements that have reached yield are plotted in black. In this figure it can be seen that the soil just below the foundation in the top clay layer reaches yield as one would expect. Following this the yielded soil elements representing the middle layer extend laterally as they get squeezed out. The soil elements in the stiff, bottom clay layer below this, by and large, do not reach yield.

The above example demonstrates the use of FE analyses in the design of a shallow foundation located on layered soil strata. Clearly it provides additional information to the geotechnical engineer that helps in visualizing the

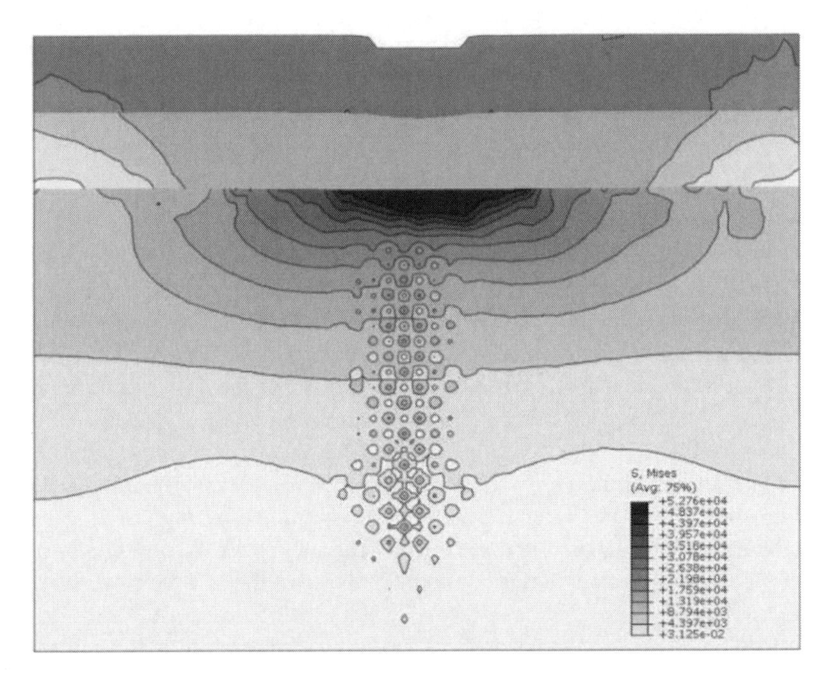

Figure 2.4 Von Mises stress contours.

Figure 2.5 Yield of soil elements.

soil yielding, failure mechanism mobilised under a given loading, and the stresses induced in the soil strata. Such additional information can help the engineer recognize the assumptions made in the design and use the results from the FE analyses with good engineering judgment.

2.3 NEED FOR PHYSICAL MODELLING

Although numerical modelling is quite powerful and can be applied to problems with no obvious closed-form solutions, it is important to check if the numerical procedures are capturing the correct failure mechanism. For the problem of shallow foundation on layered soil strata discussed in Section 2.2 above, the failure mechanism that involved "squeezing out" of the soft clay layer is quite obvious and agrees with the engineer's intuition. However, when the boundary value problems are more complex, this can be more difficult. To demonstrate this point let us consider the following problem. Let us assume a 3-m-diameter tunnel with its tunnel axis 3 m below the soil surface as shown in Figure 2.6(a). The axis of the tunnel is offset from the center line of the foundation by 1 m. The crown of the tunnel enters the soft clay layer in the middle by 0.5 m. Let us assume that the tunnel is lined with steel plates of 25.4 mm thickness. It would be interesting to know the capacity of the footing in the presence of this tunnel. Clearly it is possible to run an FE analysis of this problem in a similar vein as before.

The FE analysis using ABAQUS is run on the mesh shown in Figure 2.6(a) by incrementing the bearing pressure of the footing as before. In Figure 2.3 the increasing value of Q_u is plotted against the normalised settlement as a percentage identified by the legend "Footing with tunnel." In this figure it can be seen that the footing response is stiffer and stronger in the presence of the tunnel compared to the "Footing only" case. At the nominal strain of 2 percent the value of Q_u reaches 90 kPa compared to 60 kPa for the

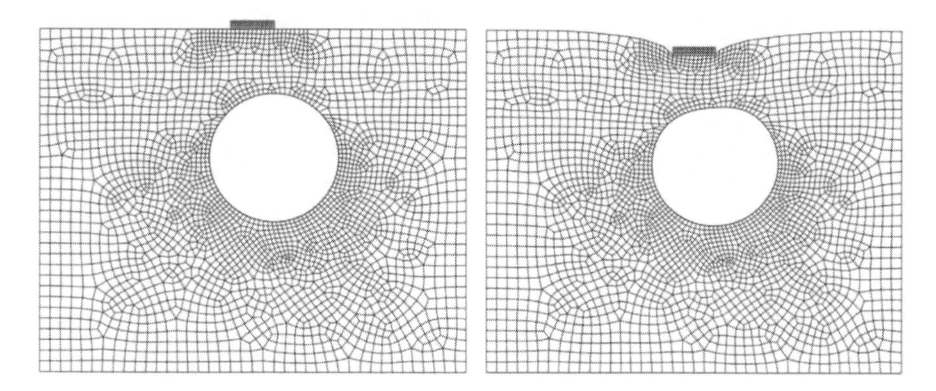

Figure 2.6 (a) Finite element discretization of the tunnel problem. (b) Deformed mesh (magnified by 10).

footing only case. This result is rather surprising and may not be the one an engineer would have anticipated. The deformed mesh when the footing reaches 2 percent strain is plotted in Figure 2.6(b). The shape of the tunnel clearly changes and is no longer circular. The presence of the tunnel also makes the deformations in the soil asymmetric, as one would expect. Despite the deformation of the tunnel, it seems it is able to offer more "support" to the footing, as the tunnel has steel-plated lining.

It is possible to plot the Von Mises stress contours for this problem as shown in Figure 2.7. Clearly these stress contours are quite different to those seen in Figure 2.4 due to the presence of the tunnel. As before, we can plot the soil elements that have reached the yield as shown in Figure 2.8. We can compare this figure to the one in the previous "footing only" analysis shown in Figure 2.5.

From this analysis it appears that the presence of the tunnel actually reduces the region of soil that reaches a yield. Also more of the soil elements to the left of the footing axis reach a yield compared to the right. This asymmetry may be attributed to the presence of the tunnel whose axis is offset from the footing axis. It appears that the presence of the tunnel that deforms allows the soil yield to be more localised to regions in the immediate vicinity of the footing. The presence of the tunnel also made the footing response much stiffer and stronger as shown in Figure 2.3. It is as if the softer clay layer in the middle has been strengthened due to the steel-lined tunnel.

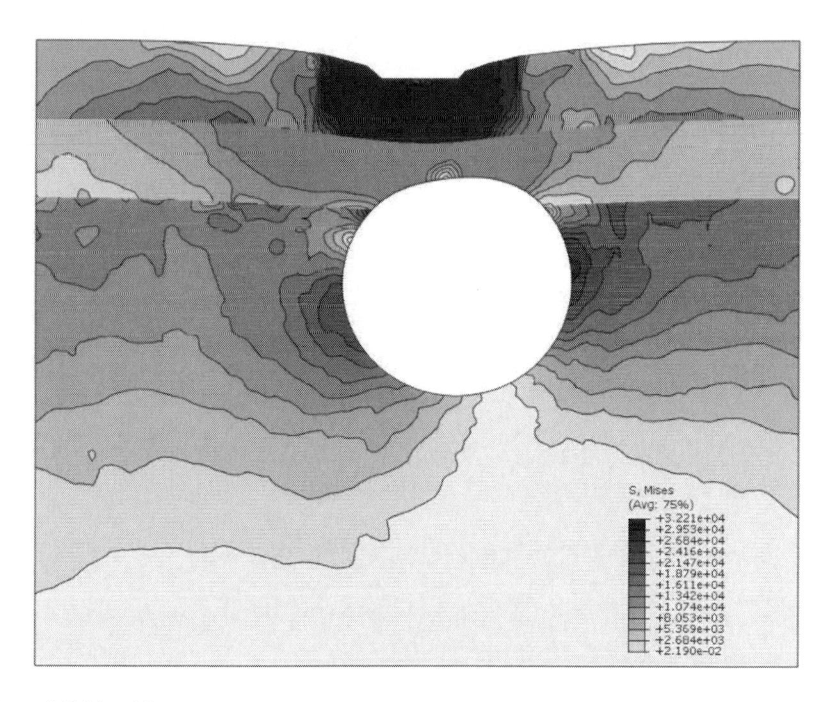

Figure 2.7 Von Mises stress contours.

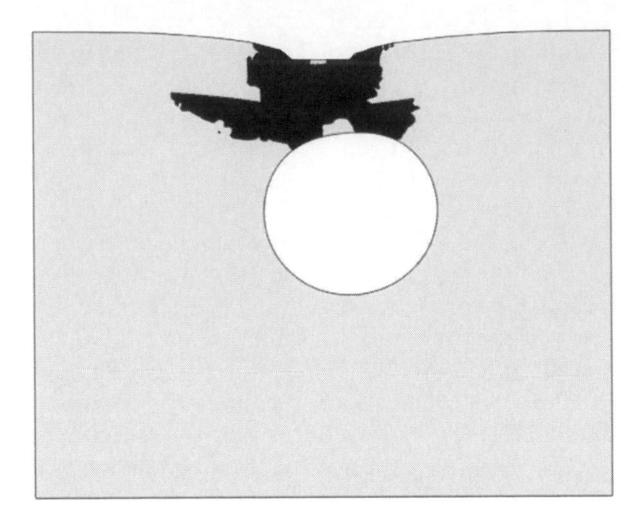

Figure 2.8 Yield of soil elements.

Although the FE analyses are able to offer solutions to complex problems of this nature, how do we know that these solutions are accurate and reliable? One can expect that with increased bearing pressure on the footing the soft clay layer in the middle will need to be squeezed out, past the crown region of the tunnel. And therefore the shear stresses generated between this clay and the tunnel lining and any slip at this interface play a very important role. In the FE analyses the interface between the tunnel and soil layers is modeled as a surface-to-surface contact in the analysis shown above. Such an assumption does not allow any localised slip to occur. How much impact does this have on the mechanism observed and overall results?

The accuracy and reliability of FE analyses such as the one discussed above can be verified by conducting physical model tests in a geotechnical centrifuge. The whole boundary value problem can be modeled by constructing a reduced scale model and testing it in the increased gravity created in the centrifuge. The results of the centrifuge tests can be used to verify the accuracy of the failure mechanism captured by the FE analyses and also the footing response obtained with and without the presence of the tunnel. Such an approach will help us validate the FE analysis and determine its reliability for use in such complex problems.

2.4 BENEFITS OF CENTRIFUGE MODELLING

Centrifuge modelling has become established as a reliable research tool that helps to clarify geotechnical behaviour in a wide variety of boundary value problems. For complex problems in the oil industry such as offshore jack-up platforms, spud-can foundations, and so on, centrifuge modelling was used as a tool to develop a better understanding of the problem and to help develop design methodologies for specific problems. In addition, centrifuge modelling also played a crucial role in clarifying the soil mechanics at a fundamental level in a number of problems. However, the main strength of centrifuge modelling is its ability to capture the correct failure mechanisms in a boundary value problem. This arises mainly due to its ability to capture the correct stress-strain behaviour of the soil by invoking the prototype stresses and strains in the centrifuge models. Broadly speaking, some of the benefits of centrifuge modelling can be listed as:

- Ability to investigate complex problems by constructing small-scale physical models and testing them in the enhanced gravity field of a geotechnical centrifuge.
- Use of miniature instruments within the model that can record the soil behaviour before, during, and after a failure event is simulated.
- Geotechnical laboratories can create models accurately and carry out repeatable tests to increase confidence in the results obtained and the behaviour observed.
- Soil models with well-known and well-controlled stress history can be prepared.
- Extreme and/or rare loading events such as earthquake loading can be simulated aboard a centrifuge using specially developed and customized actuators.
- Complex construction sequences can be modelled on centrifuge models in-flight, to simulate the correct stress history of the problem.
- As physical models are tested, either true plane strain models or fully three-dimensional models can be developed as required.

On the other hand, centrifuge tests are specialized and it is often not possible (or necessary) to conduct all possible variations of a given problem in terms of loading and soil properties that can occur in the field. So parametric studies with a large number of centrifuge tests are impractical and can be expensive. Centrifuge modelling can therefore be used to test a set of well-defined problems and the data used to understand the failure mechanisms that are developing in that class of problems. This understanding can then be used to develop simplified design methodologies or used to verify the predictions obtained from numerical methods.

2.5 BENEFITS OF NUMERICAL MODELLING

As demonstrated in Sections 2.2 and 2.3 above, numerical modelling in the form of FE analyses can be quite a powerful tool in analysing complex geotechnical problems, especially when closed-form analytical solutions are not possible or are too cumbersome. The use of FE analyses has many benefits, some of which are outlined here.

- The FE method is able to handle complex geometries easily.
- The two-phase (and in some cases three-phase) nature of soil can be captured and fully coupled analyses between the soil and pore water phases can be conducted.
- Different levels of complexity in the material behaviour can be implemented. For example, one can start with simple elastic analyses and then carry out elasto-plastic analyses and model the yielding in soil.
- Sophisticated constitutive models that capture the nonlinear soil behaviour under extreme loading events such as cyclic loading imposed by earthquakes are available.
- Although plane strain or axisymmetric analyses are common, FE codes can analyse true three-dimensional problems. Automatic mesh refinement and automatic time stepping schemes can reduce the analysis times and the computational effort required significantly.
- Visualization of the results from FE analyses such as deformed meshes and stress contours (see Sections 2.2 and 2.3) provides an excellent opportunity for engineers to make judgments on the response of the system and its stability. It can help understand the failure mechanism that is developing in the problem.
- Another advantage of FE analyses is that parametric studies can be undertaken easily and the sensitivity and influence of each parameter can be established on the performance of the system.

It must be pointed out that, while the finite element and finite difference methods are the most popular numerical methods currently used by geotechnical engineers, there are several other numerical methods currently available, such as the boundary element method, discrete element method, mesh-free formulations, and so on. Their use may increase in the future with the availability of robust codes.

The main issue with numerical methods is that their ability to capture the correct failure mechanisms that can develop in a given boundary value problem is not always guaranteed.

2.6 COMPLEMENTARITY OF CENTRIFUGE MODELLING AND NUMERICAL MODELLING

In the above sections the benefits of physical modelling in the form of centrifuge testing and numerical modelling using the finite element method are outlined. The two techniques can be combined to form a very powerful set of methods that can help geotechnical engineers who have to deal with quite complex problems. As an example of this let us consider earthquake loading. Clearly performance of key infrastructure during and following a major earthquake event is a concern in the seismic regions around the world. Earthquake loading is a random and unpredictable event and very little information is available on the performance of buildings or other civil engineering infrastructure that was collected during the earthquake itself. However, failures and good performance of certain structures are well documented through post-earthquake missions such as those organized by EEFIT in the United Kingdom or GEER in the United States. Let us consider the case of an anchored quay wall that is subjected to dynamic loading from an earthquake. We wish to study the performance of the quay wall under such loading. This problem is of considerable interest given the failures of quay walls in ports and harbors observed in many earthquakes around the world.

Specialist FE codes have been available for more than 25 years to investigate earthquake problems in geotechnical engineering. Codes like Swandyne and Dynaflow have been mentioned earlier that are able to solve the fully coupled soil-fluid systems using Biot's formulation. These FE analyses have to be carried out in the time domain to capture the nonlinear behaviour of soil. It is possible to analyse the performance of an anchored quay wall following earthquake loading using one such FE code. Similarly it is possible to conduct dynamic centrifuge tests in which a physical model of an anchored quay wall can be subjected to earthquake loading. One of the advantages of doing the centrifuge testing, as explained earlier, is that it captures the correct failure mechanism. If we model the same problem numerically using an FE code, we can calibrate the performance of the numerical analysis against the centrifuge test results. If good comparison is observed and we are confident that the correct physical mechanism is being captured by the numerical analysis, the confidence in the FE code increases. Such validated codes can be used to investigate much more complex geometries that are close to physical reality in the field or conduct parametric studies to investigate the sensitivity of various parameters.

Such a comparative study on a large scale was first conceived in late 1980s by Dr. Cliff Astill and was carried out as a National Science Foundation-funded project called VELACS. Many centrifuge modellers and numerical analysts took part in this project. More details of the project and the results from the project are described by Arulanandan and Scott (1993). This project established the complementarity of both centrifuge and FE techniques in studying a wide range of problems in earthquake geotechnical engineering. A similar project that aims to investigate the dynamic interaction between tunnels and their surrounding soil strata using both centrifuge testing and FE analyses is currently being conducted by Universita' degli

Studi di Napoli Federico II in Italy. It is expected that such a VELACS-style project will lead to a better understanding of the tunnel behaviour during earthquakes and provide well-validated numerical procedures that can be used to study tunnels subjected to earthquake loading.

2.6.1 Comparison of centrifuge data with results from FE analysis

As mentioned before, quay walls have failed in many of the recent earthquakes, such as the 1994 Northridge earthquake (Madabhushi 1995), the 1995 Kobe earthquake (Soga, 1998), the 1921 Ji-Ji earthquake (Madabhushi, 2007), and the 2001 Bhuj earthquake (Madabhushi, Patel, and Haigh, 2005). Anchored quay walls are particularly vulnerable as their stability depends on the anchor stiffness and strength. Zeng (1990) has conducted a series of centrifuge tests investigating this problem. A typical cross-section of the centrifuge models tested is shown in Figure 2.9a. The finite element discretization used to analyse this problem is shown in Figure 2.9b. The

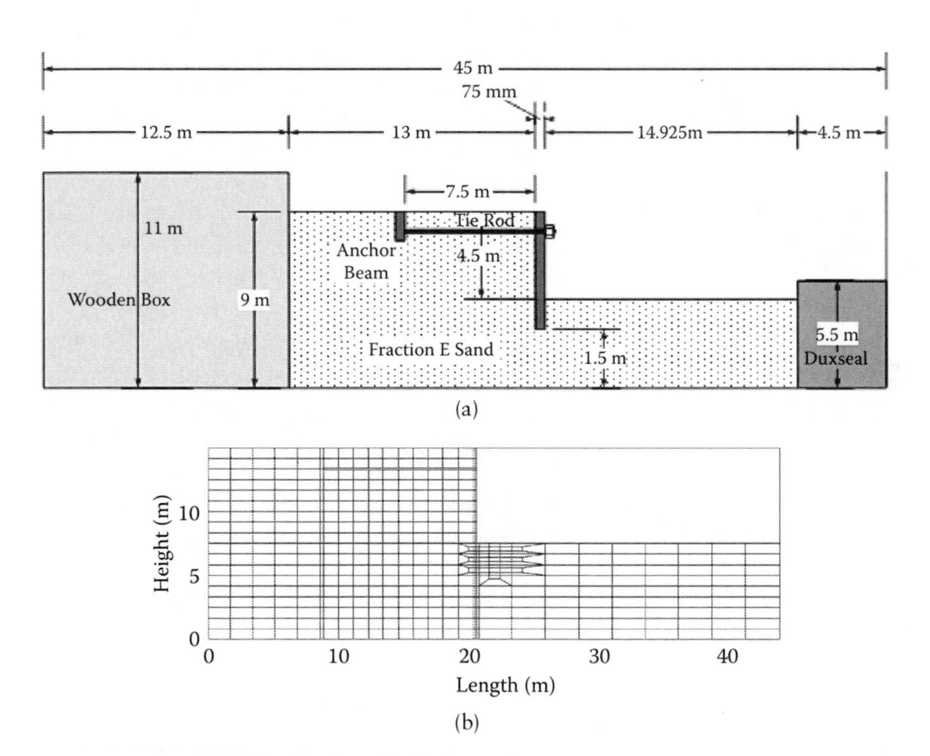

Figure 2.9 (a) Cross-section of a centrifuge model of anchored quay wall (Zeng, 1990). (b) Finite element discretization of the anchored quay wall problem.

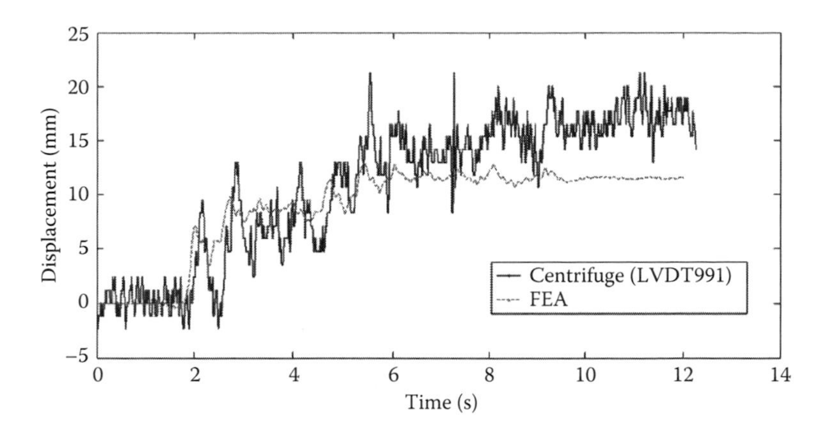

Figure 2.10 Comparison of wall displacement in the centrifuge test and FE analysis.

centrifuge model was subjected to earthquake loading using the Bumpy Road actuator (Kutter, 1982). The same earthquake loading was applied to the FE mesh at the base nodes. The FE analysis was carried out using the Swandyne code (Chan, 1989). The simple Mohr-Coulomb constitutive model was used for the soil in this analysis. More details of this problem can be found in Cilingir and Madabhushi (2011a).

It is now possible to compare the lateral displacement at the tip of the retaining wall in the centrifuge test and that predicted by the FE analyses using Swandyne code during an earthquake loading event as shown in Figure 2.10. Clearly the FE code is able to predict the lateral displacement suffered by the wall during the earthquake loading reasonably well although the residual displacement after the earthquake loading is slightly underestimated by the FE analysis.

Similarly the bending moments generated in the wall before, during, and after the earthquake loading measured in the centrifuge tests using strain gauges on the wall can be compared to the bending moments obtained from the FE analysis. These are presented in Figure 2.11 and show that the FE analysis is able to capture the bending moment distribution along the length of the wall reasonably well.

2.6.2 Deformed shape and horizontal stresses

As in the problem of the footing over layered soil strata explained in Sections 2.2 and 2.3 before, the FE analyses can give additional information with respect to the deformed mesh and the stresses in the soil. For this problem, the deformed mesh is presented in Figure 2.12 plotted with a magnification of 100 on deformations. In this figure the deformation

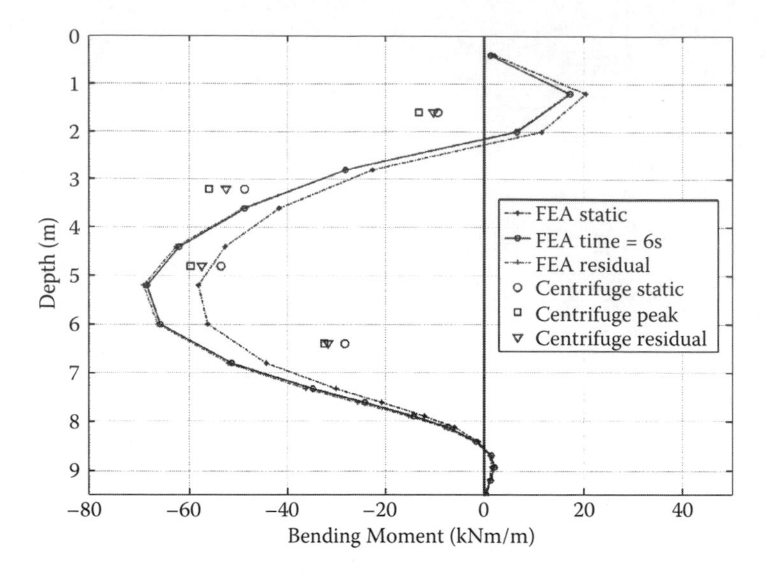

Figure 2.11 Bending moment distribution in the anchored retaining wall.

suffered by the anchored quay wall following the earthquake loading can be seen quite clearly. Based on the experimental observations the movement of the anchor beam (see Figure 2.9) and the subsidence of the soil behind the anchor beam were also seen to match correctly. In Figure 2.13 contours of the residual horizontal stress in the soil following the earthquake loading are presented. The mobilisation of the active soil wedge behind the quay wall and the passive pressures building up before the anchor beam and in front of the toe of the retaining wall can be seen in Figure 2.13.

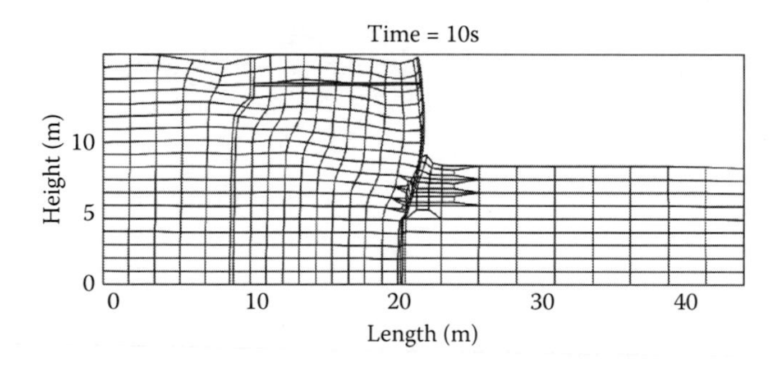

Figure 2.12 Deformed finite element mesh.

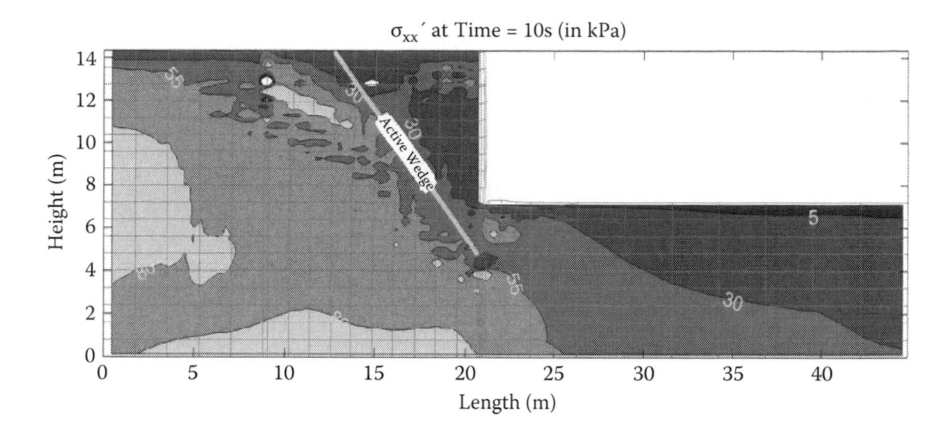

Figure 2.13 Residual horizontal stress contours.

2.7 SUMMARY

In the above sections the problem of an anchored quay wall subjected to earthquake loading was considered. Such a problem is quite complex and cannot be solved using analytical solutions. The dynamic centrifuge test has given valuable experimental data in terms of wall displacements and bending moments induced in the wall. In fact more detailed information on the acceleration fields within the soil model are also obtained and can be found in Zeng (1990), although not discussed here. The experimental data was used to identify the response of the wall, anchor beam, and tie, and the FE analysis was adjusted accordingly. The final results from the FE analysis were able to capture the observed experimental behaviour accurately. This gives good confidence in the results of the FE analysis and therefore this code can be used to solve these types of problems. Additional complexity in terms of geometrical variations in the wall or soil strata can be introduced easily in the FE analysis. Further, parametric studies can be carried out to investigate the influence of each component of the anchored tie wall system. For example, the influence of tie stiffness on the deformation mechanism was investigated by Cilingir and Madabhushi (2011b). Similarly the influence of the retaining wall stiffness was investigated by Cilingir and Madabhushi (2011b).

The centrifuge modelling technique and the FE analysis can be considered complementary to one another and can be used in conjunction to form a very powerful set of tools to investigate quite complex problems such as those encountered in earthquake geotechnical engineering. The complementarity between centrifuge modelling and finite element analysis is further discussed by Madabhushi and Chandrasekaran (2005, 2006) for solving an excavation behind retaining walls of different height. Similar examples can be extended to other geotechnical problems such as effects of construction of tunnels below existing buildings or staged excavations in front of a retaining wall and its effects on buildings behind the retaining wall. In these cases the centrifuge tests are used to capture the failure mechanisms developed in the basic geometry of the problem and the FE analyses are then calibrated against the centrifuge data. The validated FE analyses can then be used with confidence to solve more complex problems encountered in the field.

Chapter 3

Uniform circular motion

3.1 INTRODUCTION

In centrifuge modelling we deal with a rotating soil body mounted on a geotechnical centrifuge to represent a scale model of a given prototype we are trying to model. In Figure 3.1 the Turner Beam centrifuge at the Schofield Centre of the University of Cambridge is shown. The diameter of this centrifuge is nominally 10 m with a working radius of 4.125 m. This is a balanced beam centrifuge that carries the scaled model at one end and a counterweight at the other, each mounted on a swing platform. The maximum payload that can be carried is approximately 1000 kg. As the centrifuge starts to spin about the central vertical axis, the swing platforms rotate about a pivot until they become horizontal. This is illustrated in Figure 3.2 for one end of the centrifuge carrying a container with the soil model. The "swing-up" is achieved for the Turner beam centrifuge at about 10 g when the centrifuge is spinning at 45 RPM. When the centrifuge speed is increased beyond 45 RPM the model is subjected to increasing levels of centrifugal acceleration that act normal to the soil surface. It must be noted that the normal earth's gravity continues to act on the soil model perpendicular to the centrifugal acceleration. The maximum centrifugal acceleration that can be applied by the centrifuge to the model is 150 g, that is, 150 × earth's gravity. The definition of centrifugal acceleration is explained in detail in Section 3.3.

Often the scaled models tested in a centrifuge may contain moving parts such as actuators applying loads on model foundations placed on the soil or the whole model being shaken as in the case of earthquake modelling (see Section 2.6). These actuators operate while the centrifuge is spinning. It is therefore important for us to establish the mechanics that govern the uniform circular motion. In this chapter the basic mechanics of circular motion are explained as relevant to centrifuge modelling. For more details on this topic please refer to Halliday, Resnick, and Walker (2004) and Hibbeler (2007).

Figure 3.1 The 10 m diameter Turner Beam centrifuge at the University of Cambridge.

3.2 UNIFORM CIRCULAR MOTION

A body is said to be in uniform circular motion if it travels around a circle or circular path at a constant speed. Let us consider a solid sphere travelling around in a circular path of radius r as shown in Figure 3.3. The sphere is travelling at a uniform speed of v. However, its velocity $\overline{v}$ is constantly changing as the sphere changes its direction of travel as it goes round the circular path. In uniform circular motion the acceleration of the particle occurs because of change in direction of the velocity although the magnitude of velocity (speed) remains constant.

Imagine that at a given instant the sphere's location is described by an angle θ as shown in Figure 3.3. The sphere is rotating about the center of the circle with an angular speed $\dot{\theta}$. We wish to discover the acceleration acting on this sphere due to uniform circular motion. Let us define a set of unit vectors i, j, and k as shown in Figure 3.3. The angular velocity of the sphere is $\dot{\theta}\, k$. It must be noted that in a Cartesian system of coordinates, the unit vectors i, j, and k are always fixed in direction, so their time derivatives are zero.

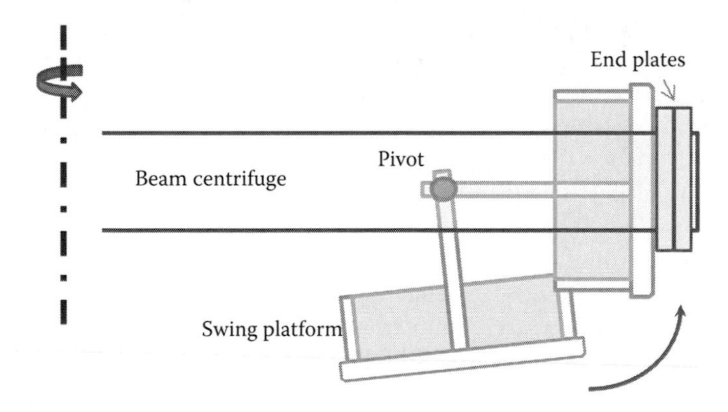

Figure 3.2 Swing-up of the centrifuge model at the end of the arm.

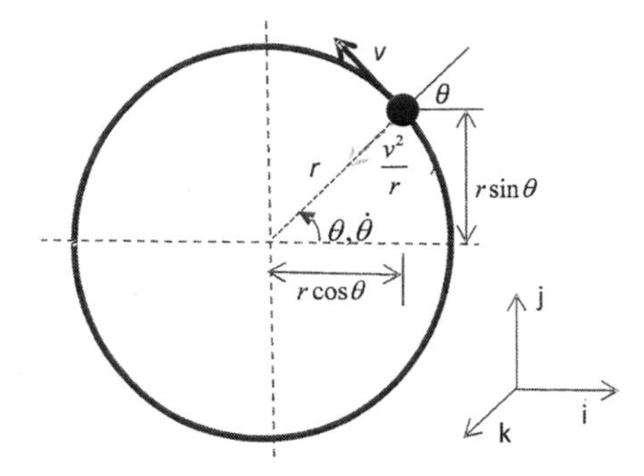

Figure 3.3 Uniform circular motion of a sphere.

The position vector of the sphere at a given instant shown in Figure 3.3 can be written as

$$\bar{r} = r\cos\theta \; i + r\sin\theta \; j \tag{3.1}$$

We can obtain the velocity of the sphere by differentiating the position vector in Equation 3.1 as

$$\bar{v} = \frac{d\bar{r}}{dt} = r\,(-\sin\theta)\dot{\theta}\; i + r(\cos\theta)\dot{\theta}\; j \tag{3.2}$$

However, we know that the linear speed of the sphere v is related to the angular speed $\dot{\theta}$ as

$$v = r\,\dot{\theta} \tag{3.3}$$

So substituting these, Equation 3.2 can be rewritten as

$$\bar{v} = (-\,v\,\sin\theta)\; i + (v\,\cos\theta)\;\; j \tag{3.4}$$

This is clearly correct as $-v\,\sin\theta$ and $v\,\cos\theta$ are the horizontal and vertical components of the velocity vector $\bar{v}$.

In order to get the acceleration of the sphere, we can differentiate the velocity vector $\bar{v}$.

$$\bar{a} = \frac{d\bar{v}}{dt} = -v(\cos\theta)\dot{\theta}\; i + v\;(-\sin\theta)\dot{\theta}\; j \tag{3.5}$$

Substituting for the angular speed $\dot{\theta}$ from Equation 3.3 into Equation 3.5, we can write:

$$\bar{a} = -\frac{v^2}{r}(\cos\theta)i - \frac{v^2}{r}(\sin\theta)j = \left(-\frac{v^2}{r}\right)(\cos\theta\ i + \sin\theta\ j) \tag{3.6}$$

The magnitude of this acceleration is therefore $\frac{v^2}{r}$ and the minus sign indicates that it acts toward the center of the circular path. This acceleration is termed centripetal acceleration and acts on any body that undergoes a uniform circular motion.

In a centrifuge that is rotating at a uniform speed, the soil body is carried in a model container at the end of the centrifuge arm. The centrifuge end plates will apply this centripetal acceleration to the model container. The direction of the centripetal acceleration is always toward the central vertical axis of the centrifuge about which it rotates.

3.3 BASIC DEFINITIONS: CENTRIPETAL AND CENTRIFUGAL FORCES

The word centripetal is derived from the Latin words *centrum* meaning center and *petere* meaning to seek. The centripetal acceleration derived in Section 3.2 acts on a body undergoing uniform circular motion and is always directed toward the center of the circular path. According to Newton's first law, acceleration on any body must be caused by a force. The force that causes centripetal acceleration is termed the centripetal force. Newton's second law describes this force as being equal to the product of the mass of the body and the acceleration acting on it. Using this and following on from Equation 3.6, the centripetal force is defined as:

$$F = m\frac{v^2}{r} \tag{3.7}$$

A common example of centripetal force is given by considering a car going around a corner. Imagine a person sitting on the back seat of the car in the middle. As the car starts to go round, it is subject to centripetal acceleration toward the center. This is exerted as a frictional force on the tyres of the car. The body of the person would tend to continue in a straight line while the seat supporting him would be under centripetal acceleration. If the friction between the person and the seat is too little, then the seat will slide from underneath the person. This will continue until the person slides and reaches the car door and then he feels squeezed as the car goes round the corner. This is because of the centripetal force exerted by the car door on the person.

Let us consider another example of a sphere of mass m attached to the end of a string. If we hold the other end of the string and start to spin it around, the sphere is subject to centripetal acceleration as discussed in Section 3.2. This centripetal acceleration is caused by a centripetal force as defined by Equation 3.7. However, the string attached to the sphere experiences a tension that can be felt by the hand holding the other end. The tension in the string applies the centripetal force on the sphere pulling it toward the center of rotation. This force must be balanced by another force, otherwise the sphere would move toward the center. This can be explained by means of a "fictional" or "inertial" force that acts on the sphere radially outward. This fictional force is termed centrifugal force. The reason we use the term fictional force must be explained here. If we consider a noninertial frame of reference, then we need to introduce "additional forces" to explain the motion of a body. For example, consider a ball falling freely toward the surface of the earth. If an observer on the surface of the earth is looking at the ball, it will not appear to fall straight down. This is because the observer would have moved horizontally due to the rotation of the earth in the time it takes for the ball to fall to the surface. We can account for this by considering a fictional "horizontal" force (called the Coriolis force, explained in Section 3.5) acting on the ball. More details of this can be found in Borowitz and Bornstein (1968) and Rothman (1989).

The word centrifugal is derived from the Latin words *centrum* meaning center and *fugere* meaning to flee. By Newton's third law the centrifugal force and the centripetal force must be equal and opposite. The centrifugal force causes a centrifugal acceleration that acts radially outward. In the example of the car going around a corner, we can say that the tires of the car will exert centrifugal forces on the road acting radially outward. Centrifugal forces are exploited in a wide variety of engineering devices such as centrifugal pumps and centrifugal clutches.

In a geotechnical centrifuge, we are interested in the centrifugal accelerations created in the soil model as it spins around. Referring back to Figure 3.2, the soil model will be subjected to centrifugal accelerations that act on it radially outward. This causes the body forces in the soil model to increase. The higher the speed of rotation of the centrifuge, the larger will be the centrifugal acceleration and hence the body forces within the soil model. The container carrying the soil model is supported on the end plates as shown in Figure 3.2. These end plates will apply a centripetal force acting radially inward on the model container.

3.4 USE OF POLAR COORDINATES IN UNIFORM CIRCULAR MOTION

We have utilised the Cartesian coordinate system with unit vectors i, j, and k for deriving the centripetal acceleration in Section 3.2. This coordinate system is rather cumbersome when dealing with uniform circular

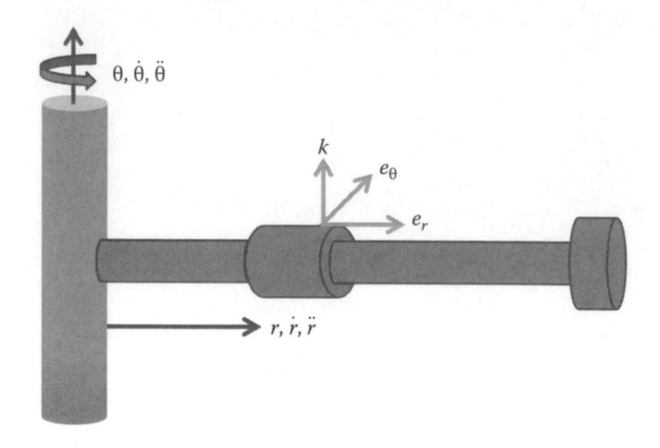

Figure 3.4 Rotating shaft with a sliding sleeve.

motion of a body particularly when the body has parts that can undergo other motions. For example, let us consider the case of a rotating shaft shown in Figure 3.4. Let us suppose that the rotation of the vertical shaft is defined by θ and it has an angular velocity of $\dot{\theta}$ and an angular acceleration of $\ddot{\theta}$. The vertical shaft has a horizontal bar attached to it rigidly. There is a frictionless sleeve on the horizontal bar that is able to undergo a sliding motion along the horizontal bar. The position of the sleeve at any time can be defined by r and its velocity and acceleration are defined as $\dot{r}$ and $\ddot{r}$, respectively. Let us say we are interested in the velocity and acceleration components of the sleeve due to the combined action of rotation and sliding. Such a problem can be quite cumbersome to solve using the Cartesian coordinate system. We can simplify the problem by attaching the unit vectors to the body of interest, that is, the sliding sleeve, and allow them to rotate with the body. Such a system is called the polar coordinate system, in which all the motions can be defined in terms of the coordinates (r, θ) and their time derivatives.

The key aspect of the polar coordinate system is the treatment of the unit vectors that are attached to the body of interest. Let us define a unit vector e_r in the radial direction along which the sleeve is sliding. Orthogonal to this unit vector, let us define a second unit vector e_θ in the tangential direction to the sleeve. These two unit vectors are attached to the sleeve and rotate with it about the vertical axis with an angular velocity of $\dot{\theta}k$.

The position vector of the sleeve can be written as:

$$\overline{OR} = r\ e_r \tag{3.8}$$

The velocity of the sleeve can be obtained by differentiating Equation 3.8 as:

$$\bar{v} = \frac{d}{dt}(r\ e_r) = \dot{r}\ e_r + r\ \dot{e}_r \tag{3.9}$$

As the unit vector e_r is rotating with an angular velocity of $\dot{\theta}k$, its time derivative is not zero. We can evaluate this time derivative as a cross product of the angular velocity and the unit vector.

$$\dot{e}_r = \dot{\theta}k \times e_r = \dot{\theta}\ e_\theta \tag{3.10}$$

Similarly, the time derivative of the unit vector e_θ rotating with an angular velocity of $\dot{\theta}k$ is evaluated as

$$\dot{e}_\theta = \dot{\theta}k \times e_\theta = -\dot{\theta}\ e_r \tag{3.11}$$

Substituting Equation 3.10 into Equation 3.9, the velocity of the sleeve is obtained as

$$\bar{v} = \dot{r}\ e_r + r\dot{\theta}\ e_\theta \tag{3.12}$$

Thus, the sleeve will have a radial velocity component of $\dot{r}$ and a tangential velocity component of $r\dot{\theta}$. The magnitude of the sleeve velocity vector is

$$\sqrt{(\dot{r})^2 + (r\dot{\theta})^2}.$$

We shall now evaluate the acceleration components of the sleeve. We can obtain these by differentiating the velocity vector shown in Equation 3.12.

$$\bar{a} = \frac{d}{dt}(\bar{v}) = \frac{d}{dt}(\dot{r}\ e_r + r\dot{\theta}\ e_\theta)$$

$$\bar{a} = \ddot{r}\ e_r + \dot{r}\ \dot{e}_r + \dot{r}\dot{\theta}\ e_\theta + r\ddot{\theta}\ e_\theta + r\dot{\theta}\ \dot{e}_\theta \tag{3.13}$$

As before, recognizing that the unit vectors in the polar coordinate system are rotating with an angular velocity of $\dot{\theta}k$ and using Equations 3.10 and 3.11, we can rewrite Equation 3.13 as:

$$\bar{a} = \ddot{r}\ e_r - r\dot{\theta}^2\ e_r + r\ddot{\theta}\ e_\theta + 2\dot{r}\dot{\theta}\ e_\theta \tag{3.14}$$

This equation shows that acceleration has two components in the radial and tangential directions. In the radial direction the acceleration depends on the radial acceleration of the sleeve ($\ddot{r}$) and the centripetal acceleration ($-r\dot{\theta}^2$). The latter agrees with what we have obtained in Equation 3.6 using the Cartesian coordinate system, when we substitute for

$$\dot{\theta} = \frac{v}{r}$$

In the tangential direction the acceleration depends on the angular acceleration ($\ddot{\theta}$) and an additional component ($2\dot{r}\dot{\theta}$). This additional component is called the Coriolis acceleration and is named after the French scientist Gaspard-Gustave de Coriolis. This acceleration component is a function of both radial and angular velocities. Coriolis acceleration affects any moving parts that are contained within a centrifuge model.

3.5 CORIOLIS FORCE AND EULER FORCE

In the previous section we have established all the acceleration components in the polar coordinate system. We can obtain the corresponding force components by using Newton's second law. In Section 3.3 we discussed the centripetal and centrifugal forces. Using the same principles, we can determine the radial and tangential force components.

Let us consider the radial forces acting on a body undergoing circular motion. Note that we no longer are restricted to uniform circular motion as we are allowing for radial and angular accelerations. The radial forces on a body of mass m are given as:

$$F_t = m\left(\ddot{r} - r\dot{\theta}^2\right) \quad (3.15)$$

This force acts in the radial (e_r) direction and is the net force acting on the body and allows for centripetal force. In this context, also see the discussion of centrifugal force in Section 3.3.

Let us now consider the forces on the body in the tangential (e_θ) direction. These can be written as:

$$F_\theta = m\,(r\ddot{\theta} + 2\dot{r}\dot{\theta}) \quad (3.16)$$

The first force component in Equation 3.16 arises due to angular acceleration and is called the Euler force. The second force component is due to the Coriolis acceleration introduced in Section 3.4 and is called the Coriolis force. In centrifuge modelling the Euler forces are usually quite small as the angular acceleration of the centrifuge is quite small and only exists at the starting process of the centrifuge while it picks up angular speed and until the desired g level is reached. It is also possible to reduce the Euler forces by adjusting the ramp-up speed of the centrifuge motor controller. However, if the stability of a centrifuge model is important during this acceleration phase, then the Euler force can be calculated as:

$$F_{Euler} = m\ r\ddot{\theta} \tag{3.17}$$

The Coriolis force arises from any movement that occurs within the centrifuge model. For example, if we try to construct an embankment by raining sand in flight or if we drop a ball from the center of the centrifuge to study projectile motion and impact on the soil, the moving object will have a radial velocity ($\dot{r}$) and an angular velocity ($\dot{\theta}$) due to the rotation of the centrifuge. The Coriolis force can be calculated as:

$$F_{coriolis} = 2m\ \dot{r}\dot{\theta} \tag{3.18}$$

3.6 SUMMARY

Many of the problems that are modeled using geotechnical centrifuge testing require knowledge of circular motion. In this chapter we have considered the basic principles of uniform circular motion that are relevant to centrifuge modelling. Initially the Cartesian coordinate system was used and later the polar coordinate system was introduced. The latter is more useful in centrifuge modelling. In both systems the velocity and acceleration vectors are obtained by differentiating the position vector. Coriolis acceleration was introduced, which can influence any moving objects within a centrifuge model. The forces that result from the angular accelerations in the radial and tangential directions were also explained.

Chapter 4

Principles of centrifuge modelling

4.1 INTRODUCTION TO CENTRIFUGE MODELLING

Testing of small-scale models is used widely in civil engineering. For example, in fluid mechanics we build a scaled model to study the flow of water in open channels. In such problems we ensure certain criteria are satisfied both in the model and the prototype (for example, the Reynolds number is the same in both model and prototype). Structural engineers also use scaled models to investigate the behaviour of a complex prototype (an example would be estimation of the dynamic behaviour of a cable stayed bridge during severe crosswinds; such experiments will be carried out in a wind tunnel). In geotechnical engineering we need small-scale models for testing in a laboratory especially when we are dealing with soils of unknown stress-strain behaviour (i.e., constitutive model unknown). Full-scale testing in the field is often very expensive and sometimes even unfeasible, for example when simulating earthquake loading on a large dam or retaining wall as illustrated in Section 2.6.

In soil mechanics we have to resort to something more complicated when testing small-scale models. If we build a scale model of an embankment, for example, the stress exerted by our model embankment on the foundation will be very small. The behaviour of this model, compared to a prototype that may be N times larger than the model, will be totally different. This is because the stress-strain behaviour of soil is known to be highly nonlinear. This is explained with the aid of Figure 4.1.

Let us consider the stress-strain behaviour of sand tested in a shear box apparatus and plot the shear stress against shear strain as illustrated in Figure 4.1. If we consider dense sand, the stress-strain curve will reach a peak stress, after which the dense sand will suffer strain softening before reaching its critical state at large strains as shown in Figure 4.1. Similarly if we consider loose sand then the stress-strain curve will be smooth until it reaches the critical state at large strains. For both dense and loose sands, the initial stiffness is high at small strains, as indicated by the tangents to

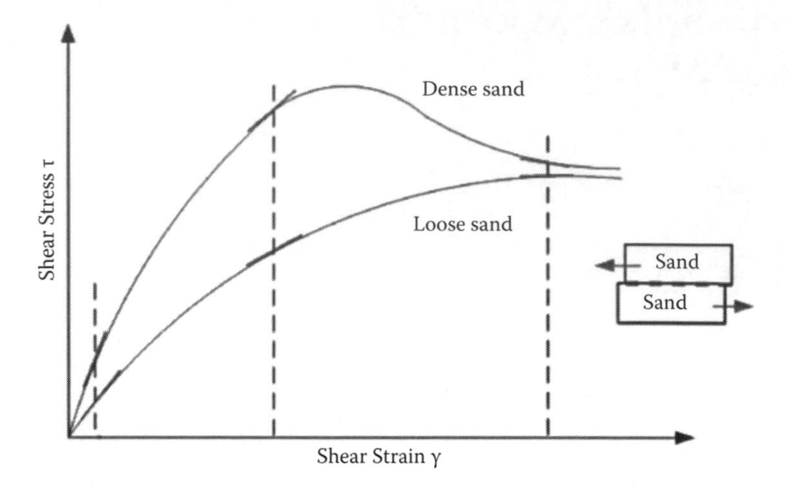

Figure 4.1 Stress-strain relationship of sand tested in a shear box apparatus.

the stress-strain curves in Figure 4.1. At large stresses and large strains, the stiffness of the sand reduces. Similarly at very large strains, that is, close to the critical state, the stiffness reduces to quite small values.

If we make small-scale models and test them in the laboratory, then the soil will respond with large stiffness as the stresses and strains in these models will be small. As a consequence of this, the observed deformations in the soil model such as settlements will be small. However, the same soil below a large prototype structure that exerts larger stresses will respond with a lower stiffness and hence the settlements will be larger. It is therefore not conservative to predict the behaviour of a prototype by testing small-scale models in a laboratory to solve geotechnical problems, due to the nonlinear stress-strain behaviour of soils.

A $1/N$ scale model of a prototype will exert only a $1/N$ fraction of the stresses compared to the stresses exerted by a prototype. You can satisfy that this is the case by solving example problem 1 at the end of this chapter. We use centrifuge modelling to get around this problem; that is, by creating prototype stresses and strains in our model we can capture the true behaviour of the soil.

4.2 PRINCIPLES OF CENTRIFUGE MODELLING

The basic premise in centrifuge modelling is that we test a $1/N$ scale model of a prototype in the enhanced gravity field of a geotechnical centrifuge. The gravity is increased by the same geometric factor N relative to the normal earth's gravity field (referred to as 1 g).

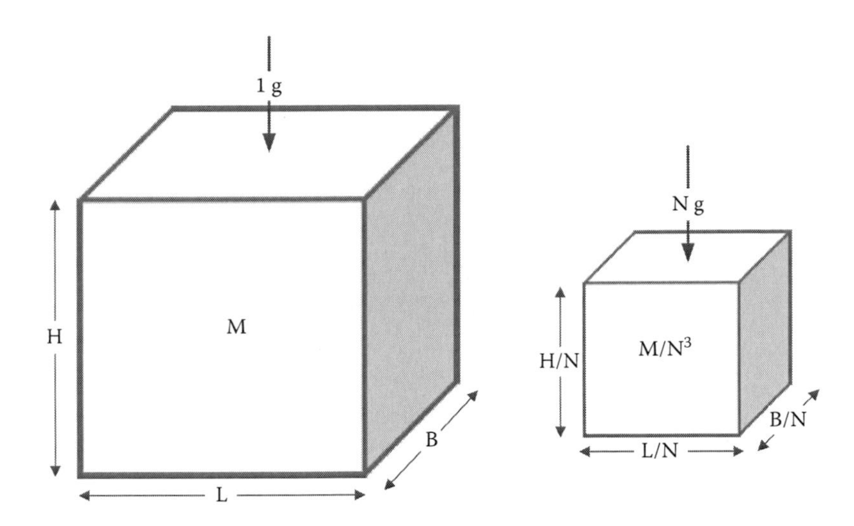

Figure 4.2 Principle of centrifuge modelling.

This can be illustrated using a simple example. Let us consider a block structure of mass M and with dimensions $L \times B \times H$ sited on a horizontal soil bed as shown in Figure 4.2. The average vertical stress exerted by this block on the soil can be easily calculated as:

$$\sigma_v = \frac{M\ g}{L \times B} \tag{4.1}$$

Similarly the vertical strain induced in the soil for a characteristic length α as:

$$\varepsilon = \frac{\delta\alpha}{\alpha} \tag{4.2}$$

Now let us consider a scale model of this block in which all the dimensions are scaled down by a factor N as shown in Figure 4.2. As all the dimensions are scaled down by a factor N, the mass of this scaled down block will be M/N^3. Let us now imagine that this scale model of the block is placed in the increased gravity field of $N \times$ earth's gravity.

If we now recalculate the vertical stress underneath this scale model of the block, we can see that

$$\sigma_v = \frac{\frac{M}{N^3} \times Ng}{\frac{L}{N} \times \frac{B}{N}} = \frac{M\ g}{L \times B} \tag{4.3}$$

Thus, the vertical stress below this scale model of the block is the same as that below the larger block obtained in Equation 4.1. Similarly, if we consider strains in the soil:

$$\varepsilon = \frac{\delta\alpha/N}{\alpha/N} = \frac{\delta\alpha}{\alpha} \tag{4.4}$$

we can see that the prototype strain in Equation 4.2 is recovered, as the changes in displacements and the original length are both scaled by the same factor N.

We increase the "gravity" acting on our scaled model by placing it in a geotechnical centrifuge, introduced in Chapter 3. The centrifugal acceleration will give us the "Ng" environment in which the scaled model will behave in an identical fashion to the prototype in the field. Simplifying the equations derived in Section 3.4, we can relate the angular velocity of the centrifuge to the required "g" level. When the centrifuge is rotating with an angular velocity of $\dot{\theta}$, the centrifugal acceleration at any radius r is given by:

$$\bar{a} = r\,\dot{\theta}^2 \tag{4.5}$$

We wish to match this centrifugal acceleration to be the same geometric scale factor as the one we used to scale down our prototype by N.

$$N\,g = r\,\dot{\theta}^2 \tag{4.6}$$

The centrifugal acceleration changes with the radial distance from the axis of rotation of the centrifuge as indicated in Equation 4.6. We will normally arrange the speed of the centrifuge such that the model at the desired radius (e.g., a typical point in the model such as its centroid) will experience the desired centrifugal acceleration Ng. This will give us the angular velocity $\dot{\theta}$ with which we have to rotate our centrifuge. For example, for the Turner Beam Centrifuge at Cambridge discussed in Chapter 3 the nominal working radius is 4.125 m. If we need to create a centrifugal acceleration of 100 g on a centrifuge model, then using Equation 4.6, we can calculate the angular velocity as:

$$\dot{\theta} = \sqrt{\frac{100 \times 9.81}{4.125}} = 15.42 \;\; rad/s$$

$$\dot{\theta} = 147.3 \;\; RPM \tag{4.7}$$

So by spinning the centrifuge at 147.3 RPM, we can create the required 100 *g* of centrifugal acceleration. For other centrifuges we can use the same approach of calculating the angular velocity depending on the working radius from the axis of the centrifuge to the representative point within the soil model. More details on the selection of representative point within the centrifuge model will be discussed later in Section 6.2 of Chapter 6.

4.3 CONCEPTS OF A FIELD STRUCTURE PROTOTYPE AND THE CENTRIFUGE MODEL

While attempting centrifuge modelling we often use the terms prototype and centrifuge model as we did in Section 4.2. In this section we wish to clarify what we mean by these terms and explore the relationship between them. Let us consider a real multistoried tower structure in the field that is founded on a sand deposit as shown in Figure 4.3(a), subjected to earthquake motion through bedrock accelerations. We can term this the full-sized field structure. We can now imagine a reduced-scale model of this, which is an exact 1/*N*th scale model of the field structure as shown in Figure 4.3(b). This 1/*N*th scale model placed in an ideal Ng gravity field will behave exactly like the full-sized field structure. However, in centrifuge modelling it is often difficult or impractical to build the exact 1/*N*th scale model. We often resort to simplifying these models in such a way that they capture the essential behaviour of the full-sized field structure. Referring to Figures 4.3b and 4.3(c), we can see that the reduced-scale model in Figure 4.3(c) is simplified so that it will capture the rocking behaviour of the multistoried tower structure under earthquake loading. Thus, we are only considering the lumped mass of the structure at the top and equivalent flexural stiffness of the columns in our model structure in Figure 4.3(c). This simplified model is what we would term our centrifuge model.

If we scale up our centrifuge model as shown in Figure 4.3(d) using the same scaling laws, we would end up with the prototype structure. Comparing this prototype structure with the full-sized field structure we started with in Figure 4.3(a) allows us to see the effect of simplifications we made in arriving at the centrifuge model. Clearly in this example, we lost a lot of detail of the structure such as beam-column joints, openings for windows, etc. However, we arrived at the centrifuge model to capture the rocking of the whole structure under earthquake loading and hence this detail was considered not important in this problem. For other problems, these assumptions may not be valid. For example, if we were interested in testing the performance of the beam-column joints under the earthquake loading, then clearly we need to incorporate that detail into the centrifuge model, by carefully designing the scaled models of beams and columns that have appropriately scaled stiffness and representative joint details. So the

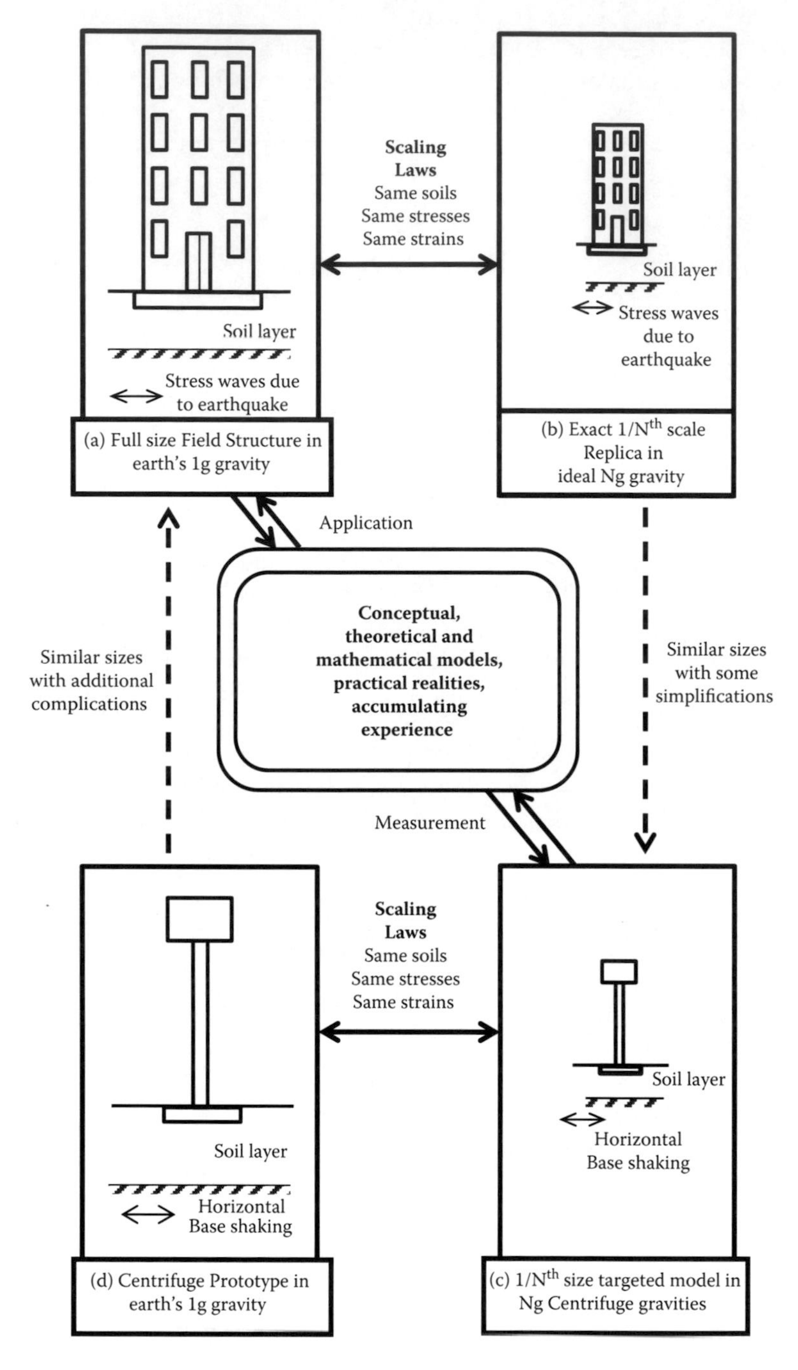

Figure 4.3 Relationship between the field structure and centrifuge model.

centrifuge models we develop depend on the specific problem we wish to investigate. It is very important at the time of idealizing the centrifuge model that we are not losing any important features we are planning to study in our centrifuge experiments. In Figure 4.3 we can see that we require a set of scaling laws to relate the behaviour observed in the centrifuge model to the full-sized field structure. The scaling laws are derived based on well-known mathematical models as discussed next in Section 4.4.

4.4 SCALING LAWS IN CENTRIFUGE MODELLING

Scaling laws are relationships that relate the behaviour of the centrifuge model and the prototype. These are required to relate the observed behaviour of the scale model in the centrifuge experiment to the behaviour of a prototype. If we can predict the behaviour of a prototype we can infer the behaviour of the field structure. This of course assumes that the prototype and the field structure are closely related and all the essential features of the field structure are present in the prototype.

4.4.1 Derivation of simple scaling laws

Some of the scaling laws come directly from the principle of centrifuge modelling outlined in Section 4.2. For example, by observing the settlement of a structure in the centrifuge model we would like to predict the settlement of the prototype. The scaling law for this is quite straightforward. As we have scaled down all the length dimensions of the prototype by a factor of N in the centrifuge model, the scaling law for settlement would be N, that is, settlements in the centrifuge model are N times smaller than in the prototype. Similarly area and volume in the centrifuge model will be related to the prototype by factors of N^2 and N^3, respectively. These can be useful if, for example, we were to model a landslide problem in the centrifuge. The area and volume of the soil mass involved in the landslide in a centrifuge model can be related to the field-scale landslide soil movements using these scaling laws. These simple scaling laws can be written formally as:

$$\frac{Displacement_{model}}{Displacement_{prototype}} = \frac{1}{N} \tag{4.8}$$

$$\frac{Area_{model}}{Area_{prototype}} = \frac{1}{N^2} \tag{4.9}$$

$$\frac{Volume_{model}}{Volume_{prototype}} = \frac{1}{N^3} \tag{4.10}$$

In Section 4.2 we have already proved that the stresses and strains are identical in the centrifuge model and the prototype. So we can write the scaling law for stress and strain as:

$$\frac{Stress_{model}}{Stress_{prototype}} = 1 \tag{4.11}$$

$$\frac{Strain_{model}}{Strain_{prototype}} = 1 \tag{4.12}$$

$$\frac{Acceleration_{model}}{Acceleration_{prototype}} = \frac{Ng}{1g} = N \tag{4.13}$$

Similarly the scaling law for mass is

$$\frac{Mass_{model}}{Mass_{prototype}} = \frac{1}{N^3} \tag{4.14}$$

These scaling laws are all derived based on the principles of centrifuge modelling considered in Section 4.2.

4.4.2 Scaling law for time of consolidation

Consolidation of soil is a diffusion process that occurs when excess pore pressures are generated in the soil due to application of rapid loading. With passage of time these excess pore pressures decrease and the effective stress in the soil increases. The void ratio of the soil changes allowing for the settlement to take place. The theory of consolidation originally attributed to Terazaghi (see Terazaghi, Peck, and Mesri 1963), governs the dissipation of the excess pore pressures within the soil with time. The governing equation for consolidation in three dimensions can be written as:

$$\frac{\partial u}{\partial t} = C_v \left(\frac{\partial^2 u}{\partial x^2} + \frac{\partial^2 u}{\partial y^2} + \frac{\partial^2 u}{\partial z^2} \right) \tag{4.15}$$

where u is the excess pore pressure. In this section we wish to investigate the scaling law that relates the time for consolidation in a centrifuge model to that in the prototype. The degree of consolidation we wish to achieve in the centrifuge model is the same as that of the prototype soil. The degree of consolidation is linked to the time factor T_v, which depends on the coefficient of consolidation C_v of the soil, the drainage distance d for the pore fluid to escape, and the time t for achieving the required degree of consolidation. The time factor T_v is related to these three parameters as:

$$T_v = \frac{C_v \ t}{d^2} \tag{4.16}$$

Let us consider Equation 4.16 for the centrifuge model and the prototype as shown in Equation 4.17.

$$\frac{[T_v]_{model}}{[T_v]_{prototype}} = \frac{\left[\frac{C_v \ t}{d^2}\right]_{model}}{\left[\frac{C_v \ t}{d^2}\right]_{prototype}} \tag{4.17}$$

The left-hand side of Equation 4.17 must be equal to 1 if we wish to achieve the same degree of consolidation in the centrifuge model as that in the prototype. Also if the same soil is tested in the centrifuge as that in the prototype then we can assume that the coefficient of consolidation C_v is the same in the centrifuge model and the prototype. Taking into consideration these factors, Equation 4.17 reduces to:

$$\frac{\left[\frac{t}{d^2}\right]_{model}}{\left[\frac{t}{d^2}\right]_{prototype}} = 1$$

$$\frac{t_{model}}{t_{prototype}} = \frac{[d_{model}]^2}{[d_{prototype}]^2} \tag{4.18}$$

However, using the scaling law for length shown in Section 4.4 we know that the drainage paths in the centrifuge model and the prototype are related by a factor of N as shown in Equation 4.19.

$$\frac{d_{model}}{d_{prototype}} = \frac{1}{N} \tag{4.19}$$

Substituting Equation 4.19 into 4.18, we can get the scaling law for the time of consolidation as:

$$\frac{t_{model}}{t_{prototype}} = \frac{1}{N^2} \tag{4.20}$$

This scaling law for time of consolidation suggests the consolidation of soil in a centrifuge model occurs N^2 times faster compared to the prototype. This is quite simple to visualize physically. The distance the pore water has to travel in the centrifuge model has to be reduced by a factor of N compared to an equivalent prototype. The pressure head driving the seepage flow is the same in the prototype and the centrifuge model but is applied over a distance scaled down by a factor N. These two combine to result in speeding up the consolidation time in the centrifuge by N^2.

The scaling law of consolidation time is a significant advantage in the modelling of many soil mechanics problems. Let us consider an example. Let us suppose that a road embankment is to be constructed on a 10-m-thick soft clay deposit. Using the soil properties from the site, let us say we calculated that the clay will take 5 years to reach 95 percent consolidation. In such problems, the common question would be "what is the magnitude of settlement that the embankment will suffer due to the consolidation of the soft clay?" We could model this problem in a centrifuge at 100 *g* using a model embankment or equivalent surcharge placed on a soft clay layer that is 100 mm thick. In the centrifuge we could model the time of consolidation to reach the same 95 percent consolidation using the scaling law in Equation 4.20. So we could scale the consolidation time in this centrifuge test in 4.38 hours and measure the settlement of the embankment at this time. Using this we can predict the settlement that the prototype embankment will undergo using the scaling law in Equation 4.8.

The scaling law for time of consolidation makes the centrifuge an ideal tool to investigate consolidation problems. Of course, in the above example we were only considering a simple soft clay layer for which easy analytical solutions are available to predict the settlements. However, more complex geometries with multiple layers of varying soils or the presence of inclusions such as tunnels or pipelines or embankments constructed on soft clay layers can easily be modeled in a centrifuge using a relatively short period of centrifuge running time. With the help of such centrifuge tests, predictions can be made with regard to settlements that an embankment will undergo after 10 or 20 years of consolidation after construction. Thus, in a way we can imagine the centrifuge to be a geotechnical "time machine," in that reliable predictions on future behaviour of geotechnical structures can be made.

4.4.3 Scaling law for seepage velocity

The scaling law for time of diffusion processes is important in many problems. This was shown for the case of consolidation of a clay layer. In this context we may also consider the seepage velocity. Seepage velocity is sometimes called Darcy's velocity and it describes the overall speed of the fluid movement through a porous medium. Darcy's law states that the seepage velocity v is directly proportional to the hydraulic gradient driving the flow i and the constant of proportionality is called the hydraulic conductivity K. Thus, for laminar flow, Darcy's law can be written as:

$$v = Ki \tag{4.21}$$

Comparing the seepage velocity in the centrifuge model and the prototype, we can write:

$$\frac{v_{model}}{v_{prototype}} = \frac{[Ki]_{model}}{[Ki]_{prototype}} \tag{4.22}$$

If we are using the same soil in the centrifuge model and the prototype, then we should expect the same hydraulic conductivity K for the soil in both. The hydraulic gradient, however, is defined as the change in pressure head over a given distance, and this is different in the model and the prototype.

$$\frac{i_{model}}{i_{prototype}} = \frac{\left[\frac{dp}{ds}\right]_{model}}{\left[\frac{dp}{ds}\right]_{prototype}} \tag{4.23}$$

However, the pressure head in the centrifuge model and the prototype will be the same, although they occur over a much smaller distance in the centrifuge model. The distances scale by a factor N as seen in Equation 4.8.

$$\frac{i_{model}}{i_{prototype}} = \frac{[ds]_{prototype}}{[ds]_{model}} = N \tag{4.24}$$

This scaling law for hydraulic gradients suggests that the centrifuge models will have much higher hydraulic gradients than in the prototypes. If we substitute Equation 4.24 in Equation 4.22, we can get the scaling law for seepage velocity as:

$$\frac{v_{model}}{v_{prototype}} = \frac{1}{1} \times N = N \tag{4.25}$$

So the scaling law for seepage velocity is N, which indicates that the seepage velocity in the centrifuge model will be relatively high. This result is quite important as we must make sure that the flow is still laminar in the centrifuge models despite this increase in fluid velocity. Further, it is also important to distinguish the scaling law for seepage velocity from the scaling law for velocities in dynamic events. This will be considered separately in Section 4.4.4.

4.4.4 Force, work and energy

Considering the basic definition of Newton's second law of motion, a force F acting on a body of mass m will cause an acceleration a, such that:

$$F = m\,a \tag{4.26}$$

In the context of centrifuge modelling, let us consider the force that must be applied in a small-scale model that is subjected to a centrifugal acceleration of Ng.

$$\frac{F_{model}}{F_{prototype}} = \frac{[ma]_{model}}{[ma]_{prototype}} \tag{4.27}$$

Using the scaling laws derived earlier for mass and acceleration in Equations 4.13 and 4.14, we can write that:

$$\frac{F_{model}}{F_{prototype}} = \frac{1}{N^3} \times \frac{N}{1} = \frac{1}{N^2} \tag{4.28}$$

So we have derived a scaling law for force. Equation 4.28 suggests that the forces required in a centrifuge model are relatively small. For example, a 1 MN force in a 100 g centrifuge test scales down to be only 100 N. This is another advantage of a centrifuge test, as we can easily make actuators to load piles, retaining walls, etc., and forces that need to be applied by these actuators are relatively small, yet they simulate very large forces in the prototypes.

The basic definition of work done W is the product of a force F moving through a distance d. Using the scaling laws for force in Equation 4.28 and the distance in Equation 4.8, we can derive the scaling law for work done in a centrifuge model and in an equivalent prototype as:

$$\frac{W_{model}}{W_{prototype}} = \frac{[Fd]_{model}}{[Fd]_{prototype}} = \frac{1}{N^2} \times \frac{1}{N} = \frac{1}{N^3} \tag{4.29}$$

As with the force before, this scaling law for work done suggests that the work done in a centrifuge model is relatively small compared to that in a prototype. This is also advantageous for centrifuge modellers.

In physics, work done and energy spent are equivalent. So we should be able to recover the same scaling law for energy as we have derived for work done in Equation 4.29. Let us consider the definition of potential energy E_p normally expressed as energy lost by a falling mass m through a height h, as:

$$E_p = mgh \tag{4.30}$$

Using the scaling laws for mass in Equation 4.14, acceleration in Equation 4.13, and distance in Equation 4.8, we can write the scaling law for energy.

$$\frac{[E_p]_{model}}{[E_p]_{prototype}} = \frac{[mgh]_{model}}{[mgh]_{prototype}} = \frac{1}{N^3} \times \frac{N}{1} \times \frac{1}{N} = \frac{1}{N^3} \tag{4.31}$$

This confirms that we can recover the same scaling law of $1/N^3$, whether we consider work done or energy spent. The scaling law for energy can have profound implications for military applications. For example, effects of blasts can be modeled on buildings using quite small charges in a centrifuge model to simulate blasts with extensive energy release in reality. Let us say we want to model a blast event that has an energy release of 1 GJ. We could model this event in a 100 g centrifuge test as:

$$\frac{[E_p]_{model}}{[E_p]_{prototype}} = \frac{1 \times 10^9}{100^3} = 1 \times 10^3 \text{ J} \tag{4.32}$$

Thus, we only need to have a blast event in the centrifuge model that releases 1 kJ of energy. This is approximately equivalent to having an explosion from 0.24 grams of TNT. Thus centrifuge modelling can offer a very effective way of investigating the effects of explosions on civil engineering structures without the need to conduct these studies at full scale, which can be both expensive and damaging to the environment.

Equally, we could have considered kinetic energy in the centrifuge model and a prototype and again that should have the same scaling law for energy. By considering the kinetic energy E_k for the model and the prototype:

$$\frac{[E_k]_{model}}{[E_k]_{prototype}} = \frac{\frac{1}{2}[mv^2]_{model}}{\frac{1}{2}[mv^2]_{prototype}} = \frac{1}{N^3} \times 1 = \frac{1}{N^3} \tag{4.33}$$

This suggests that the scaling law for velocity is 1. It must be noted that this scaling law for velocity is valid for all dynamic events. However, the scaling law for diffusion velocities such as seepage of groundwater will be different.

The scaling laws derived thus far are shown in Table 4.1 for "slow" events and "dynamic" events, following Madabhushi (2004). In Chapter 14 we discuss in detail the differences between these two types of events. The scaling laws for the rest of the parameters can be derived following Schofield (1980, 1981).

Table 4.1 Scaling laws

	Parameter	*Scaling law model/prototype*	*Units*
General scaling laws (slow events)	Length	1/N	m
	Area	$1/N^2$	m^2
	Volume	$1/N^3$	m^3
	Mass	$1/N^3$	$Nm^{-1}s^2$
	Stress	1	Nm^{-2}
	Strain	1	–
	Force	$1/N^2$	N
	Bending moment	$1/N^3$	Nm
	Work	$1/N^3$	Nm
	Energy	$1/N^3$	J
	Seepage velocity	N	ms^{-1}
	Time (consolidation)	$1/N^2$	s
Dynamic events	Time (dynamic)	1/N	s
	Frequency	N	s^{-1}
	Displacement	1/N	m
	Velocity	1	ms^{-1}
	Acceleration/Acceleration due to gravity (g)	N	ms^{-2}

4.5 MODELLING OF MODELS

Modelling of models is a technique used in centrifuge modelling to ensure that the scaling laws derived earlier are valid. We resort to modelling of models when we are unsure of the scaling law for any given parameter.

We always begin centrifuge modelling with a *real structure* in mind. We simplify this into a *prototype* which we can incorporate in the physical scaled-down version we call the *centrifuge model*. We have already seen the specific relationship of these in Figure 4.3. Now if we imagine a prototype structure for a given problem, we can always scale it down using two different geometric scaling factors. This can be illustrated with the example of an embankment on soft clay.

Consider a 5-m-high embankment on a soft clay layer that is 4 m thick. The soft clay layer itself is underlain by bedrock. The two scale models are shown in Figure 4.4 as Model 1 and Model 2, tested at 75 *g* and 100 *g*,

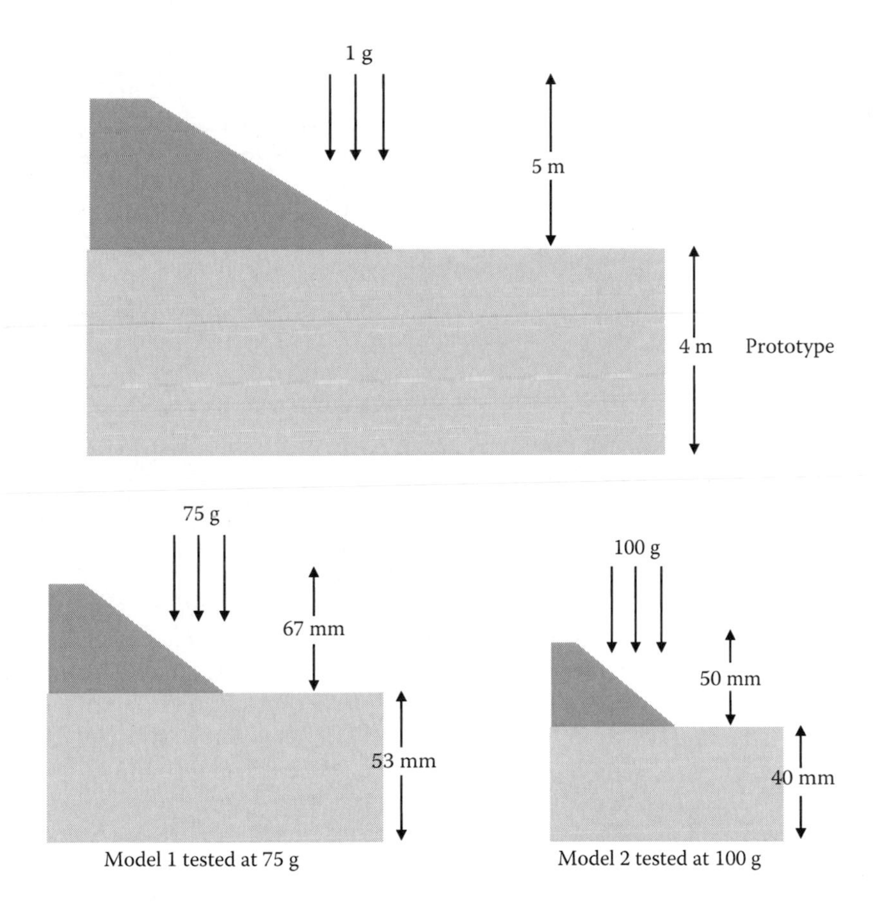

Figure 4.4 Modelling of models for an embankment on soft clay layer.

respectively. It can be easily seen that when these model dimensions are converted into prototype dimensions using the scaling law in Equation 4.8, we arrive at the same prototype dimensions (shown on the top of Figure 4.4).

If we consider any one parameter in these models, then it must scale up to give the same result. Let us say that we wish to consider the ultimate settlement of the embankment and we record this in both the centrifuge models at 75 *g* and 100 *g*, respectively. We then scale up the ultimate settlement recorded in each test to get the ultimate settlement of the prototype embankment; that is, ultimate settlements observed in both these models will be different absolute values but when scaled up by the respective *g* levels, they should yield the same value, which will be the ultimate settlement of the prototype structure. If this is the case then we can be sure that the scaling law for settlements is correct.

Similarly, the pore water pressures observed at homologous points in both the models must have the same values and must correspond to the pore water pressures expected in the prototype at an equivalent depth.

The technique of testing scaled models at different *g* levels with the aim of verifying the scaling laws is termed modelling of models. The modelling of models technique is used whenever we are trying to model a new physical phenomenon in the centrifuge and the scaling laws for one or more parameters cannot be derived easily.

4.5.1 Centrifuge testing of shallow foundations

Graduate and research students at the University of Cambridge carry out centrifuge tests on shallow foundations as part of the 5R5 module on "Advanced Experimental Methods in Geo-Mechanics." These are fairly simple centrifuge tests, which are carried out in an 850-mm-diameter steel tub that carries a horizontal layer of uniform sand of known relative density. Circular footings of different diameters are loaded vertically using a two-dimensional actuator (Haigh et al. 2010). This actuator is described in detail in Chapter 10. The results from these centrifuge tests are available on the website www.tc2teaching.org and the data can be downloaded freely (Madabhushi et al. 2010). The website also has more details about these centrifuge tests and is constantly updated with news of ongoing centrifuge tests and data from more recent centrifuge tests conducted by students as part of the 5R5 module. A full description of these tests is covered in Chapter 10. The vertical load applied and the settlements suffered by the model footing are monitored in these tests. Here this data will be used to demonstrate the principle of modelling of models.

The schematic view of the cross-section of the centrifuge model is presented in Figure 4.5. The footings are tested at two different *g* levels. The 80-mm-diameter footing is loaded when the centrifugal acceleration was 40 *g*. Similarly the 40-mm-diameter footing was loaded when the

Color Figure 5.1 A view of the Turner beam centrifuge at Cambridge.

Color Figure 5.3 *A view of the University of California, Davis Centrifuge Facility. (Photo courtesy of Dr. Dan Wilson, UC, Davis.)*

Color Figure 5.4 A view of the U.S. Army Corps of Engineers Centrifuge Facility at Vicksburg, Mississippi. (Photo courtesy of Ms. Wipawi Vanadit-Ellis.)

Color Figure 5.6 A view of the National Centrifuge Facility at the Indian Institute of Technology, Bombay. (Photo courtesy of Professor B.V.S. Viswanadham.)

Color Figure 5.8 Cambridge 2-m-diameter drum centrifuge with central tool table.

Color Figure 5.9 A view of the three-legged offshore jack-up rig tested on the Cambridge drum centrifuge.

Color Figure 5.10 A view of the Cambridge mini-drum centrifuge.

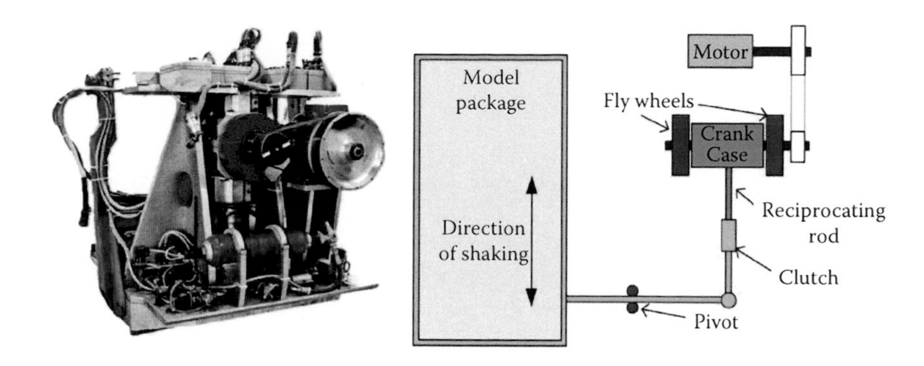

Color Figure 14.1 Schematic diagram and a view of the SAM earthquake actuator.

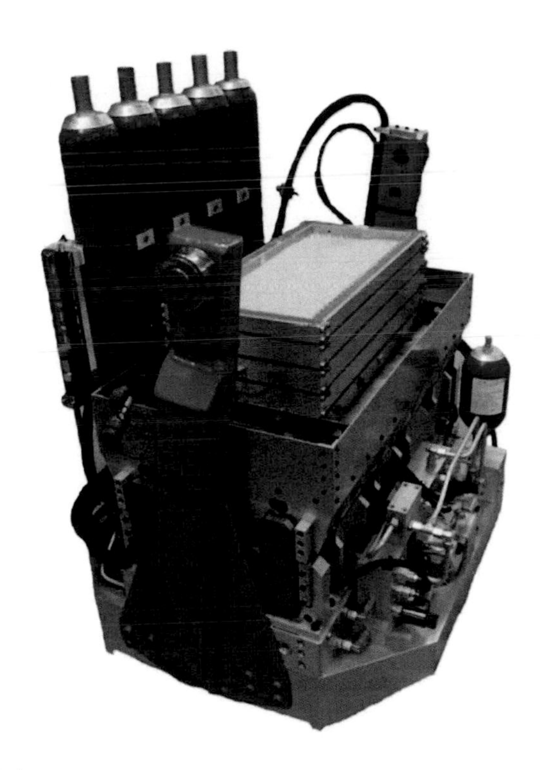

Color Figure 14.4 A view of the new servo-hydraulic earthquake actuator.

Color Figure 14.8 The IFSTTAR earthquake actuator. (From J.L. Chazelas, S.P.G. Madabhushi, and R. Phillips (2007). In: Proc. IV International Conference on Soil Dynamics and Earthquake Engineering. Thessalonica, Greece. With permission.)

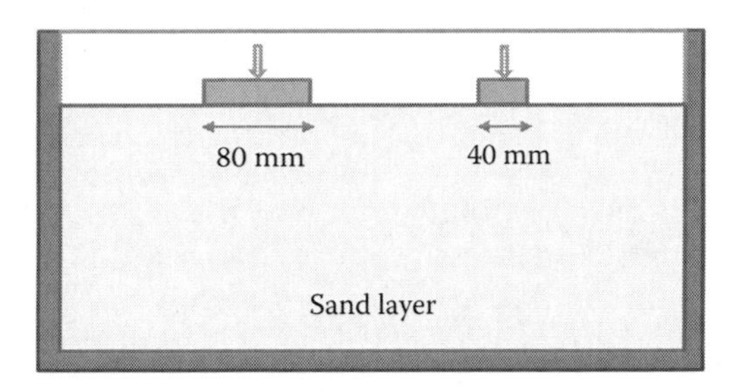

Figure 4.5 Cross-section of the centrifuge model of shallow foundation tests.

centrifugal acceleration was 80 *g*. Both these models represent a circular footing of 3.2 m diameter at the prototype scale.

4.5.2 Bearing capacity

The vertical load applied onto the circular footings can be converted into a bearing pressure by dividing the load at any given instant with the area of the footing being tested. Similarly the vertical settlement can be normalised with the diameter of the footing to obtain a representative strain in percent. In Figure 4.6 the evolution of the bearing pressure with increasing strain is plotted for both the centrifuge tests for the two footings tested at 40 *g* and 80 *g*, respectively. In this figure it can be seen that both centrifuge tests have resulted in very similar stress-strain behaviour, that is, the same soil stiffness is mobilised in each test.

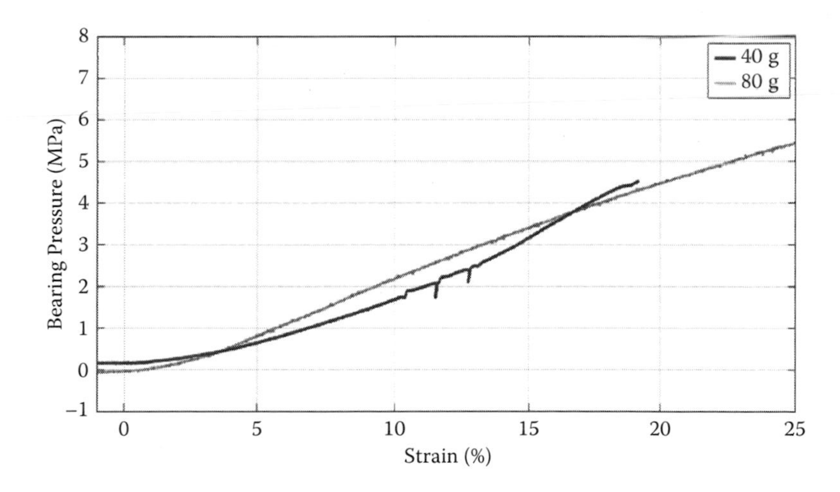

Figure 4.6 Modelling of models: comparison of a a 40 *g* and 80 *g* centrifuge test on shallow foundations.

This data serves as an example of how modelling of models can be used to confirm the scaling laws. In this case the bearing capacity of the shallow foundation can be taken to be at 10 percent of representative strain. Then both centrifuge tests suggest a nominal bearing capacity of about 2 MPa at this strain level. It must be pointed out that although both centrifuge tests give very similar results there are some differences in the data; that is, the solid and dotted lines in Figure 4.6 are not identical. This is to be expected due to experimental errors and small changes in density of soil below the foundations. More carefully constructed centrifuge models can yield even better results.

4.6 SUMMARY

In this chapter the fundamental principles of centrifuge modelling were explained. While centrifuge modelling is a powerful tool available to model complex problems in civil engineering, the underlying principles of centrifuge modelling are quite straightforward. The need for centrifuge modelling in dealing with scale models that contain nonlinear elastoplastic materials such as soils was explained. Testing of scale models of prototypes in the enhanced gravity field of a geotechnical centrifuge was considered. The definitions of a centrifuge model and the prototype and the relationships between these are presented. Scaling laws are required to relate the observed behaviour of the centrifuge model and the anticipated behaviour of a prototype. Derivation of simple scaling laws for stress and strain was presented along with the scaling laws for length, area, volume, etc., that can be directly induced. Scaling laws for consolidation time and seepage velocity were derived by considering the underlying principles of consolidation. Similarly scaling laws for work and energy could be derived based on simple physical principles.

Finally in this chapter the concept of modelling of models was presented. This is a powerful technique that can be used to develop and verify scaling laws for unknown parameters and carry out internal consistency checks for known scaling laws. An example of shallow foundation was considered and data from the centrifuge tests was shown to confirm the scaling laws.

4.7 EXAMPLES

1. A clay embankment in the field has a height of 20 m at the crest and is found on a dense sand layer of thickness 25 m, which is underlain by bedrock. Calculate the vertical stress at the mid-depth of the sand layer in the same vertical plane as the crest of the embankment. Assume suitable unit weights for the clay and dense sand. Now assume

that you have constructed a 1/100 scale model of the embankment and the sand layer. Repeat your calculation for the vertical stress at the mid-depth of your model sand layer. Compare the vertical stress in the model and the prototype. In this example consider the water table to be at the same level as the top of the sand layer.

2. Rework the vertical stress under the 1/100th scale model embankment described in problem 1 with the gravity increased by a factor of 100 and prove the principle of centrifuge modelling.
3. Let us say we have a centrifuge with radius r = 4.125 m. You have built a centrifuge model with a geometrical scaling factor of 100 and whose centroid is 0.125 m above the base. Determine the angular velocity of the centrifuge that will correctly recreate the prototype stresses and strains in your centrifuge model.
4. Consider a soft, 20-m-thick clay layer. This clay layer takes 12 years to reach 98 percent consolidation. Calculate how long it will take to consolidate a centrifuge model of this clay layer in:
   i) a 50 g centrifuge test and
   ii) a 100 g centrifuge test.
5. In a centrifuge model of the above problem, a thin sand seam of negligible thickness is introduced at the mid-depth of the clay layer. Calculate how this changes the consolidation times in each of the above centrifuge tests.
6. Design a centrifuge model that can be tested at 100 g, which represents a blast event that is equivalent to a 30 Megaton event.
7. Suggest suitable scale models at 50 g and 100 g to simulate the construction of a 10-m-high embankment on soft clay that is 12 m thick and is underlain by bedrock.
8. In the above problem, how would you estimate the ultimate settlement of the prototype embankment? How would this compare with the ultimate settlements recorded in each of the centrifuge models? You may assume suitable soil properties for soft clay.
9. Estimate the pore water pressure 2 hours after the start of each of the centrifuge tests in problem 7. Should these values be the same in both models? Explain the reasons for any deviations.

Chapter 5

Geotechnical centrifuges

Some design considerations

5.1 INTRODUCTION TO GEOTECHNICAL CENTRIFUGES

Chapter 2 established the need for centrifuge modelling in solving geotechnical problems. In Chapter 4 the basic principles of centrifuge modelling were presented along with simple scaling laws. From a modelling point of view we can define a geotechnical centrifuge as a machine that can impart radial centrifugal accelerations to small-scale physical models of a given class of problems being investigated, such as retaining walls or tunnels in difficult soil strata. As explained in Chapter 4, we will organize the centrifugal acceleration to be increased by the same factor as the one we used to scale the prototype to obtain our physical model.

Geotechnical centrifuges need to be robust machines that can carry hundreds of kilograms of soil models. The radius of these machines is usually between 1 m and 5 m. Soil is a bulky material and therefore we would anticipate that our physical models will have considerable mass. It is normal to call the physical model (excluding any swing platform) the payload. The geotechnical centrifuge design must be such that these machines can tolerate some amount of out-of-balance forces that may arise from inaccurate measurement of the mass of the model or due to change of mass of the soil model during centrifuge flight. The latter can happen, for example, due to evaporation of moisture from the soil or unexpected leakage of pore water from the soil during a long centrifuge test. In addition to mechanical robustness, geotechnical centrifuges often require additional services. For example, we may need electrical power on our centrifuge models to move actuators or to power instrumentation. Increasingly more and more sophisticated instruments are being used in centrifuge models to measure the soil behaviour during flight. For this we would require electrical slip rings that can carry either electrical power or data. Recent innovations in optical fiber technology have enabled modern centrifuges to have an optic fiber slip ring that enables data to be carried from the centrifuge to the control room via a high-speed optical switch that can support Ethernet connections. In addition to these, geotechnical centrifuges also require a certain number of fluid slip rings. These are used for movement of fluids from outside into the centrifuge during flight, allowing, for example, changes of water table in the soil models or to replenish moisture lost due to evaporation. Specially designed fluid slip rings can also be used to provide hydraulic power to actuators on the centrifuge models by using hydraulic fluids at high pressure.

Thus, geotechnical centrifuges are bespoke machines that are built specifically to test physical models of geotechnical structures. They require special systems such as electrical and fluid slip rings to support the experimental activities envisaged in any geotechnical problem. There are different types of geotechnical centrifuges available, such as beam centrifuges and drum centrifuges. These are explained next.

5.2 BEAM CENTRIFUGES

A beam centrifuge consists of horizontal structural beams that carry the payload at one end and a counterweight at the other end. The horizontal beams are all attached together either by welds or bolts and act as a single structural beam. There is usually a vertical shaft that supports the horizontal beams. The vertical shaft is mounted on bearings so that it is able to spin freely along with the horizontal beams. The power to drive the centrifuge is derived from electrical motors that are normally housed below the centrifuge chamber. Normal operations of a geotechnical centrifuge would require it to spin at different angular velocities in different experiments so that the geometric scaling factor matches the centrifugal acceleration as explained in Chapter 4. In order to achieve this speed variation a constant-speed alternating current (AC) motor is connected to the centrifuge through a magnetic coupling and a gear box. The speed of the centrifuge can be changed by varying the field strength of the magnetic coupling. Modern centrifuges tend to achieve the speed variation through an inverter that can vary the speed of rotation of a standard AC motor. Another alternative is to use a direct current (DC) motor, but this requires an elaborate electrical system that can convert a standard three-phase AC to provide DC.

In normal practice the soil model that constitutes the payload is placed on a swing platform at one end of the horizontal beam. The mass of the model is balanced by a counterweight that is placed on an identical swing platform on the other end of the beam. The counterweights are adjusted in each centrifuge test to balance the soil model being tested. Beam centrifuges can have further subtle differences. For example, the counterweight that balances the soil model can be mounted directly on the main horizontal beam, thus removing the need to have a swing platform for the counterweight. Such an arrangement comes with the added benefit of automatic counterbalancing systems that are able to monitor the out-of-balance forces in the beam centrifuge and move the counterweight accordingly to minimize these forces. On the other hand, having two identical swing platforms allows the users to switch the testing ends at which soil models are placed or use both ends of the centrifuge to test soil models of matching masses.

5.2.1 Examples of beam centrifuges

There are over 85 geotechnical centrifuges currently operating worldwide. A significant number of these are beam centrifuges. The older machines tend to have swing platforms on both ends while the newer machines have a single swing platform and a counterweight mounted on the main beam itself. A few examples of these machines are given below. A more exhaustive list of centrifuges worldwide can be found at the website www.tc204.org.

5.2.1.1 The Turner beam centrifuge at Cambridge

The Turner beam centrifuge was designed by Philip Turner and was built in the workshops of the Department of Engineering at the University of Cambridge. It became operational in the late 1970s. Schofield (1980) describes the specifications of this machine and the operation of this centrifuge in detail. It has a nominal diameter of 10 m and the payload capacity is 1 ton at an operational *g* level of 150 times earth's gravity. It is therefore classed as a 150 g-ton machine. A view of this centrifuge is presented in Figure 5.1.

The beam centrifuge is powered by a 260-kW, three-phase electric motor that is coupled to the beam centrifuge through a magnetic coupling and a beveled gear box that drives a vertical shaft passing through the center of the beam. Speed control is achieved by adjusting the field strength on the magnetic coupling. The two ends of this machine are color coded blue and red. Although both ends are nominally identical, in regular operations the red end carries the centrifuge models while the blue end carries the counter weight made from steel plates. During earthquake tests the ends

Figure 5.1 A view of the Turner beam centrifuge at Cambridge.

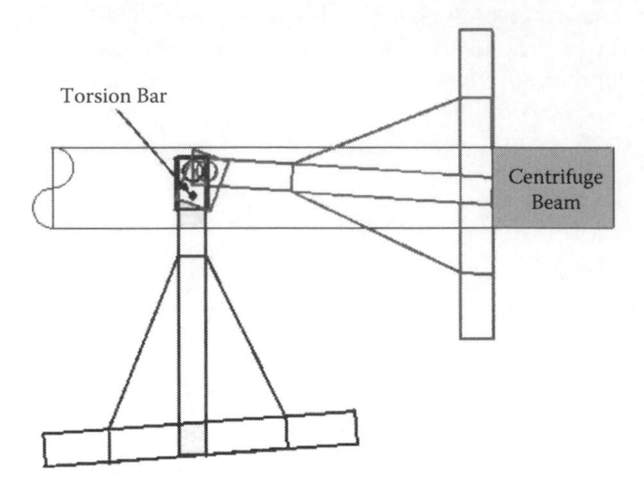

Figure 5.2 Torsion bar arrangement.

are reversed, that is, the blue end carries the earthquake actuator and the centrifuge model while the red end carries the counterweight.

The Turner beam centrifuge has many unique design features. The swing platforms on either end hang at a slight angle to the horizontal when the centrifuge is stationary. This enables easy swing-up once the centrifuge starts rotating. At about 8 *g* the swing platforms become vertical and make a positive contact with end stops. At this stage the surface of the swing platform, and hence any centrifuge model placed on it, is vertical. Another design feature of this machine is the torsion bar arrangement. The pivot of the swing is attached to a torsion bar that is offset. As a result when the *g* level increases beyond 8 *g*, the weight of the swing platform and the centrifuge model cause the torsion bar to twist elastically allowing the swing to sit back onto the face plates, as shown in Figure 5.2. Once this happens, the increasing weight of the swing platform and the centrifuge model with increasing *g* level is transferred into the main beam structure as axial tension forces. The pivots of the swings are not subjected to any further loads and therefore they can be small. This allows the pivots to be small and provides the centrifuge models with unrestricted head room to mount actuators or penetrometers on top to carry out geotechnical investigations.

5.2.1.2 The University of California, Davis, centrifuge

The centrifuge at the University of California, Davis (UC, Davis) has the largest radius and largest platform area of any geotechnical centrifuge in the United States. It is one of the top few in these categories in the world. The working radius of the centrifuge is a staggering 9.1 m. The centrifuge can carry 5 tons of payload and operate at 75 *g*, and therefore is classified

Figure 5.3 A view of the University of California, Davis Centrifuge Facility. (Photo courtesy of Dr. Dan Wilson, UC, Davis.)

as a 375 g-ton machine. A view of the UC, Davis centrifuge is shown in Figure 5.3. This centrifuge is one of the oldest centrifuges to have a single swing platform. A counter weight is placed on the main beam itself at the other end of the centrifuge.

In 1978, in response to a broad agency announcement from the National Science Foundation, UC, Davis collaborated with the National Air and Space Administration (NASA) on a successful $2..5 million proposal to construct and install the four-ton payload, 9.1-m radius, 300-*g* National Geotechnical Centrifuge at NASA's Ames Research Center. The centrifuge was first operated at NASA Ames in 1984. In 1987 the centrifuge was disassembled and hauled to UC, Davis. Using funds from Tyndall Air Force Base, Los Alamos National Laboratories, the National Science Foundation, and the University of California, the centrifuge was installed in an open-air pit two miles west of the UC, Davis campus. This permitted operations at centrifugal accelerations of 19 *g*. In 1989, an enclosure was constructed, increasing the peak achievable centrifuge acceleration to 50 *g* (Kutter et al., 1991). From 1990 to 1995, a large amount of effort was expended to fund, design, and develop a 1 m × 2 m shaking table for earthquake simulation mounted on the end of the centrifuge. In April 1995, the first proof tests of the shaker were performed at a centrifugal acceleration of 50 *g* Kutter et al. (1994). More recently the UC, Davis centrifuge has become one of the two centrifuges available as part of the George E, Brown Jr. Network for Earthquake Engineering Simulation (NEES) and was upgraded to 75 *g* operations.

5.2.1.3 U.S. Army Corps of Engineers centrifuge

The U.S. Army Corps of Engineers centrifuge is designed to carry a maximum payload of 8.8 tons on its 1.3 m × 1.3 m platform. The radius of the centrifuge arm from center of rotation to the platform base is 6.5 m. The range of gravity to which a model can be subjected is 1 to 350 *g*. A view of this centrifuge is presented in Figure 5.4. Further details of this centrifuge are described by Ledbetter (1991).

The envelope of performance for the centrifuge is as shown in Figure 5.5. Any combination of mass and gravity is permitted as long as it is within the performance envelope. At the maximum payload of 8.8 tons, the maximum gravity permitted would be 150 *g*. At the maximum gravity of 350 *g*, the maximum payload permitted would be 2 tons. This centrifuge has a capacity of 1320 g-tons. The performance envelope is presented in Figure 5.5. The U.S. Army Corps of Engineers centrifuge is presently the largest centrifuge anywhere in the world in terms of its g-ton capacity.

5.2.1.4 National Geotechnical Centrifuge Facility at Indian Institute of Technology, Bombay

Another example of a beam centrifuge is the Indian centrifuge called Sudarshan, named after the mythological wheel-shaped weapon of the Hindu god, Lord Vishnu. This centrifuge was developed indigenously within India and was installed at the Indian Institute of Technology, Bombay. The development of this centrifuge facility is described in detail by Chandrasekaran (2001).

Figure 5.4 A view of the U.S. Army Corps of Engineers Centrifuge Facility at Vicksburg, Mississippi. (Photo courtesy of Ms. Wipawi Vanadit-Ellis.)

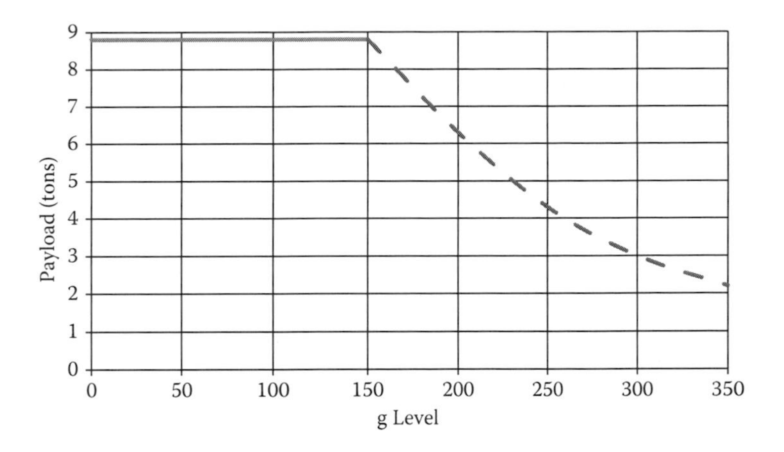

Figure 5.5 Performance envelope of the U.S. Army Corps of Engineers centrifuge.

This centrifuge has a working radius of 4.5 m and a maximum acceleration of 200 *g*. It can carry a maximum payload of 2.5 tons at 100 *g* or 0.75 tons at 200 *g*. The centrifuge is classed as a 250 g-ton machine. Unusually, this centrifuge is powered by a DC motor and therefore can accelerate to its top speed corresponding to 200 *g* in about 6 minutes. The centrifuge can accommodate soil models of 0.9 m × 1 m × 0.65 m. A view of this centrifuge is shown in Figure 5.6.

Figure 5.6 A view of the National Centrifuge Facility at the Indian Institute of Technology, Bombay. (Photo courtesy of Professor B.V.S. Viswanadham.)

5.2.2 Size implications

The examples given in the previous sections deal with relatively large-diameter centrifuges. However, it is possible to reduce the diameter of the centrifuge and increase the rotational speed of the centrifuge to achieve the same *g* level. This brings up the question: What is the "right" size for a geotechnical centrifuge? Unfortunately there is no unique answer to this question. To some extent the answer depends on the type of geotechnical problems that are being studied.

Let us illustrate this with the aid of an example. A typical slope stability problem is to determine a safe height h of a slope cut of clay. If the soil has a unit weight $\gamma = 20$ kN/m^3 and has undrained shear strength c_u, then typical limiting stability occurs when Taylor's stability number $c_u/\gamma h = 0.1$. Slopes with heights in the range of

$$100 \text{ m} > h > 4 \text{ m} \tag{5.1}$$

are safe in the ground with undrained shear strengths in the range of

$$200 > c_u > 8 \text{ kPa} \tag{5.2}$$

Consider Figure 5.7, which shows a prototype slope of height h and a centrifuge model slope of height h_m in a soil with undrained shear strength c_u. Let the radius to the center of rotation from the centrifuge model be r. Typically, the centrifuge model slope height h_m may be $r/10$. Let us say that the angular velocity of the centrifuge is $\dot{\theta}$ rad/s. Referring back to Chapter 4, the linear velocity v of the centrifuge model container will be

$$v = r\,\dot{\theta} \tag{5.3}$$

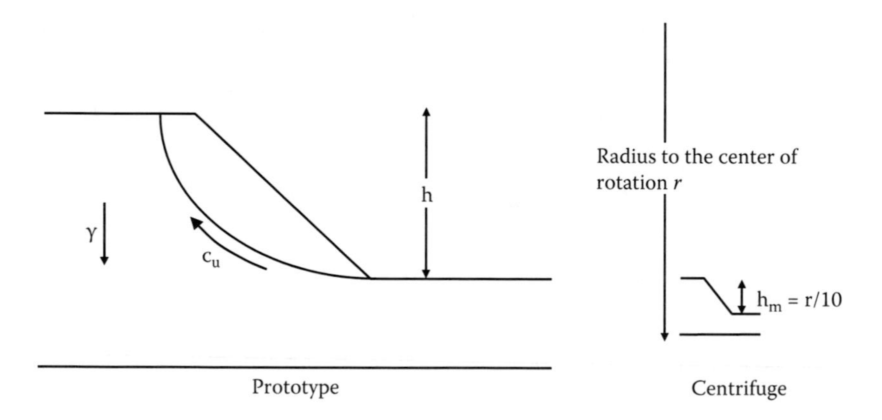

Figure 5.7 Slope stability.

and the acceleration of the model will be

$$a = Ng = r\,\dot{\theta}^2 = \frac{v^2}{r} \tag{5.4}$$

Considering the Taylor's stability number for the prototype slope, we can write:

$$\frac{c_u}{\gamma h} = 0.1 \tag{5.5}$$

Similarly, for the model slope in the centrifuge that is subjected to Ng, the Taylor's stability number may be written as:

$$\frac{c_u}{(N\gamma)\left(\frac{h}{N}\right)} = \frac{c_u}{\left(\gamma \frac{v^2}{rg}\right)\left(\frac{r}{10}\right)} = 0.1 \tag{5.6}$$

Substituting the value of unit weight of the soil and taking g as 10 m/s^2, we can write the linear velocity of the model slope as:

$$v = \sqrt{50\ c_u} \tag{5.7}$$

To make model slopes fail in flight we require that the velocities v be as listed in Table 5.1. The soil slopes considered below will approach limited conditions in flight when the linear velocity v reaches the values specified in the table.

Whatever the centrifuge radius r, we can run it faster at high speeds to make strong ground fail.

A centrifuge with a large radius allows a large model to be tested with many instruments. Further, other errors may be introduced by making the model small and flying it at high speeds, such as changes in gravity within the height of the model, curvature of the gravity field, etc. We will consider these in more detail in Chapter 6.

In the present discussion we have considered the slope stability number and derived an approximate relation between the undrained strength and the velocity with which we should fly the centrifuge to observe failure of the centrifuge model slope. In every geotechnical application in which we wish to employ centrifuge modelling, we must carefully consider the failure mechanism that might be expected and make reasonable back-of-the-envelope

Table 5.1 Limiting velocity to induce slope failure

Strength c_u (kPa)	8	72	200
Height h (m)	4	36	100
Velocity v (m/s)	20	60	100

calculations to ensure that the centrifuge being used is of appropriate size and that we are expecting the right things to be observed in the centrifuge test.

5.3 DRUM CENTRIFUGES

In the beam centrifuge we are only utilising a relatively small portion of the centrifuge pit. Considering the balanced beam centrifuge at Cambridge discussed in Section 5.2.1.1, it is possible to fly two centrifuge models on either end of the centrifuge. If one can imagine centrifuges with two beams welded perpendicular to one another we can fly four packages simultaneously. If we keep increasing the number of beams spinning on a common spindle, we will soon think of a drum in which the entire perimeter is used to carry the soil model. Drum centrifuges have the great advantage of having uniform soil models with identical soil parameters, which can be used to perform many tests (like jack-up rig platforms, spud can footings, etc.) and perform a parametric study. Further, parts of the circumference can be partitioned to make smaller soil models, in which case the drum centrifuge reverts to being a beam centrifuge.

5.3.1 Examples of drum centrifuges

The concept of drum centrifuges has taken off and there are many working examples of such centrifuges around the world. The most notable of these are the large drum (2 m diameter) and the mini-drum (0.85 m diameter) centrifuges at Cambridge, the drum centrifuge at ETH, Zurich, and the Centre for Offshore Foundation Systems (COFS) drum centrifuge at the University of Western Australia, Perth. At Cambridge, the use of the large 2 m centrifuge is discontinued but the mini-drum centrifuge has been recently upgraded to a diameter of 1.2 m and is very active in offshore pipeline research.

5.3.1.1 Two-meter-diameter drum centrifuge at Cambridge

The concept of the beam centrifuge was proposed by Professor A.N. Schofield and the first 2-m-diameter drum centrifuge was commissioned at Cambridge in the early 1980s. It was designed to reach a centrifugal acceleration of 500 *g*. The high *g* level allows for quick consolidation of clay samples in the drum, when required. A picture of this centrifuge is presented in Figure 5.8. In this figure the soil sample can be seen in a channel section that runs along the perimeter of the drum. The central space is used for a tool table and various tools such as actuators can be mounted on the tool table. In Figure 5.8 a single spud-can foundation is being loaded on the soil surface using an electrical actuator. Once the test has been completed at a site, the spud-can could be withdrawn by retracting the actuator, and

Figure 5.8 Cambridge 2-m-diameter drum centrifuge with central tool table.

then the tool table could be rotated relative to the outer ring channel and loaded at another site. In this way, parametric studies can be carried out at different locations along the perimeter with identical soil conditions. In fact, the length of the soil sample can be 0.6 km at prototype scale in a 100 *g* centrifuge test.

Many research projects have been conducted using the 2 m drum centrifuge. One other example is its use in establishing the stability and bearing capacity of an offshore jack-up platform. A detailed model of the three-legged jack-up foundation was fabricated as shown in Figure 5.9. As described earlier, this can be tested at different locations along the perimeter of the drum centrifuge with different combinations of vertical, horizontal, and moment loadings acting on the foundation system.

The drum centrifuge configuration, however, has some disadvantages. The size of this centrifuge means that it takes nearly 1.5 tons of soil in the channel section. Emptying the centrifuge after a test takes a long time. Another issue is making the soil samples. These needed to be prepared in flight, by pouring the soil from the tool table with special hoppers. In order to minimize the problems with the use of the large drum centrifuge, a new mini-drum centrifuge was designed and fabricated at Cambridge. This is described next.

5.3.1.2 Cambridge mini-drum centrifuge

The original mini-drum centrifuge had a diameter of 800 mm. The channel section was approximately 250 mm wide and 220 mm deep and could accommodate 175 kg of soil. A view of the mini-drum centrifuge is presented in Figure 5.10. The main feature of this centrifuge is that it can

Figure 5.9 A view of the three-legged offshore jack-up rig tested on the Cambridge drum centrifuge.

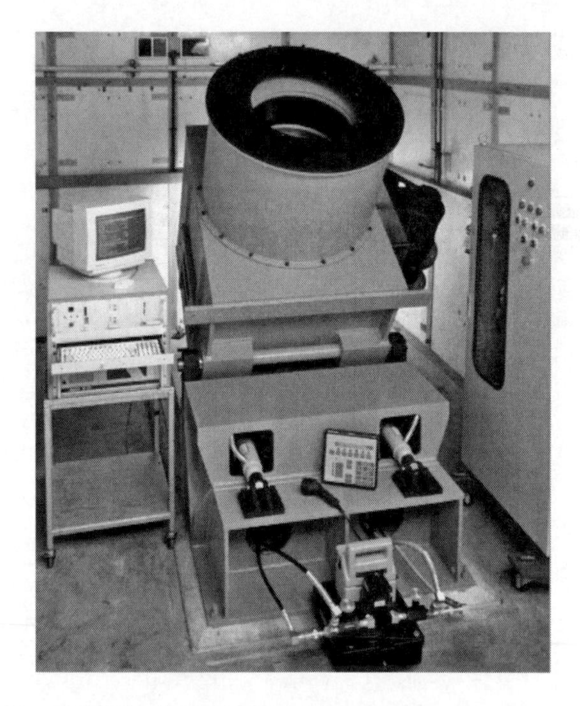

Figure 5.10 A view of the Cambridge mini-drum centrifuge.

rotate with its axis vertical or horizontal. A hydraulic pump is used to control the tilt and change it from a horizontal to a vertical configuration as seen in Figure 5.10. This facility allows model preparation by simply pouring the soil into a partitioned region of the channel with the centrifuge axis horizontal. Following this we can start spinning the centrifuge at low speed and then rotate the centrifuge into a vertical position and increase the *g* level to the required value. Another feature is the twin-axis system that allows the central tool table to be fixed or free to rotate relative to the channel section. This allows changing of the equipment that is mounted on the central tool table without the need for stopping the centrifuge and subjecting the soil to unload/reload cycles.

For example, at the start of a test the sand-pouring equipment can be mounted on the central tool table as shown in Figure 5.11. This is used to pour a uniform layer of sand into the channel section. The sand-pouring equipment is motorized so that it can move laterally and pour sand across the width of the channel. Once this is done, the central tool table can be stopped while the outer channel section continues to spin. A safety shield is lowered that covers the rotating channel section and protects the researchers accessing the central tool table. The researchers can then remove the sand-pouring equipment and place a new actuator that can apply radial loads onto foundations. Once the tool has been changed into a loading actuator, foundation loading tests can be conducted in different regions along the perimeter of the mini-drum.

The ability to change the equipment midway through a centrifuge test without the need to stop the machine allows for very versatile testing programs to be executed.

Figure 5.11 Sand-pouring kit mounted on the central turntable.

5.3.1.3 Drum centrifuge at ETH Zurich

Following the use of drum centrifuges at Cambridge, other centers around the world started to acquire drum centrifuges. The most prominent of these are discussed here briefly, although this list is not exhaustive.

ETH Zurich started operations on its 2.2-m-diameter drum centrifuge in 2001 (Springman et al., 2001). This is one of the two largest drum centrifuges anywhere in the world. It has a payload capacity of 2 tons and can reach a maximum of 440 *g*, making it an 880 g-ton machine. The channel section that carries the soil is 300 mm deep and 700 mm wide. A view of the centrifuge and the channel section with a sand layer is shown in Figure 5.12.

Many tools have been developed for this centrifuge. A spreader tool that can be used to level the sand bed is seen in Figure 5.12.

The ETH drum centrifuge has been used in many research projects, such as an investigation into the varved, lacustrine soft clay deposits in Switzerland, holistic soil-structure interaction, and advanced modelling of ground improvement techniques for soft clays. This centrifuge facility was one of the earliest to use the tactile pressure measuring devices to directly measure earth pressures in geotechnical problems.

5.3.1.4 Drum centrifuge at the University of Western Australia

The Centre for Offshore Geotechnics (COFS) at the University of Western Australia in Perth houses a 1.2-m-diameter drum centrifuge. This machine

Figure 5.12 A view of the geotechnical drum centrifuge at ETH Zurich. (Photo courtesy of Professor S.M. Springman.)

Figure 5.13 A view of the central turntable of the COFS drum centrifuge at the University of Western Australia. (Photo courtesy of Professor D.J. White.)

is capable of reaching a maximum of 500 *g*. A view of this machine is shown in Figure 5.13. Since its inception this machine has been very productive and has been used in the investigation of a number of problems related to offshore geotechnics. One of the main features of this machine is the precision control of the central turntable relative to the outer channel section. This allows for any actuator that is mounted on this table to move to a specific location relative to the outer channel. Once this is achieved a loading test using the actuator on the turntable can be conducted before moving to a new location. It is also possible to change the tools mounted on the turntable as described earlier.

5.3.2 Classes of experiments suitable for drum centrifuges

One of the advantages of a drum centrifuge is the relatively large soil models that can be tested. Virtually identical soil samples are present all along the perimeter of the channel of the drum allowing for multiple tests to be conducted at different locations. For example, foundations on layered soil strata can be tested for various combinations of horizontal, vertical, and moment loading. Because the soil strata along the perimeter are identical, results from a series of such tests in a single centrifuge test can yield invaluable information on the response of the foundations. Parametric studies of this nature can be carried out with relative ease in a drum centrifuge.

Another advantage of the drum centrifuge is that the model does not have any lateral boundaries as it is continuous around the channel. This can lend itself to model certain problems that are very difficult to model in a beam centrifuge. For example, propagation of a tsunami wave on the seabed can be modeled in a drum centrifuge. A displacing block arrangement can set off a solitary wave that can travel on the model seabed, that is, along the perimeter of the channel section. This wave can travel around the drum centrifuge several times until it is finally damped out. Wave gauges can be used to measure the wave height and travel times at different locations. Multiple waves can also be generated using a wave paddle arrangement, but such experiments require special arrangements to prevent the reflection of the waves when they reach the backend of the paddle.

5.4 SUMMARY

In this chapter the two types of geotechnical centrifuges, namely beam and drum centrifuges, are introduced. Examples of each type are presented. Choosing a particular type of centrifuge or even a centrifuge of a given size depends on the problem that is to be investigated. This was emphasized by considering an example of a slope stability problem. The size of the centrifuge both in terms of the payload capacity and the radius are important. Large-diameter machines can lead to minimizing some of the errors associated with centrifuge modelling. These aspects are considered in detail in Chapter 6.

Chapter 6

Errors and limitations in centrifuge modelling

6.1 INTRODUCTION

Like any modelling technique centrifuge modelling has some limitations. It is well known that the variation of gravity field within the centrifuge model, radial nature of the gravity field, and particle size effects all lead to some errors in centrifuge modelling. It is important to understand these limitations and ensure that the problem we are investigating is not adversely affected by them. In this chapter we consider some of these limitations in some detail. Where possible we will develop a quantitative measure for the errors involved. The success of centrifuge modelling worldwide in the last two decades, to a large extent relied on using centrifuge modelling to investigate relevant problems as well as using an appropriately sized centrifuge. It is widely acknowledged that centrifuge modelling is well suited for capturing the true failure mechanism in a given problem as explained in Chapters 1 and 2.

6.2 VARIATION IN GRAVITY FIELD

The earth's gravity field may be assumed to be constant with the height of a real structure in the field such as a building or an earth dam. In a similar vein we generally ignore the change of acceleration due to gravity with depth, for example in determining the effective stress under an earth dam and similar problems we encounter in the field.

However, in a centrifuge we are testing small-scaled models in a gravity field, which is a linear function of the distance from the center of rotation. We have already seen the following equation in Chapter 4, which links the g level required to the centrifugal acceleration.

$$Ng = r\,\dot{\theta}^2 \qquad (6.1)$$

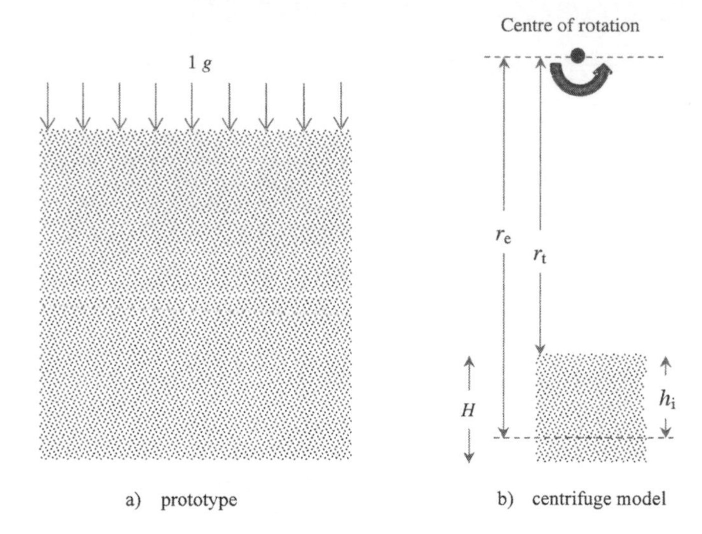

Figure 6.1 Prototype and centrifuge model of a horizontal soil layer.

Thus, the gravity field varies linearly with radius r from the center of rotation and as a square of the angular velocity $\dot{\theta}$ of the centrifuge. Consequently the error due to change in radius r in a centrifuge model may not be insignificant. Unlike the 1 – g gravity field experienced by our "real structures," which does not change with depth, the Ng gravity field in the centrifuge varies linearly with the depth of the model. We may choose to organize the angular velocity $\dot{\theta}$ of the centrifuge to give the required gravity field of Ng at a particular depth within the model. Consequently, the top regions of the centrifuge model above this depth will experience an under-stress (less than nominal g) and the lower regions will experience an over-stress (more than nominal g). In this section, we wish to quantify this error due to the variation in gravity field.

Let us consider a horizontal layer of soil of thickness H and its equivalent centrifuge model as shown in Figure 6.1. Let us say, referring to Figure 6.1(b), that the radius from the center of rotation to the top surface of the centrifuge model is r_t. The vertical stress at any depth z in the centrifuge model is given by

$$(\sigma_v)_{model} = \int_0^H \rho\dot{\theta}^2 (r_t + z)\,dz \tag{6.2}$$

$$= \rho\dot{\theta}^2 H\left(r_t + \frac{H}{2}\right) \tag{6.3}$$

where ρ is the mass density of the soil layer.

The vertical stress in the centrifuge model and the prototype are organized to be identical at a nominal radius of r_e by choosing the appropriate value of angular velocity $\dot{\theta}$ of the centrifuge. This radius r_e corresponds to a particular depth in the centrifuge model, h_i, and it can be seen in Figure 6.1(b) that

$$r_e = (r_t + h_i) \tag{6.4}$$

Also following Equation 6.1 we can now write:

$$Ng = r_e\dot{\theta}^2 \tag{6.5}$$

We can visualize the under-stress and over-stress in our horizontal soil layer with the aid of Figure 6.3. In this figure the linear increase of the vertical stress in the prototype horizontal soil layer is shown by the dashed line. In the equivalent centrifuge model the vertical stress will increase as a quadratic given by Equation 6.2 as indicated by the solid line in the figure. These two lines cross at a depth h_i below the surface of the centrifuge model so that the vertical stresses in the prototype and the centrifuge model match at this depth. Above this depth, the vertical stress in the centrifuge model falls below those in the model, that is, a region of under-stress. Similarly, the vertical stress in the centrifuge model exceeds the prototype stress below this depth and may be called a region of over-stress. The maximum under-stress occurs at a depth of $h_i/2$. Similarly, the maximum over-stress occurs at the base of the centrifuge model, that is, a depth of H. These are marked in Figure 6.2. It must be pointed out that in Figure 6.2 the magnitudes of under-stress and over-stress in the centrifuge model are exaggerated for clarity.

It is possible to minimize the error in stress distribution by knowing the relative magnitudes of the over-stress and the under-stress in the centrifuge model. The ratio of maximum under-stress R_{under} in the model, which occurs at the model depth of $h_i/2$, to the prototype stress at that depth can be written as:

$$R_{\text{under}} = \frac{\frac{h_i}{2}\rho g N - \frac{h_i}{2}\rho\dot{\theta}^2\left(r_t + \frac{h_i}{4}\right)}{\frac{h_i}{2}\rho g N} \tag{6.6}$$

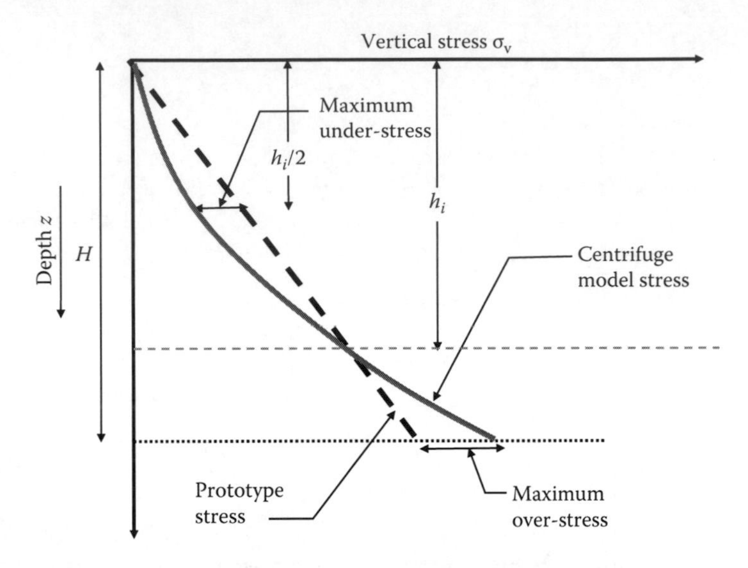

Figure 6.2 Under-stress and over-stress in a centrifuge model.

Substituting for Ng from Equation 6.5 into Equation 6.6, we get:

$$R_{\text{under}} = \frac{\frac{h_i}{2}\rho\dot{\theta}^2 r_e - \frac{h_i}{2}\rho\dot{\theta}^2\left[r_t + \frac{h_i}{2} - \frac{h_i}{4}\right]}{\frac{h_i}{2}\rho\dot{\theta}^2 r_e} \tag{6.7}$$

$$= \frac{r_e - \left[r_e - \frac{h_i}{4}\right]}{r_e} \tag{6.8}$$

$$R_{\text{under}} = \frac{h_i}{4r_e} \tag{6.9}$$

Similarly, we can calculate the over-stress ratio R_{over} for the over-stress which occurs at the base of the model H. This over-stress ratio R_{over} can be obtained as:

$$R_{\text{over}} = \frac{\rho\dot{\theta}^2 H\left(r_t + \frac{H}{2}\right) - \rho g N H}{\rho g N H} \tag{6.10}$$

Substituting for Ng, we get:

$$R_{\text{over}} = \frac{\rho\dot{\theta}^2 H\left(r_t + \frac{H}{2}\right) - \rho H\ r_e\dot{\theta}^2}{\rho H\ r_e\dot{\theta}^2} \tag{6.11}$$

$$R_{\text{over}} = \frac{\left(r_t + \frac{H}{2}\right) - \left(r_t + \frac{h_i}{2}\right)}{r_e} \tag{6.12}$$

$$R_{\text{over}} = \frac{H - h_i}{2.r_e} \tag{6.13}$$

One way of minimizing the error due to the variation of g level with depth within a centrifuge model is to ensure that the ratio of maximum under-stress and the maximum over-stress are equal. Thus, equating the two ratios R_{under} and R_{over} we get:

$$h_i = \frac{2}{3}H \tag{6.14}$$

Substituting this back into Equations 6.9 and 6.13, we can see that

$$R_{\text{under}} = R_{\text{over}} = \frac{H}{6.r_e} \tag{6.15}$$

and also that

$$r_e = r_t + \frac{2H}{3} \tag{6.16}$$

Using these equations, we can see that there is an exact correspondence in stress between the model and prototype at two-thirds model depth and that the effective radius must be measured from the central axis to one-third above the base of the model. The measure of the maximum error is given by the under-stress ratio and the over-stress ratio.

In the Cambridge beam centrifuge the models are typically 300 mm high. The radius to the swinging platform is 4125 mm giving us an effective radius of 3925 mm. This gives us an over-stress and under-stress error of 1.27 percent. This error is quite small and is acceptable within the context of any experimental testing. Similar calculation should be made for any centrifuge facility that is being used to ensure that the error due to variation in the gravity field with depth is small.

6.3 RADIAL GRAVITY FIELD

Apart from being constant, the earth's gravity field can be assumed to be parallel in the context of civil engineering structures. In other words the curvature of the earth is not considered within the scale of a civil engineering structure, even if it is a large dam or a long bridge. In the centrifuge, the centrifugal acceleration field which provides the high *g* is radial by definition emanating outward from the center of rotation of the centrifuge. Obviously, the error due to the radial gravity field varies from one centrifuge machine to another. The larger the diameter of the centrifuge the smaller this error gets. This point is clarified in Figure 6.3.

In this section let us try and quantify the error due to the radial gravity field. Let us consider two centrifuges, one with a large diameter and the

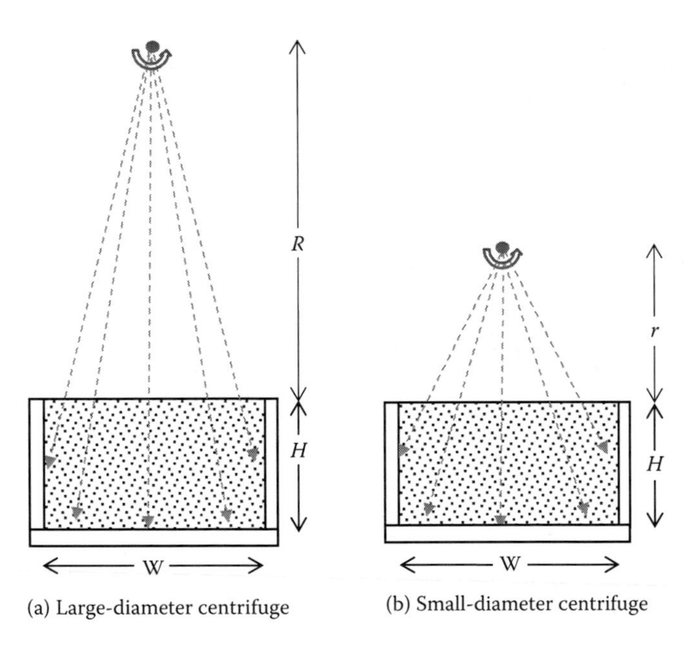

Figure 6.3 Radial gravity field in a large- and a small-diameter centrifuge.

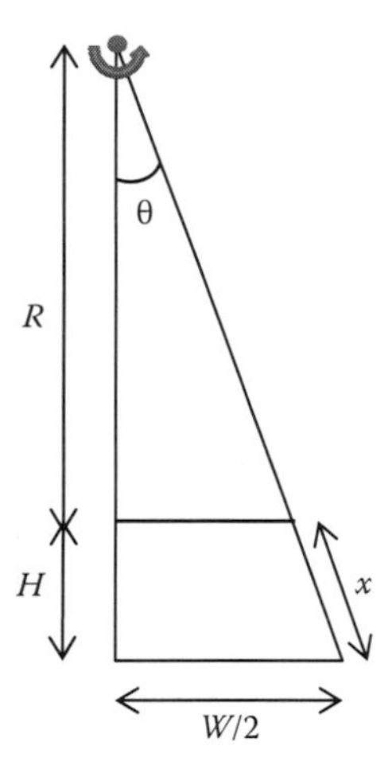

Figure 6.4 Quantification of radial error.

other with a small diameter. Let their radii be R and r, respectively. If we assume that we are flying the same centrifuge model in both centrifuges, we can see in Figure 6.3 the error introduced due to radial gravity fields. The error is a minimum at the middle of the model and increases as we go along to either edge of the centrifuge model. In a way this may not affect our results much, as we always consider the middle section of the model, for example, while you are placing a structure (such as a model tunnel or a model embankment).

It is obvious from Figure 6.3 that the error due to the radial gravity field is worse in a small-diameter centrifuge compared to a large-diameter one. In this section, we wish to quantify this error. One way to do this is by comparing the length of the inclined gravity ray at one edge of the model to the length of the vertical ray H (which will be the case in an ideal parallel gravity field).

In Figure 6.4, we can see that

$$\tan\theta \cong \theta = \frac{W}{2(R+H)} \tag{6.17}$$

and that

$$\cos\theta = \frac{H}{x} \tag{6.18}$$

The error due to the radial gravity field may be calculated as:

$$\varepsilon_{radial} = \left(\frac{x - H}{H}\right) \times 100 \tag{6.19}$$

For the Cambridge beam centrifuge the radius to the swinging platform is 4.125 m. If we take that the approximate height H of the model is 300 mm, and the width W of the model to 900 mm then we have:

$$R = 4.125 - 0.3 = 3.825 \text{ m}$$

giving us:

$$\text{Tan}\theta = \frac{0.9}{2 \times 4.125} \rightarrow \theta = 6.226° \quad \text{and}$$

$$x = \frac{H}{\cos\theta} = \frac{0.3}{\cos 6.226} = 0.30178$$

Using these values, we get:

$$\varepsilon_{radial} = \left(\frac{0.30178 - 0.3}{0.3}\right) \times 100 = 0.5933 \text{ \%}$$

For a reasonable-sized centrifuge such as the Cambridge machine and with a reasonable sized model, the error due the radial gravity field is ~0.6 percent. Obviously, this is a very small percentage. One of the parameters in the design of the centrifuge is to minimize the above error by choosing the diameter of the centrifuge and limiting the maximum dimensions of the model that can fly on it.

In some centrifuges such as the Schaevitz machine at the University of California, Davis with a radius of 1 m, the modellers have adopted to minimize this error by making their centrifuge models curved. Thus, a horizontal plane in a prototype is represented by a curved plane in the centrifuge model, the degree of curvature being obtained from the radius of the centrifuge. This, however, means complicated model-making procedures, which may be justified while using smaller centrifuges.

6.4 PARTICLE SIZE EFFECTS

In the centrifuge model we are scaling the prototype dimensions by the gravity scaling factor N. However, we are not scaling down the particle sizes by the same factor. One of the most common questions asked about the centrifuge modelling technique is why are the particles themselves not scaled down, that is, why is there no scaling down of the sizes of the particles? This is an appropriate question and needs be addressed carefully. One way to do this is to consider the following argument.

If we are trying to model gravel in the prototype with an average particle size of 40 mm in a 100 *g* centrifuge test, then the model gravel will have average size particles of 0.4 mm, that is, it should be modeled using relatively fine sand. If we continue the same argument for scaling down particles, then fine sand with average size particles of 0.4 mm in the prototype must be represented by fine clayey silts with a particle size of 0.004 mm in the model. In this argument we are of course ignoring the mineralogy of the soil particles and their affinity to water retention. This would of course be a mistake. Scaling down particles of soil in this way will lead to use of soils in the model that are totally different in their classification by normal soil mechanics definitions. With our knowledge of soil mechanics we can very definitely say the stress-strain behaviour of clay is different from that of fine sand, so it would be erroneous to use clay in centrifuge models to represent sands in the prototype. In order to capture the true behaviour of the soil in the prototype we need to use the same soil as the prototype, which has the same stress-strain relationship. This enables us to observe the correct deformations as the soil mobilises appropriate stiffness for the strains induced.

The result of the above argument is that, while we are scaling down the dimensions of the overall prototype, the soil particles themselves are not being scaled down in size. In other words we are still considering the soil to be a continuum. This assumption is true in soil mechanics in general. We are used to treating soil as a continuum although it is quite clearly particulate in its nature. As an example, we can consider the problem of bearing capacity of shallow foundations. Terazaghi's bearing capacity factors were derived based on the assumption that the soil is a continuum. In fact they have their origins in the plastic deformation of metals under indentation loads. The continuum assumption of soil is also true in the numerical analysis using the finite element method. The finite element mesh in most cases is much coarser than the soil particles but we assume that the continuum assumptions hold good when we carry out such analyses.

The continuum assumption is only valid within some limits. When the effect we want to see in the centrifuge model approaches the particle sizes, the continuum approach breaks down and under those circumstances the validity of centrifuge modelling ceases. For example, let us say that we are trying to test a pile embedded in a sand layer say for measuring the ultimate pile capacity. Let us say that the prototype pile has a diameter of 0.5 m and the average size of sand particles is about 0.5 mm. The ratio of pile diameter to average particle size is 1000. In a 50 *g* centrifuge test, the model pile diameter will be 10 mm and the same sand is used. The ratio of pile diameter to average particle size in this case will be about 20. If the pile diameter is at least 10 to 15 times the typical particle size of the sand grains then the continuum approximations hold good. The pile will have a sufficient number of particles of soil with which to interact. However, if we increase our *g* level to say 500 *g* and bring down the diameter of the pile to 1 mm, then the ratio of the pile diameter to average particle size will drop to 2. In other words the pile will only have two sand particles to interact with. This is akin to using a sewing needle embedded in the soil to model the pile. In this case obviously our centrifuge model will result in erroneous predictions. It is therefore important to ensure that the critical dimensions of insertions such as piles, tunnels, shafts, etc., into the soil are much larger than the average particle size of the soil in a centrifuge test.

Some researchers like Tatsuoka et al. (1990) believe that the ratio of particle size to the shear band width is the important parameter that determines the soil behaviour at failure. The size of the shear band may be important while considering the particle size effects in the centrifuge. The shear bands that form below a model foundation or next to a model retaining wall in a centrifuge must have a sufficient number of particles within the thickness of the shear band. This must be the same in the prototype as in the centrifuge model for the error due to particle size effects to be a minimum.

6.5 STRAIN RATE EFFECTS

In the centrifuge we are modelling the prototype stresses and strains accurately as seen in Chapter 3. However, it may not always be possible to model the strain rates observed in the prototype, in a centrifuge model. Consequently, any soil behaviour that is a strong function of the strain rate will be modeled incorrectly under those circumstances. Some good examples of this are the secondary consolidation in clays and creep phenomenon. In general the strain rate effects need to be investigated more thoroughly using centrifuge modelling techniques. Techniques like modelling of models explained in Chapter 4 can be used to better understand the phenomena where strain rate effects are important. Research is also needed to establish the scaling laws for such cases. Some effort in this direction has been recently directed at investigating foundations on clays that are subjected to time-dependent loading at different rates.

Study of the strain rate effects in a centrifuge is part of ongoing research and we anticipate building up our knowledge in modelling the strain rate dependent phenomena with time.

6.6 CORIOLIS ACCELERATIONS

In the centrifuge we have a rotating acceleration field. If we have in our centrifuge model any radial movement relative to this acceleration field then the moving part will experience Coriolis acceleration as explained in Chapter 3 (see Section 3.5).

The radial gravity field is caused by the inertial acceleration of the model that is given by:

$$a_{\text{inertial}} = r\dot{\theta}^2 = \dot{\theta}V \tag{6.20}$$

where V is the linear velocity at any instant of time. If we assume that the moving component of the centrifuge model (for example, during pouring of sand to build an embankment or movement of the centrifuge package under earthquake loading) has a velocity $\dot{r}$, then the Coriolis acceleration is given as:

$$a_{\text{coriolis}} = 2\dot{r}\dot{\theta} \tag{6.21}$$

We can measure the error due to Coriolis acceleration as a ratio between the inertial acceleration a_{inertial} and the Coriolis acceleration a_{coriolis}. Thus, this error is given by:

$$\varepsilon_{\text{coriolis}} = \frac{a_{\text{coriolis}}}{a_{\text{inertial}}} = \frac{2\dot{r}\dot{\theta}}{V\dot{\theta}} \times 100 = \frac{2\dot{r}}{V} \times 100 \tag{6.22}$$

If we assume that in any given centrifuge test, the velocity of a moving component in the centrifuge model is 5 percent of the centrifuge's linear velocity, then the error due to Coriolis acceleration is calculated as 10 percent, using Equation 6.22. This is a significant error compared to the previous errors such as those due to the variation in gravity field with depth or radial gravity field. Therefore, whenever a moving part is present in a centrifuge model, the error due to Coriolis acceleration must be taken into account.

Coriolis acceleration causes objects falling radially onto a centrifuge model to follow a curved path. Imagine that we wish to drop a steel ball

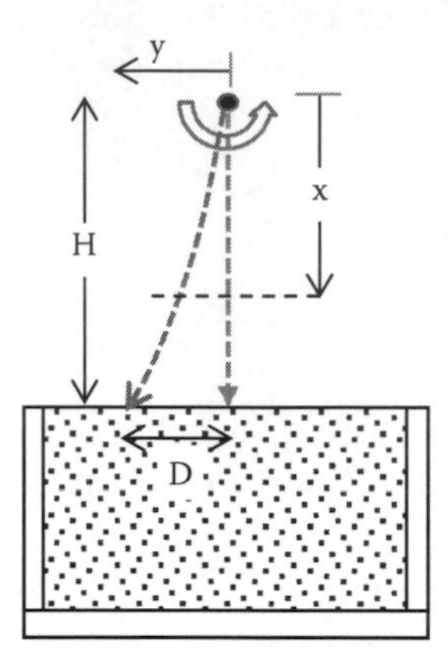

Figure 6.5 Lateral drift due to Coriolis acceleration.

onto a centrifuge model constituting a horizontal sand layer from a height H above the sand layer as shown in Figure 6.5.

Let us assume that the steel ball is initially at rest relative to the centrifuge model and that secondary effects such as air resistance are negligible. Let us suppose that the centrifuge is rotating at a constant angular velocity of $\dot{\theta}$. When the steel ball is released it will have no radial velocity but it will have the same angular velocity $\dot{\theta}$ as the centrifuge. As it falls towards the soil surface it will accelerate due to the centrifugal acceleration of $Ng = r\dot{\theta}^2$. In other words its velocity $\dot{r}$ increases until it hits the soil surface and is brought to rest. In this period it will suffer a Coriolis acceleration that acts on the steel ball in a direction perpendicular to its fall. This acceleration will pull it away from its initial straight path and cause it to fall along a curved path as shown in Figure 6.5. Further, the magnitude of the Coriolis acceleration increases as the linear velocity of the falling ball increases. Let us evaluate this by considering the situation when the steel ball has travelled a distance of x from its start position.

At $t = 0$, the steel ball starts at rest, that is, it has no velocity relative to the centrifuge model. When it falls by a distance x, its velocity in the vertical direction can be calculated using Newton's laws of motion as:

$$v_x = Ngt \tag{6.23}$$

Referring to Section 3.5, we can write the Coriolis accelerations as:

$$a_y = 2Ngt\dot{\theta} \tag{6.24}$$

We know that this Coriolis acceleration will act in the y direction on the steel ball. We can obtain the velocity of the steel ball in the y direction, by integrating Equation 6.24.

$$v_y = \int 2Ng\dot{\theta}t\,dt = Ng\dot{\theta}t^2 + C_1 \tag{6.25}$$

We know that the steel ball had no initial velocity relative to the centrifuge model, that is, $v_y = 0$ when $t = 0$. This implies that $C_1 = 0$.

We can obtain the lateral displacement of the steel ball by integrating Equation 6.25.

$$S_y = \int Ng\dot{\theta}t^2 dt = Ng\,\dot{\theta}\,\frac{t^3}{3} + C_2 \tag{6.26}$$

We know that the steel ball had no initial displacement relative to the centrifuge model, that is, $S_y = 0$ when $t = 0$. This implies that $C_2 = 0$. Therefore:

$$S_y = \frac{1}{3}Ng\dot{\theta}t^3 \tag{6.27}$$

The time t taken by the steel ball to fall a height H vertically can be calculated as:

$$H = \frac{1}{2}\,Ngt^2 \tag{6.28}$$

Substituting for t from Equation 6.28 into Equation 6.27 and noting from Figure 6.5 that

$$x = H \;\rightarrow\; S_y = D;$$

we get that

$$D = \frac{1}{3}Ng\,\dot{\theta}\left(\frac{2\,H}{Ng}\right)^{\frac{3}{2}} \tag{6.29}$$

Simplifying Equation 6.26 we can write the offset distance D due to Coriolis acceleration as:

$$D = \frac{2\sqrt{2}}{3}\,\dot{\theta}\left(\frac{H^{\frac{3}{2}}}{\sqrt{Ng}}\right) \tag{6.30}$$

Let us consider that we are dropping the steel ball that is initially at rest relative to the centrifuge model from a height of 300 mm above the soil surface in a 100 g test. If the radius of the centrifuge is 4.125 m (as in the case of the Cambridge beam centrifuge), the angular velocity $\dot{\theta}$ is calculated as:

$$\dot{\theta} = \sqrt{\frac{Ng}{r}} = \sqrt{\frac{100 \times 9.81}{4.125}} = 15.42 \text{ rad/s}$$

The offset distance D due to Coriolis acceleration for this case can be calculated as:

$$D = \frac{2\sqrt{2}}{3} \times 15.42 \times \left(\frac{0.300^{\frac{3}{2}}}{\sqrt{100 \times 9.81}}\right) = 0.07627 \text{ m or } 76.27 \text{ mm.}$$

This distance D can be a significant proportion of the width of the model container. It must be pointed out that the variation of g as the steel ball falls towards the soil surface has been neglected in this calculation. This assumption is acceptable as long as the height of fall H is small compared to the radius of the centrifuge.

Coriolis acceleration becomes important only when the velocity of the moving objects within the centrifuge model is high. This is usually the case for centrifuge models subjected to earthquake loading or where falling objects are involved, such as shooting projectiles into the soil models or raining sand onto clay for rapid construction of model embankments. Although Coriolis acceleration cannot be eliminated, the error due to this factor can be quantified. Further, in cases such as the raining sand problem, the hopper outlets can be offset to account for the shift in dropping zone due to Coriolis acceleration. Such practical measures can help minimize the effects of Coriolis acceleration.

6.7 SUMMARY

Each modelling technique has its own advantages and limitations. In this chapter we have discussed some of the limitation in centrifuge modelling. In some cases such as variation of gravity field within the soil model being tested or the effect of radial gravity field, we were able to quantify these errors and show that for a reasonable-sized centrifuge they are in fact quite small. In the case of particle size effects we have discussed the necessity to maintain a reasonable ratio between the soil particles and any inclusions we may place into the soil body such as piles, tunnels, etc. It was also recognized that where strain rate effects are important, further research is needed to establish the relevant scaling laws perhaps by resorting to modelling of models. Where moving objects are involved within a centrifuge package, Coriolis acceleration effects must be considered carefully. Using basic mechanics it is often possible to calculate the effect of Coriolis acceleration on the moving objects.

Chapter 7

Centrifuge equipment

7.1 MODEL CONTAINERS

The first bit of equipment that is required before embarking on a centrifuge testing program is a well-designed model container. Several aspects need to be considered before designing a new model container or choosing one from an existing stock. First, there are structural aspects of the container to consider. The model container must safely withstand the large pressures generated within it by both soil and water.

Using the scaling laws derived in Chapter 4, we can estimate these pressures. For example, a soil sample that is 400 mm high will exert a pressure of 600 kPa in a 100 *g* centrifuge test. In addition, if the soil sample is saturated we will get an additional hydrostatic pressure of 400 kPa at the base of the model container, making the total internal pressure acting on the model container 1 MPa. This pressure will create large stresses in the walls of the container and they must have sufficient thickness to carry these pressures safely. While designing model containers it is prudent to use large factors of safety against first yield; normally a factor of 5 is recommended. Further, the model container is often required to support head works that are mounted on the top, which can take the form of an actuator or gantries to support other instrumentation, electronic junction boxes, and so on. Again a mass of 100 kg supported as head works can exert a vertical load of 10 kN on the model container.

Second, the model containers are often required to be watertight especially when testing saturated samples. This requires all the hydraulic fittings, pipe work, and valve arrangements to be rated for high-pressure operations. Unwarranted leaks during a centrifuge flight can result in serious imbalance of the centrifuge. Often such leaks can lead to rapid drainage from the soil sample due to the high gravity. Loss of a mere 10 litres of water (for example, about 1 percent of the volume of the model container) can result in a 10-kg mass difference between the centrifuge model and the counterweight. In this regard, a movable counterweight system, which modern centrifuges are equipped with, can protect the centrifuge from damage due to imbalance. However, it is always better to prevent the water leaks in the first place.

Depending on the type of prototype being modeled, different centrifuge model containers can be used. These are considered in the following sections.

7.1.1 Containers for one-dimensional models

If the prototype being modeled or the event of interest is occurring in one direction only, we can use simple one-dimensional (1-D) model containers. Normally for many 1-D problems we have good analytical methods that give us good solutions. Consolidation of clay layers is a good example of this, where using the traditional Terazaghi's theory we can determine ultimate settlements. However, let us say we are interested in consolidation of clay layers interspersed with thick sand layers wherein the immediate settlement in sand layers and consolidation settlement in clay layers has to be determined. In such cases we may want to carry out centrifuge testing of the problem. A 1-D model container can be a very useful and straightforward choice for such a problem. One such model container is shown in Figure 7.1. The wall thickness for this container is 30 mm and the maximum sample height is 380 mm. A piston can fit into the model container that facilitates preconsolidation of clay samples prior to centrifuge testing. More details on preconsolidation of samples are presented in Section 7.2.2. Further base drainage is enabled by means of a ring main that runs along the perimeter.

One research project that was conducted with a 1-D model container was in the area of environmental geotechnics. Potter (1996) investigated the migration of contamination in a consolidating clay layer. The analytical solution for such a problem is quite complex as one needs to solve two sets of partial differential equations, one for consolidation and the

Figure 7.1 A one-dimensional model container with base drainage.

other for contaminant transport, in a coupled fashion. Centrifuge testing provided unique data for this complex problem against which numerical analyses were validated. For this problem use of a 1-D model container was adequate as a consolidating clay layer was created and a contaminant was introduced at the top surface of the model and its propagation with depth was studied.

7.1.2 Containers for two-dimensional plane strain models

Many civil engineering problems can be idealised as plane strain conditions. Examples of this include wide earth dams, tunnels, embankments, quay walls, or retaining walls, where the height of the structure is relatively small compared with its width. By assuming plane strain conditions we are of course saying that all the vertical cross-sections passing through the structure are identical; in other words, no lateral strains (ε_z) are permitted. In order to model the plane strain problems, specially designed two-dimensional (2-D) model containers are utilised in centrifuge modelling. The design of such plane strain model containers is governed by certain special requirements. The first requirement is on lateral strain. The strain in the lateral direction ε_z of the model container needs to be restricted to a very small value (typically <0.01 percent) to maintain plane strain conditions. This requirement usually controls the design of these model containers. The second requirement is that the friction between the soil in the centrifuge model and the container walls must be a minimum, so that the soil movement is not restricted by any boundary effects. Also the plane strain model containers usually have thick, Perspex sides to allow for in-flight viewing or high-resolution photography of the model cross-section in-flight. Apart from this the plane strain model containers also need to have high-pressure hydraulic fittings to allow pore fluid connections.

An example of a 2-D plane strain model container is presented in Figure 7.2. This package is called the King's package after the first collaborative project in the 1980s between King's College London and Cambridge University. In the case of the King's package the long walls of the package are stiffened by additional ribs, which reduce the lateral deflections, and the thickness of the Perspex window is about 40 mm. Figure 7.2 shows the centrifuge model consisting of a landfill liner made from Kaolin clay and a model geo-membrane. Also the model leachate retained within the landfill can be seen. Having a Perspex-sided model container is of immense value in this type of test as we can directly see any leakage of leachate into the ground outside of the landfill liner. Cameras mounted outside the model container show the leakage of the leachate in-flight and are used to detect the failure of the landfill liner system (Thusyanthan, Madabhushi, and Singh, 2006).

Figure 7.2 A two-dimensional model container with a centrifuge model of a landfill liner.

Similar model containers to the King's package have been developed at various centrifuge centers around the world. At Cambridge a more modern 2-D container named the Take box is being currently used. This model container allows for the construction of more versatile clay models. A variation of the 2-D model container is the so-called environmental chamber. This container is totally sealed from external air in the centrifuge chamber and the relative humidity, temperature, rainfall intensity, etc., within the model container can be controlled by the centrifuge modeler. More details of this model container are described by Take and Bolton (2011).

7.1.3 Containers for three-dimensional models

The most general class of problem in civil engineering is three-dimensional (3-D) in nature, because of either the geometry or loading. For example, a pile foundation that is loaded axially can be a 2-D axisymmetric problem. However, if lateral loads are applied to the pile then it turns into a 3-D problem. On the other hand, if we want to model a tunnel with a vertical shaft rising to the ground surface in a centrifuge, the model will naturally be 3-D. In order to capture the 3-D aspects of the problem cylindrical model containers are used. Different centrifuge centers use different diameters and heights for these containers. A typical size of these containers could be 850 mm in diameter and 425 mm in height (as used at the Schofield Centre). The wall thickness of these 850 mm tubs is normally about 19.05 mm (0.75 in.). Their construction involves having a tube section of the required height and welding a base and top rims to form supporting lips. Another way is to obtain a steel sheet of required thickness, rolling it into a tube and welding the edges. This method of manufacture would require some machining to remove excess welding and to ensure circularity of the tub.

The top and bottom rims also need welding. Additional lifting features are also added, such as spaces to connect eye-bolts. Similarly drainage connections at the base and provision for inserting instruments such as pore pressure transducers (PPTs) into the model container need to be well designed so that no leakage of fluid occurs at high gravities. Sometimes stand pipes are used to maintain the water in the model container at the desired level. These stand pipes are mounted on the bottom rim and are structurally attached to the tub.

In Figure 7.3(a) the problem of a monopile foundation used to support an offshore wind turbine is presented. The monopiles can be quite large, with lengths of up to 30–40 m and with diameters of 4–7 m. Driving of these monopiles and then subjecting them to lateral loads due to wind and waves has been the topic of recent research at Cambridge (Lau et al., 2013). Due to the presence of lateral loads the problem has to be modeled as a 3-D problem. In Figure 7.3(b) the model monopile that was driven into a Kaolin clay layer is presented. This test was carried out at 100 *g*, so the diameter of the soil block in prototype scale is 85 m. Similarly, the depth of the soil block is about 32.5 m. The monopile diameter tested was 4 m. The monopile was installed in-flight to mobilise the correct stresses at the base and sides of the monopile. Following this it was subjected to lateral load cycles and then to a push-over test to obtain the ultimate lateral capacity. The figure shows the pile after the centrifuge test. The figure also shows the instrumentation cables that connect the eight strain gauges that monitor bending moment in the pile under the action of the lateral loading.

Figure 7.3 (a) A monopile foundation offshore the United Kingdom. (Photo courtesy of Dong Energy.) (b) A centrifuge model of the monopile foundation driven into Kaolin clay.

7.2 MODEL PREPARATION TECHNIQUES

Centrifuge models can be made either by using undisturbed samples obtained from the field or by reconstituted soil samples in the laboratory. While the former are occasionally used for important projects where the cost of obtaining fairly large undisturbed soil samples is justified, more often reconstituted soil samples are used. One of the advantages of using reconstituted samples is that the soil properties and soil's stress history are well controlled. This means that models with very similar if not identical properties can be made repeatedly and used in parametric studies. Undisturbed samples on the other hand can be expensive to obtain and sample disturbance during removal and transportation is a worry. When clayey soils are present standard techniques have been developed for obtaining undisturbed samples required for triaxial testing. However, for centrifuge modelling much larger sample sizes are required and some thought is required for obtaining these. For sandy soils, techniques such as in situ freezing are often employed but again the sample sizes required means that these can be difficult and expensive to obtain. Despite these difficulties, it must be recognized that certain aspects of field samples such as fabric anisotropy can be replicated in a centrifuge model by using undisturbed samples.

Several model preparation techniques have been developed to create repeatable centrifuge models. In the following sections the techniques employed for sand and clay samples are presented.

7.2.1 Air pluviation of sand samples

While preparing sand samples air pluviation has been the most established technique. Uniformly graded silica sands are often used for sand models. The sand is placed in a hopper suspended above the model container. The nozzle of the hopper can be adjusted to change the size of the orifice. Also the drop height of the sand can be changed. Sand samples of different relative densities are obtained by controlling the drop height and the flow rate through the nozzle. In this manual method of sand sample preparation the researcher has to move the nozzle to cover the area of the model container. This means that there is some uncertainty in the actual relative density obtained and the uniformity of the sample. Further, researchers are required to wear protective face masks during sand pouring to prevent fine silica particles from entering their lungs. To eliminate these problems, an automatic sand pourer was developed at Cambridge. Figure 7.4 presents a schematic diagram of the sand pourer. The main aim of this device is to have a 3-D traverse of the hopper over the model container. Each of the traverse axes is computer-controlled. In addition to this there are high- and low-level indicators on the traversing hopper to show the level of the sand. Figure 7.5 shows the completed automatic sand pourer.

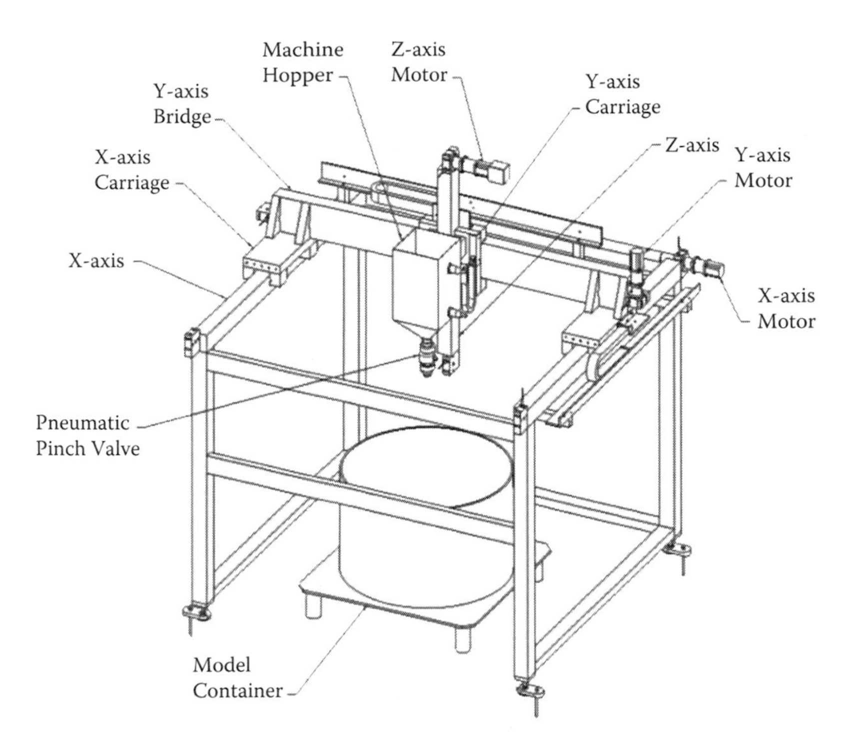

Figure 7.4 Schematic diagram of the automatic sand pourer.

Figure 7.5 A view of the automatic sand pourer.

The automatic sand pourer has been designed to stop when the sand in the hopper gets to the low level; then it goes to a stationery mother hopper and gets a feed of sand up to the high-level indicator. Following this it can recommence pouring of the sand from exactly the same spot where it stopped. The flow rate through the nozzle is controlled by a pneumatic valve. Sieves are also used below the nozzle to help spread the sand uniformly. Details of the fabrication of the automatic sand pourer are described by Madabhushi, Haigh, and Houghton (2006). The device was initially calibrated for preparing sand samples with relative densities of 60 percent or greater by Zhao et al. (2006). More recently Chian, Stringer, and Madabhushi (2010) have extended the range of the automatic sand pourer to prepare loose samples of relative density of 30 percent.

It is possible to create calibration charts to relate the nozzle diameter used and the drop height of the sand to the relative density of the sand layer. One such chart developed by Chian, Stringer, and Madabhushi (2010) is presented in Figure 7.6. Using this chart sand samples of desired relative density can be prepared. In order to maintain a constant drop height as the model is being constructed, the sand pourer incrementally moves the hopper after each layer is deposited.

It must be pointed out that the computer control allows soil samples of different shapes to be poured. The sand pourer can be stopped at desired locations to allow placement of instruments such as accelerometers, pore pressure transducers, and so on. Once the model has been poured to the required height, some leveling of the top surface is normally required.

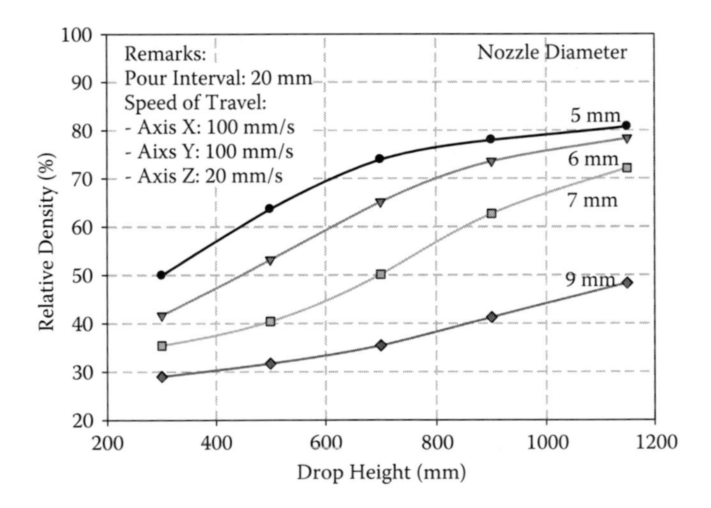

Figure 7.6 Calibration charts for the automatic sand pourer.

7.2.2 Saturation of sand models

Dry sand models prepared using the methods described in Section 7.2.1 may be required to be saturated. If the sand models have to be saturated with normal water, this can be achieved relatively easily by applying a vacuum at the sand surface and admitting the water from the base of the model in a controlled fashion. Application of a vacuum ensures that there will not be any air pockets that remain in the sand body and prevent full saturation. Alternatively permeable drains can be created at the far boundaries of the centrifuge model and water can be admitted into the drains, facilitating the saturation of the soil body.

However, in the case of earthquake geotechnical engineering, where the soil models are to be subjected to rapid loading from the earthquakes, the scaling laws require that the pore fluid viscosity must be scaled up so that the rate of excess pore pressure generation matches the rate of dissipation. The details of these scaling laws are discussed in Chapter 14.

In this section we will see the techniques used to saturate the sand models with highly viscous fluids with a viscosity of 50 or 100 times that of water. One option for the viscous pore fluid is to use silicone oil, which can be obtained in a range of viscosities. Another option is to prepare the viscous fluid by mixing methyl cellulose with water. The latter option is much more economical, especially when the centrifuge models are relatively large.

The model pore fluid used in the centrifuge tests can be prepared using deionized water. The water is mixed with methyl cellulose powder and benzoic acid. The quantity of these materials required can be estimated following Stewart, Chen, and Kutter (1998), who suggest that at a temperature of 20°C, the viscosity is related to the concentration C as a percentage of the entire solution by mass, as follows:

$$\nu_{20} = 6.92\ C^{2.54} \tag{7.1}$$

Based on Equation 7.1 the concentration C can be estimated to be 2.1784 percent to obtain a 50 cSt model pore fluid. They also suggest use of benzoic acid at 1 percent of the concentration of methyl cellulose. The purpose of benzoic acid is to prevent deterioration of the methyl cellulose with time due to bacterial action, and they found that 1 percent of C is sufficient to obtain stable solutions, in most cases.

When highly viscous fluids are used to saturate a sand model by admitting the fluid at the base, the upward hydraulic gradients need to be kept in check. Otherwise we can end up fluidizing the entire soil model. Historically high-viscosity silicone oil was used as the pore fluid and the flow of this into the model was controlled by needle valves. This required manual observation throughout the saturation process which could last 15 to 20 hours. More recently a computer-controlled saturation system called Cam-Sat was

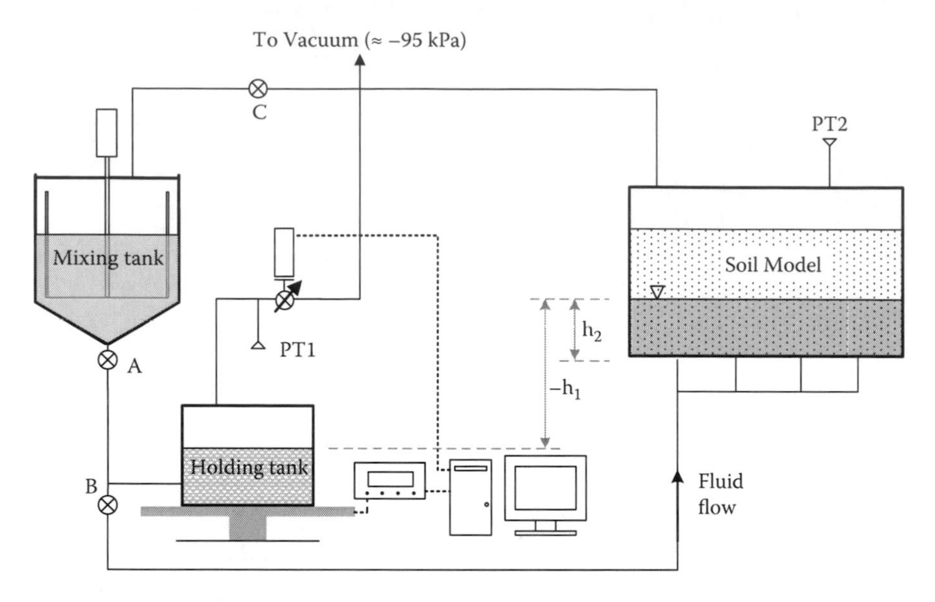

Figure 7.7 Schematic view of the Cam-Sat system.

proposed by Stringer and Madabhushi (2009). This system relies on real-time measurement of mass flux into the soil model and controls this by adjusting the differential vacuum between the viscous fluid container and the centrifuge model and controling the fluid flow rate through a valve. This system was further improved as described by Stringer and Madabhushi (2010). A schematic view of the system is presented in Figure 7.7.

This system is particularly useful when using methyl cellulose-based viscous fluids as there is no requirement to use needle valves, which can get clogged by fine particles. Further, the saturation process can run unattended and the system can even alert the researcher about the progress of saturation via email. The system was calibrated extensively for various sands and pore fluids of different viscosities as described by Stringer and Madabhushi (2009). The uniformity with which the Cam-Sat system allows the saturation of the sand model can be seen in Figure 7.8. In this figure the upward progression of the saturation front is visible in a centrifuge model. The saturation front is seen in this figure to be horizontal confirming that the soil model is saturating uniformly.

7.2.3 Consolidation of clay samples

In the absence of undisturbed samples, reconstituted clay samples can be prepared and tested in a centrifuge. These samples can have the advantage

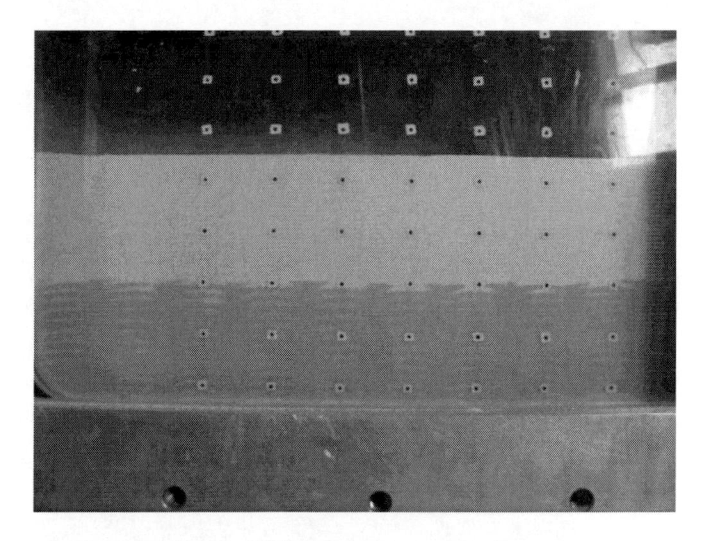

Figure 7.8 Upward progression of saturation front in a dry sand model.

of having a completely known and well-controlled stress history. The reconstituted clay samples are often made using kaolin clay. Two types of kaolin powder are normally available, namely E-grade and Speswhite kaolin. The former is slightly coarser and has higher hydraulic conductivity facilitating quicker consolidation. However, it is difficult to model truly undrained events using E-grade kaolin as some amount of drainage can take place during the loading phase (for example, of a footing on a clay layer). The properties of kaolin are very well established, with several researchers using in it their experimental programs.

It is possible to consolidate the clay layer in the centrifuge directly. However, this will take quite a long time (even with the scaling law for which time of consolidation is $1/N^2$). For example, for a 400-mm-thick clay layer to reach a normally consolidated state, we may have to run the centrifuge for 16 to 18 hours at 100 *g* depending on the properties of the clay. It would be much easier to consolidate the clay outside the centrifuge in a 1-D consolidometer which applies a load via a piston. Once the clay is consolidated and reaches its normally consolidated state, we can remove the load putting the clay into a temporary suction. The clay is then moved onto the centrifuge as quickly as possible and subjected to high gravities. The clay will then lose its suction and will return to its fully consolidated state. This process will only take a few hours. This way we can reduce the running time on the centrifuge.

The normal procedure employed in making reconstituted clay sample begins with mixing the kaolin powder with deionized water at 120 percent of its liquid limit. The mixing is carried out under a vacuum to remove

Figure 7.9 Mixing of kaolin clay under vacuum.

any air bubbles trapped in the clay. A view of the clay mixer is shown in Figure 7.9. This produces clay in a slurry form that is poured into the centrifuge model container, which may be an 850-mm tub or a plane strain model container discussed earlier. A view of the clay in the 850-mm tub is shown in Figure 7.10. Sometimes an extension is fitted to the 850-mm tub to allow consolidation of a thick clay layer. A piston is then placed on top of the clay slurry which is left to consolidate for a day or so, under the weight of the piston. Further consolidation is achieved by applying a load from the piston acting on the surface of the clay layer. The load on the piston can be created either by using high-pressure nitrogen or using a hydraulic pump. A system using the latter has been developed at Cambridge. The

Figure 7.10 Piston loaded onto kaolin clay.

Figure 7.11 A computer-controlled consolidation rig.

pressure created by the piston is computer-controlled so that any desired stress history can be applied to the soil. A view of the computer-controlled consolidation rig is shown in Figure 7.11. Another advantage of using this system is that the whole consolidation procedure can be logged continuously and can run virtually unattended.

Samples consolidated in the above fashion will give a strength profile for the clay that increases with depth. If different strength profiles are required to simulate specific site conditions, then a downward hydraulic gradient method is employed during consolidation.

7.2.4 Downward hydraulic gradient method of consolidation

The clay slurry prepared from mixing kaolin powder at 120 percent of its liquid limit is placed in a cylindrical tub as shown in Figure 7.8. A permeable membrane is placed on the top of the slurry and a sealed, impermeable piston is placed on the top of the membrane. It is possible to inject the water under pressure into the permeable membrane. There is a permeable drain at the bottom of the clay slurry through which water can be drained. The piston is jacked down in increments via high pressure created by the computer-controlled hydraulic pump. The pressure on the clay slurry is initially the weight of the piston plus the load from the pneumatic ram. Later this is increased in increments as the clay consolidates.

If the piston were to be placed on a clay slurry the consolidation pressure is uniform with depth in the model container. However, we require clay samples in which the consolidation pressure increases with depth, that is, the consolidation pressure versus depth diagram must be trapezoidal. This may be the natural case if the clay is consolidating in the field. On the lab floor, in order to increase the consolidation pressure with depth, we induce downward hydraulic gradients causing a downward seepage flow through the clay.

Normally the effective consolidation pressure at the surface of the sample needs to be smaller than at the base. The consolidation pressure required at the base is applied at the surface via the piston and the clay sample is drained to atmosphere at the base.

The consolidation pressure can be applied in two ways. The full consolidation pressure required at the base is applied via the impermeable piston. The effective consolidation pressure acting at the surface is reduced by the water pressure acting between the top piston and the drainage layer. This gives more consolidation pressure at the base compared to the top. Alternatively, we can use an unsealed piston on the top which applies the effective consolidation pressure at the top. We then pressurize the water around the piston to the pressure required at the base of the clay model.

Using this technique any required consolidation profile can be applied to the clay sample to reproduce most preconsolidation profiles, including the normally consolidated clays. By reversing the technique and applying fluid pressure at the base of the clay sample, resulting in upward hydraulic gradient consolidation, we can achieve over-consolidated surface crusts.

In employing the downward hydraulic gradient method of consolidation the following points must be kept in mind:

1. The pore pressure injected must not be more than 90 percent of the total stress applied via the piston (including the self-weight of the piston).
2. If the above condition is not met, the clay can crack and may result in nonuniform hydraulic gradients. At the beginning of the downward hydraulic gradient consolidation water flows out from the base drain but no water flows in at the top. As the parabolic isochrone reaches the top the downward hydraulic gradient is established and the clay slowly comes into equilibrium, with effective stress increasing uniformly with depth.

7.3 IN-FLIGHT ACTUATORS

Depending on the type of centrifuge test we may require the use of in-flight actuators. For example, if we wish to drive a pile during a centrifuge test or subject a pile to a lateral load we would require an in-flight actuator. By definition an in-flight actuator needs to be designed and fabricated to work in the high-gravity environment of a centrifuge. However, referring back to the scaling laws in Chapter 4, relatively small forces can represent very large prototype forces. An actuator that can apply a force of 1 kN in a 100 *g* centrifuge represents a prototype force of 10 MN.

Most actuators use electric motors as the prime movers. They can be either stepper motors or servo-motors. The movement of the actuator is controlled using these motors. Normally we can prescribe the required displacement or force to the actuators.

7.3.1 One-dimensional actuator

By definition a 1-D actuator is designed to deliver movement in one direction, for example, to push a pile or a miniature cone penetrometer (CPT) into a soil bed during flight. The simplest of these actuators could be a hydraulic jack which moves when a pressure is applied by turning on a solenoid valve. A more sophisticated 1-D actuator could have a stepper motor or a servo-motor so that the movement of the 1-D actuator is more controlled. A device such as this was fabricated at the Schofield Centre in Cambridge and is shown in Figure 7.12. This 1-D actuator can be mounted on an 850-mm tub or a plane strain model container and can deliver a maximum force of 10 kN. Its actuation can be controlled to deliver a requirement movement (displacement controlled) or a given force (force controlled). For the latter we use a load cell at the connection between the actuator and the object being driven into the soil such as a model pile. The movement of the actuator is controlled to apply the desired force at the top of the pile, thus driving the pile under force-controlled conditions.

7.3.2 Multi-axis actuators

Following the success of the 1-D actuator described in Section 7.3.1, a 2-D version was developed by Lam et al. (2012). This actuator has the ability to traverse in the horizontal and vertical directions. It can apply maximum forces of 10 kN and 5 kN in the vertical and horizontal directions, respectively. Since its inception the 2-D actuator has been widely used in many research projects at Cambridge. A view of this 2-D actuator is shown in Figure 7.13.

The 2-D actuator is controlled by a pair of aero-tech drives that are mounted on the centrifuge. These provide independent control of the x- and

Figure 7.12 A one-dimensional actuator.

y-axis movements. Further the 2-D actuator can be force- or displacement-controlled independently on each of the axis. The horizontal and vertical traverses are limited by having micro-switches which disengage the drive when the actuator reaches the end limits. Similarly the servo-motors that drive the actuator have brakes on them, so that in case of loss of power to the drives, the actuator is stopped in that position.

The 2-D actuator is a very versatile piece of equipment. It can be used to drive piles at different locations by inserting them at a given location or withdrawing and moving them to another site. This allows the researcher to conduct multiple tests in a single soil model (Li, 2010; Shepley, 2013). It can also be programmed to carry out soil excavations behind retaining walls (Elshafie, 2008; Lam et al., 2012).

In addition to the actuators described here, there are 4-D actuators that are used at the IFFSTAR centrifuge in France, the University of California, Davis and RPI (Rensselaer Polytechnic Institute) centrifuge centers in the United States, and the KAIST centrifuge center in Korea. These devices have the flexibility to move independently in x, y, and z directions and also can impart a rotation, thereby making them 4-D actuators. Ingenious ways have been developed to hold and change tools in-flight so that the same actuator can be used to carry out pile driving or miniature CPT tests.

Figure 7.13 A two-dimensional actuator.

7.4 IN-FLIGHT SOIL CHARACTERIZATION TECHNIQUES

It is important to have capabilities at any centrifuge facility to carry out in-flight soil characterization. As explained earlier in Section 7.2, most soil models are prepared at 1 *g*, whether they are sandy soils or clays. It is therefore important to characterize the soils in-flight to:

1. Check their uniformity and detect any layering that may occur during dry pluviation especially due to stop-and-start procedures to place instruments within the model.
2. Have quantitative evaluation of the soil strength and stiffness variation with depth.

For these reasons, it is important to have in-flight soil characterization much the same way as we would in the field by carrying out SPT (Standard Penetration Test) or CPT tests or using down hole/SASW (Spectral Analysis of Surface Waved) arrays and other geophysical methods. In centrifuge modelling miniature equivalents of these methods have been developed and these are discussed next.

7.4.1 Miniature cone penetrometer

Miniature CPT tests have been conducted in both clays and sands in a centrifuge for a long time. These tests are particularly straightforward in a centrifuge as the distance through which penetration must occur is relatively small (about 300 to 400 mm). The diameter of the cone penetrometer can be about 10 mm. The device can be pushed into the soil by using either a 1-D or 2-D actuator or even a simple pneumatic jack. If the rate of penetration needs to be controlled then hydraulic fluid is used below the actuator piston which is drained at the given rate through a needle valve arrangement. This reduces the penetration rate to the required value. Load cells at the tip of the CPT and the head can measure the tip resistance and the overall force of penetration. The difference of these will give shaft resistance. More sophisticated miniature CPTs can have a pore pressure transducer within the porous tip, to measure the excess pore pressure generated during the cone penetration. Also special sleeve arrangements can be designed to measure the shaft friction close to the base of the CPT. A detailed sketch of a miniature CPT is shown in Figure 7.14. This device also has a micro-electrical-mechanical systems (MEMS) accelerometer so that it can be used as a seismic CPT. This will be explained in Section 7.4.3.

Bolton et al. (1999) describe a European project that was carried out to standardize the use of CPT devices at various centrifuge centers in Europe. This was a calibration exercise that led to testing of nominally identical soil samples at different centrifuge centers in Europe and measure cone

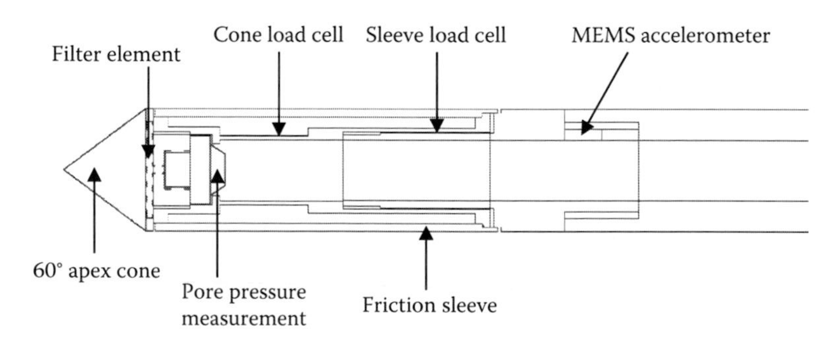

Figure 7.14 Components of a miniature cone penetrometer.

penetration resistance in each case. Teymur and Madabhushi (2003) used simplified CPT setups in dynamic centrifuge modelling to compare the cone penetration resistance at various locations within the model. A variation of miniature CPT is the T bar and ball penetrometer setups, where either a cylinder attached perpendicularly to the shaft or a ball at the end of the shaft are pushed into the soil. These devices are particularly useful in clay soils as the undrained strength can be estimated based on closed form solutions for these shapes obtained from plasticity theories.

7.4.2 Air hammer setup

Another method of characterizing the soil models in a centrifuge test is by measuring the shear wave velocity v_s. Knowing the shear wave velocity, the small strain shear modulus G_o can be obtained using the following equation:

$$v_s = \sqrt{\frac{G_o}{\rho}} \tag{7.2}$$

where ρ is the density of the soil.

Unlike in the field, it is relatively straightforward to create shear waves at the base of a centrifuge model. We can use an air hammer to create the horizontally polarized, vertically propagating S_h waves. An air hammer is a small brass tube with a metal pellet inside it. By applying high-pressure air on alternative ends, the pellet is made to accelerate and strike the end of the tube causing a shear wave to be set up at the base of the model which will propagate upward towards the soil surface as shown in Figure 7.15. The outside of the air hammer has glued sand to improve its coupling with the surrounding soil body. Accelerometers placed at different but known elevations will record the arrival times of the shear waves from which the shear

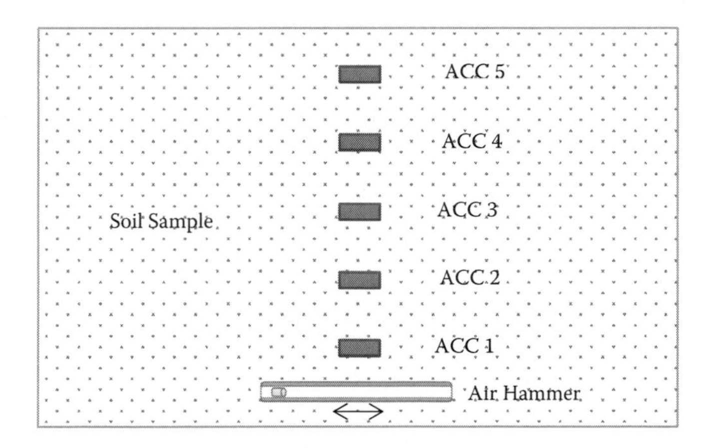

Figure 7.15 An air hammer setup.

wave velocity between adjacent accelerometers can be determined. More details of this device are described by Ghosh and Madabhushi (2002).

This arrangement is used to determine the shear wave velocity and hence the small strain shear modulus G_o in a wide variety of soil types, including saturated sands and clays. It must be pointed out that for the success of the air hammer setup it is imperative to have a very fast data acquisition system.

7.4.3 Seismic cone penetrometer

The seismic CPT combines the above two techniques of soil characterization. The miniature CPT described in Section 7.4.1 is fitted with a MEMS accelerometer as shown in Figure 7.14. An air hammer is also placed at the base of the model as shown in Figure 7.15. At the desired time during the flight, the CPT is pushed into the centrifuge model. The cone and sleeve resistances are measured normally giving us a continuous measurement of soil strength with depth. In addition, the air hammer is fired continuously creating the shear waves which are picked up by the MEMS accelerometer. As the penetration depth of the cone tip is continuously measured, the location of this accelerometer from the air hammer is known. Using this the small strain shear modulus can be calculated using Equation 7.2. Thus, using the seismic CPT we can get a continuous measure of the variation of both strength and stiffness with depth of the soil model in the centrifuge. The use of seismic CPT in the centrifuge is relatively new but it was previously used in the field as early as the 1990s by Piccoli and Smits (1991). The availability of miniature accelerometers that can fit into the body of the CPT made the fabrication of the seismic CPT possible. An example of the use of seismic CPT in normally consolidated and lightly over-consolidated clay samples of different undrained strengths is shown in Figure 7.16. In Figure 7.17 the shear stiffness measured by the arrival times of shear waves is also shown, for the case of a soil model with 50-kPa nominal undrained strength.

Theoretically, the small strain shear modulus was related to the effective confining pressure and the OCR (Over Consolidation Ratio) by Viggiani and Atkinson (1995) as:

$$\frac{G_o}{p_r} = 1964\left(\frac{p'_o}{p_r}\right)^{0.65} OCR^{0.2} \tag{7.3}$$

where p'_o is the effective confining pressure normalised to a reference pressure, normally taken as the atmospheric pressure. The experimental results can be compared to the theoretical predictions given by Equation 7.3. In Figure 7.18 the small strain shear modulus G_o obtained from this equation is compared to the experimental values obtained using an SCPT (Seismic cone penetrometer Test) device and the air hammer setup. These values compare reasonably well and all show an increase in shear stiffness with depth, although there is some scatter in the data.

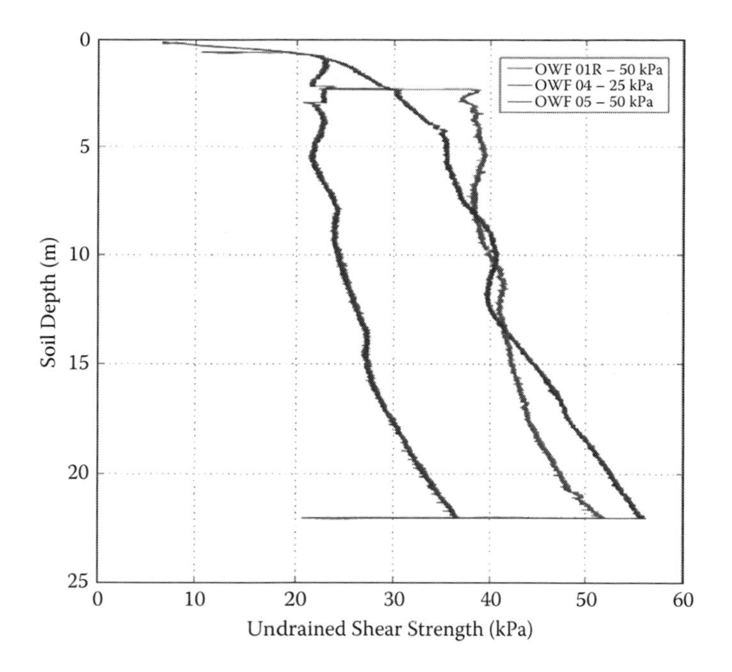

Figure 7.16 Variation of shear strength with depth in clay.

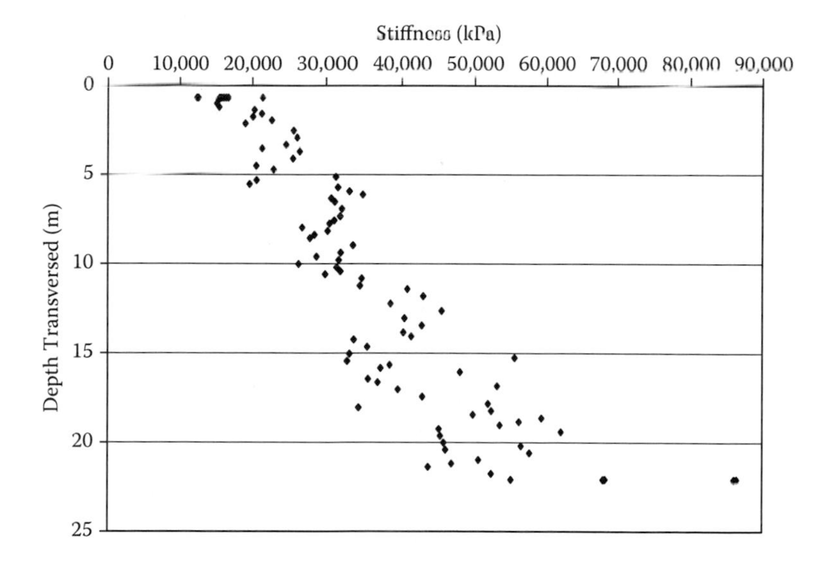

Figure 7.17 Variation of shear stiffness with depth in clay.

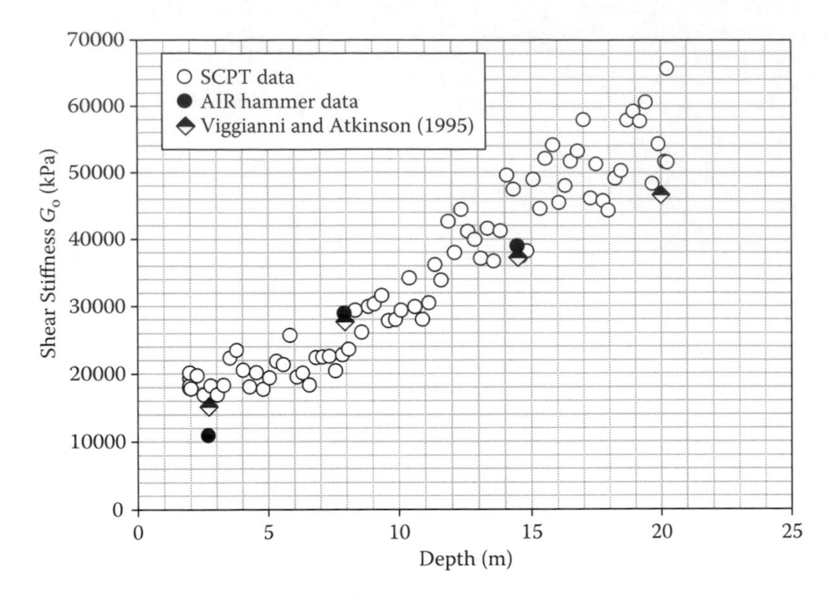

Figure 7.18 Comparison of shear stiffness obtained from SCPT and air hammer devices with theoretical predictions.

7.4.4 Bender elements

Piezo-ceramic strips expand and contract on application of an electric voltage. By sandwiching two such plates, it is possible to fabricate bender elements that will have flexural oscillations in response to an applied voltage. By fixing one end of the bender element as shown in Figure 7.19, the free end will oscillate on application of a time-varying voltage and can create shear waves in a soil sample.

The main advantage of a bender element is that it can act both as a transmitter and a receiver. A time-varying voltage signal is applied to a

Figure 7.19 A view of bender elements.

transmitter, typically 5 or 10 V at a frequency of 10 to 20 kHz. The transmitter oscillates creating a shear wave that travels to the soil and is picked up by another bender element acting as a receiver. Knowing the distance between the transmitter and the receiver the shear wave velocity is determined. Bender elements have been used for a long time in triaxial tests; for example, see Viggiani and Atkinson (1995). Figure 7.20 shows an example of the data from a bender element test in Toyura sand. A square pulse is applied to the transmitter. This creates a shear wave that travels through the soil body and is detected by the receiver. As the travel times are quite short and are often in milliseconds, the data acquisition system used in these tests must be quite fast, often collecting 50,000 samples per second.

More recently, bender elements have been used in centrifuge models as described by Lee and Santamarina (2005) and Brandenberg et al. (2006). The arrangement of bender elements in a centrifuge model can be as shown in Figure 7.21, with a stack of elements on one side of the centrifuge model acting as transmitters and another stack of bender elements on the other side acting as receivers. Kim and Kim (2011) conducted a series of centrifuge tests using this arrangement.

One of the main advantages of using bender element stacks as opposed to individual elements is that it allows for cross-hole tomography. Each bender element on the transmitting stack can be excited in turn and the shear wave pulse is received by each of the bender elements on the receiver stack. Typical data obtained during a single pulse event is shown in Figure 7.22 and the ray paths for cross-hole tomography are marked in Figure 7.23.

In this figure it can be observed that the ray paths are curvilinear in shape due to the change in shear wave velocity with changing confining pressure in a centrifuge model with depth, as prescribed by Equation 7.3.

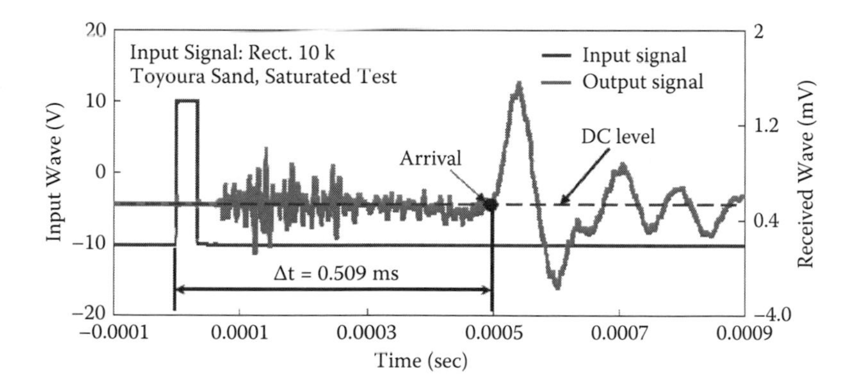

Figure 7.20 Arrival of shear waves induced by a bender element (after Kim and Kim, 2011).

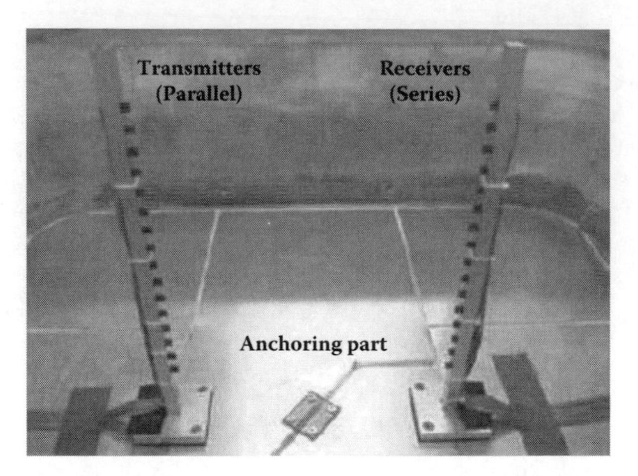

Figure 7.21 Bender element stacks for use in centrifuge tests (after Kim and Kim, 2011).

Using the shear wave velocity data obtained by conducting a cross-hole tomography, it is possible to obtain the shear velocity contours within the centrifuge model. Figure 7.24 shows the shear wave velocity contours obtained by Kim and Kim (2011) for a dry sand model. In this figure it can be seen that the shear wave velocity changes from about 140 m/s at the soil surface to about 250 m/s at the base of the soil model.

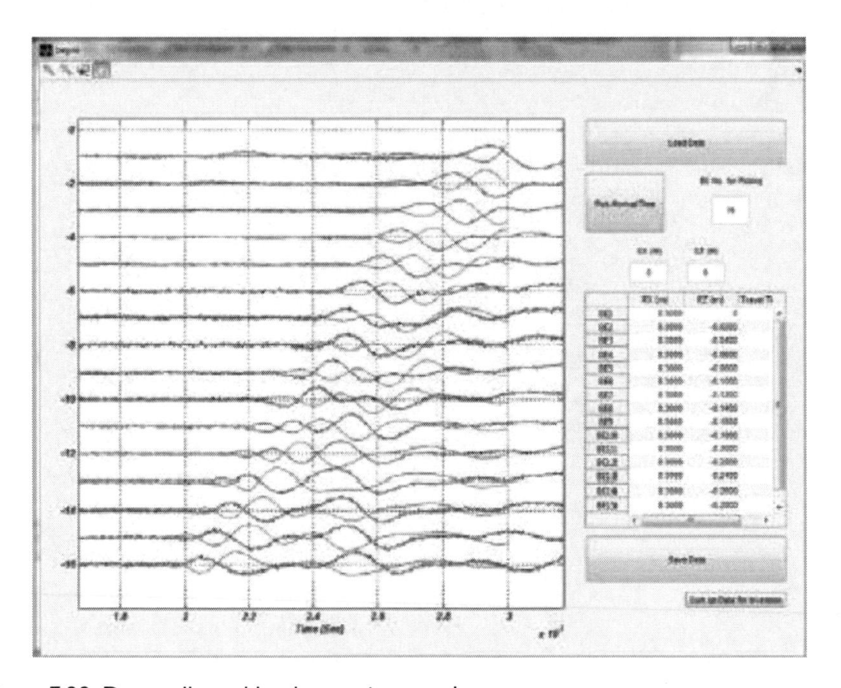

Figure 7.22 Data collected by the receiver stack.

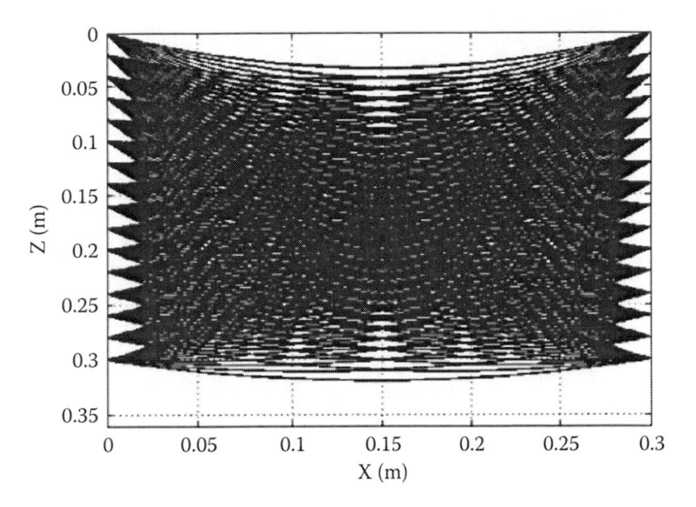

Figure 7.23 Ray paths joining transmitter and receiver bender elements (after Kim and Kim, 2011).

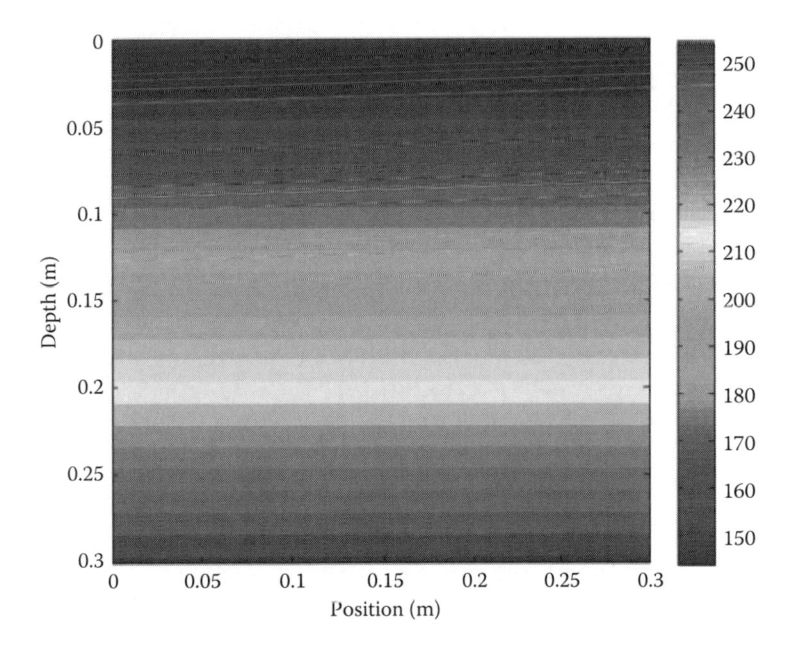

Figure 7.24 Contours of shear wave velocity in a centrifuge model (after Kim and Kim, 2011).

7.5 SUMMARY

In this chapter the various equipment that are used in centrifuge modelling have been introduced. The model containers are chosen based on the problem to be investigated and the assumptions involved such as plane strain conditions. Techniques used to prepare the centrifuge models for sand and clay samples have been discussed. In the case of sands, air pluviation and subsequent saturation are preferred. For clays, reconstituted samples prepared in a laboratory are preferred by virtue of the control that can be exercised during the preparation and the knowledge of the precise stress history of the sample. On occasions, undisturbed field samples can also be used for large projects.

During a centrifuge test various actuators are needed to perform the specific tasks such as installation of a pile or loading imposed onto a structure. These tasks are achieved by in-flight actuators. Specifically 1-D and 2-D actuators are discussed in this chapter. For all these actuators it is important to have the ability to impose either a displacement or a force onto the object placed in the centrifuge model.

Finally the importance of in-flight characterization of the soil models is discussed and various techniques that are currently available are presented in detail. Some of these techniques such as the miniature CPT are derived based on field experiences while others such as the air hammer device take advantage of the small size of the centrifuge and the accessibility to the base of the soil model, before it is constructed. Of course, such a device would be much more difficult to install under field conditions. Development of SCPT and cross-hole tomography will allow us to better characterize the soil models in the centrifuge in the future.

Chapter 8

Centrifuge instrumentation

8.1 INTRODUCTION

One of the advantages of using centrifuge modelling is our ability to instrument the models. As the testing is carried out in a laboratory environment, instrumenting the model and making in-flight measurements is a major advantage in terms of tracking the behaviour of the model before, during, and after failure. While it is possible to instrument a field structure, often such an exercise is expensive and the reliability of instruments during interesting phases of loading needs to be ensured carefully. For example, if we instrument a structure located in a seismic area, the power supply to the instruments must be ensured when the actual earthquake occurs, so that the instruments record useful data. In a centrifuge test, it is relatively straightforward to instrument a model and ensure that they provide good data during the loading event.

Much of the success of centrifuge modelling owes to the advances in electronic instrumentation over the past four decades. Especially, the miniaturization of instruments has helped centrifuge modellers as they can take advantage of the availability of these instruments that are relatively cheap and robust. While centrifuge modelling had its initial success in Soviet Russia, it is the development of the miniature instrumentation that saw its rapid adaptation first in the United Kingdom and then worldwide. It is said that development of the miniature pore pressure measuring transducer, often referred to as PPT, revolutionized centrifuge modelling. Availability of PPTs meant that the pore water pressure within the soil could be measured directly and the effective stress state of the soil could be ascertained in a centrifuge model before, during, and after a failure induced by applied loading.

In this chapter the use of various miniature instruments will be discussed. Some of the new instruments that are currently being used such as micro-electrical-mechanical systems (MEMS) accelerometers will be highlighted together with their advantages and disadvantages. Also recent refinements to traditional instruments will be presented.

8.2 TYPES OF INSTRUMENTS

In centrifuge modelling a wide variety of instruments are used depending on the parameter that needs to be measured. Classification of such instruments can be made in many ways, for example based on the physical principle used to accomplish the measurement, such as measuring strain using strain gauges or measuring the charge accumulated in a piezo-electric device. In this section we classify the instruments based on the type of measurement that needs to be made. As end users, centrifuge modellers are mainly interested in the actual measurement recorded rather than the workings of the instrument itself.

8.2.1 Pore pressure measurement

As explained earlier, one of the most important measurements that need to be taken in a centrifuge test is the pore water pressure within the saturated soil model. This allows us to establish the effective stress state within the soil model at any stage in a test. This is done using miniature pore pressure transducers (PPTs). A typical PPT uses flexible silicon diaphragms in a steel casing, which is able to bend and generate an output voltage proportional to the fluid pressure. Common types of PPTs include porous ceramic stones or sintered bronze stones, which protect the diaphragm from surrounding soil while keeping it saturated. Figure 8.1 shows a typical PPT. The main body of the PPT has a diameter of about 6 mm and the length of the PPT is about 12 mm.

A PPT requires four wires, two to supply the drive voltage and two to carry the signal back. These wires are enclosed in a flexible housing as shown in Figure 8.1. The drive voltage normally used is 0 to 5 V. The signal that is produced by the PPT in response to the applied pressure is normally only a few mV and therefore needs to be amplified before it is sent to the data acquisition system. Some types of PPTs need air at atmospheric pressure behind the flexible diaphragm. These devices measure pore water pressure relative to the atmospheric pressure. Other types of PPTs can be sealed

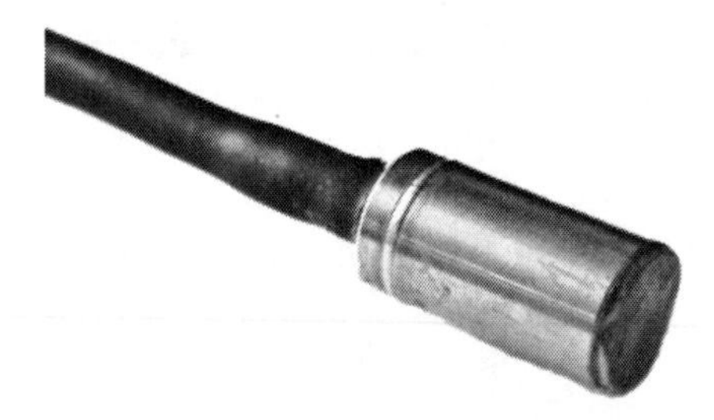

Figure 8.1 A view of a miniature pore pressure transducer.

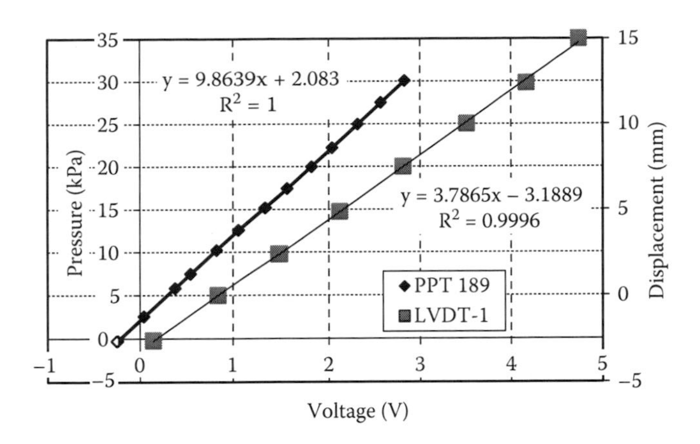

Figure 8.2 Calibration chart for a pore pressure transducer and a linearly varying differential transformer.

with 1 atmospheric pressure in a closed chamber behind the diaphragm. These devices do not need any ventilation and therefore the flexible cable can be small as it carries only four electric wires.

Before use in a centrifuge test, a PPT needs to be calibrated. This is done by applying known water pressures and measuring the voltage output from the PPT. An example of the calibration chart for a PPT (Serial No: 189, 350 millibar) is shown in Figure 8.2. From this figure the calibration factor for the PPT is obtained as 9.86 in the units of kPa/V. The data also shows that this instrument has a correlation factor R^2 of 1.0. Normally these devices are very linear within their range of operation. However, the device calibrated here has an offset of about −0.21 V at zero pressure. This offset can be important if the signal from the PPT is amplified later on. For this instrument an amplification of 10 is acceptable (which increases the offset to −2.1 V) but an amplification of 100 is not acceptable (this pushes the signals to −21 V which is greater than the range ±10 V of most data acquisition systems).

A PPT is rated as 7 bar, 1 bar, or 350 millibar, which indicates the maximum pressure it can measure. Although they can tolerate an overpressure of up to 150 percent temporarily, care must be taken not to expose a PPT to overpressure. Similarly care must be taken in the use of appropriate stones to protect the diaphragm of the instrument, which is the active part of the PPT. When used in clay a porous ceramic stone is used while in sands a sintered bronze stone is used in front of the diaphragm. These stones must be maintained in a fully saturated state, as any air bubble that is trapped in front of the active diaphragm can lead to erroneous results. This is particularly true in the case of dynamic tests such as earthquake loading, as the air bubble significantly reduces the ability of the PPT to measure a time-varying pore water pressure, that is, its dynamic response, is adversely affected.

In a centrifuge test a PPT can be placed at the desired location in the soil during the model-making process. In sand models the PPT can be simply placed carefully at the desired location and more sand is rained around the instrument. In clay models, it may be necessary to auger a hole from the side of the container, place the PPT at the desired location, and backfill the hole with wet slurry, avoiding trapping air bubbles. It is also important to plan the routing of the flexible wire. It is better to run these wires horizontally to the side boundary of the container and then take them up vertically along that boundary. We must be conscious of the possibility of additional drainage that may occur along the wire and therefore this must be kept to a minimum. By running the wires along the side boundary the central part of the centrifuge model is unaffected by any additional drainage.

PPTs are normally designed to measure positive pore water pressures. Although they can cope with measurement of suctions of up to –100 kPa and occasionally even higher suctions (Mair, 1979), their working under large soil suctions is rather unreliable. Special modifications were made to the design of the PPTs to enable them to measure large matric suctions in the soil reliably. The working of these devices, termed pore pressure tensiometer transducers (PPTTs) is discussed in detail by Take and Bolton (2003).

By and large, use of PPTs has been very successful in a wide variety of centrifuge tests. These devices are used in all the centrifuge centers around the world.

8.2.2 Displacement measurement

Displacement measurement in a centrifuge test is carried out traditionally by using contact devices such as linearly varying differential transformers (LVDTs). A view of a typical LVDT is presented in Figure 8.3. They generally consist of two detached coil windings and a rod in a cylindrical casing. The rod, whose tip is attached on the surface where the displacement

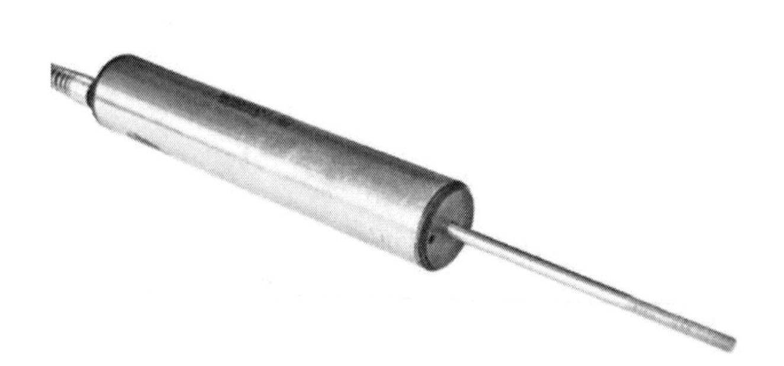

Figure 8.3 A view of a linearly varying differential transformer.

is to be measured, couples the magnetic field in one coil with the other as it moves between them. This way, the current applied on one of the coils causes induction on the other coil, the potential of which changes depending on the location of the core rod.

Prior to use, an LVDT is calibrated by applying known displacements from a screw gauge and its output is measured. A calibration chart of an LVDT is shown in Figure 8.2. From this figure the calibration factor is obtained for this instrument as 3.7865 in the units of mm/V. The data also shows that the instrument is quite linear with a correlation factor R^2 of 0.9996.

LVDTs are reliable instruments and are used in most centrifuge tests. Certain special adaptations may be required if they are used to measure the settlement of ground. Normally a small bearing pad is placed below the rod, to prevent sinking of the rod into the soil under high gravities. The dynamic response of LVDTs is limited and caution must be exercised when using these devices in earthquake or other dynamic tests.

Another device that is used to measure displacements is a rotary draw wire potentiometer. In this device a spring-loaded wire is used to attach the rotary potentiometer to the object in the centrifuge model whose displacement is to be measured. This type of device is used where the expected displacements are large. A view of this device is shown in Figure 8.4. Potentiometers of different displacement ranges from 150 mm to 400 mm can be obtained. However, their accuracy is limited to a few millimeters, depending on the ambient electrical noise in the data acquisition setup. Further, the dynamic response of the rotary potentiometers is very limited.

In some instances it is useful to have a noncontact displacement measuring device. With the advances in laser measurement technology, laser displacement transducers are now available to achieve such a noncontact

Figure 8.4 A view of the rotary draw wire potentiometer.

Figure 8.5 A view of a laser displacement transducer.

measurement. A view of the laser displacement transducer is shown in Figure 8.5. This device is usually mounted on a stationary part of the centrifuge model and is targeted toward the object whose displacement needs to be measured. The device works by sending a low-power laser beam toward the object and measuring the time it takes for the reflection to arrive back. This is converted into a displacement measurement. The dynamic response of the laser is quite good. The range of laser displacement device can be between 50 to 150 mm with sub-millimeter accuracy.

Another advantage of the laser displacement transducer is that it can measure through transparent media. For example, the level of soil surface below clear water can be measured using this device. At Cambridge a laser profiler was developed that can transit in-flight and measure the profile of the soil surface. These laser displacement devices are used widely in many centrifuge centers worldwide.

8.2.3 Acceleration measurement

Measurement of acceleration plays an important role in many centrifuge tests where dynamic loads are present, such as when earthquake, wind, or wave loading is modeled. Traditionally accelerations are measured using miniature piezo-electric devices. These work by converting the mechanical stress induced in the piezo crystal into an electric charge. The electric charge is converted into a voltage by a simple charge-coupled amplifier. A view of the piezo-electric accelerometer is shown in Figure 8.6. These devices are calibrated before use in a centrifuge test using a specially designed calibrator that can apply precisely $\pm 1\ g$ acceleration. A calibration factor for the accelerometer is obtained in the units of g/V.

Piezo-electric accelerometers are mounted in a metal body with a straight thread as shown in Figure 8.6 or a 90° thread to facilitate easy mounting onto structures. They always measure acceleration in the direction of the thread. These accelerometers can also be used directly in the soil body to measure soil accelerations at that location. The coupling between the

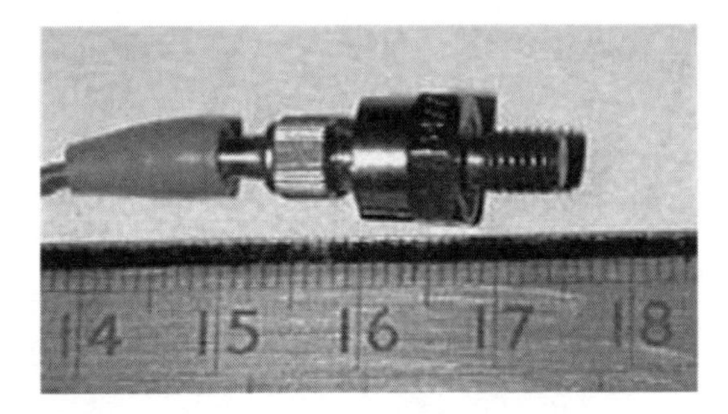

Figure 8.6 A miniature piezo-electric accelerometer.

device and the soil is normally very good and true soil accelerations can be measured. In fact attenuation of accelerations due to soil liquefaction can be measured using these devices. The frequency response of these accelerometers is very good, in the range of 5 Hz to 2 kHz. Below 5 Hz they do not give good response. Another point to remember is that the piezo devices do not record constant accelerations, that is, the acceleration due to gravity in a centrifuge model is not recorded by these devices. They require changing mechanical stresses acting on the piezo element to produce an electric charge and hence they only record time-varying accelerations.

With the revolution in mobile phone technology particularly the smart phones, the availability of cheap, miniature accelerometers has become widespread. These devices are called micro-electrical-mechanical systems (MEMS) accelerometers. They have a tiny inertial mass suspended on a spring and their displacement is used to determine the spring force and hence the acceleration of the device. MEMS accelerometers are very small, measuring only a few millimeters. A view of the MEMS accelerometer is shown in Figure 8.7. These devices are able to measure both constant and time-varying accelerations. As a result they can be calibrated by just turning the device upside down and reversing the 1 *g* component due to the earth's gravity.

The main advantage of using MEMS accelerometers is that they are very inexpensive. Also their mass and size are much smaller than the traditional piezo-electric accelerometers. Further they produce a voltage as the output and all the necessary circuitry is embedded within the device. It is possible to obtain both single-axis and three-axis MEMS accelerometers. Many centrifuge centers that carry out dynamic centrifuge tests are using MEMS accelerometers. However, they are a few issues with them that need to be highlighted. While using these devices in saturated soil models, we need to ensure that all the electrical connections to the device are properly waterproofed. Also some of the MEMS devices have low pass filters and may not pick up high frequency components. It is not possible for the end users to

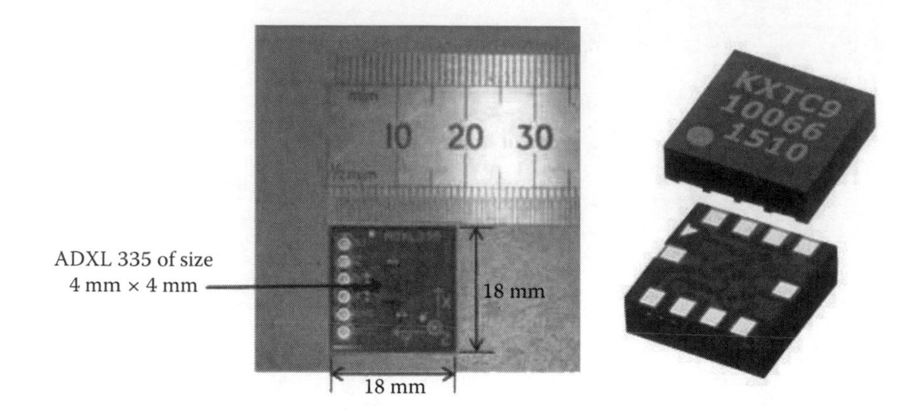

Figure 8.7 An example of a MEMS accelerometer.

control the filter characteristics. However, it is possible to obtain MEMS accelerometers that have good high frequency response. More details on filtering of signals are discussed in Chapter 9.

8.2.4 Force/load measurement

Measurement of a force or load is often required in a centrifuge test. For example, when an actuator is used to apply an axial or lateral load on a model pile, we would like to measure the magnitude of the force being applied. This is often done using a miniature load cell. Measurement of forces in a centrifuge test is straightforward as the forces in a centrifuge model are relatively small compared to the field structure. You may recall from Chapter 4 that the scaling law for force is $1/N^2$; thus, a small force measured in a centrifuge test corresponds to a large force in the prototype. An example of a miniature load cell is shown in Figure 8.8. Threading is

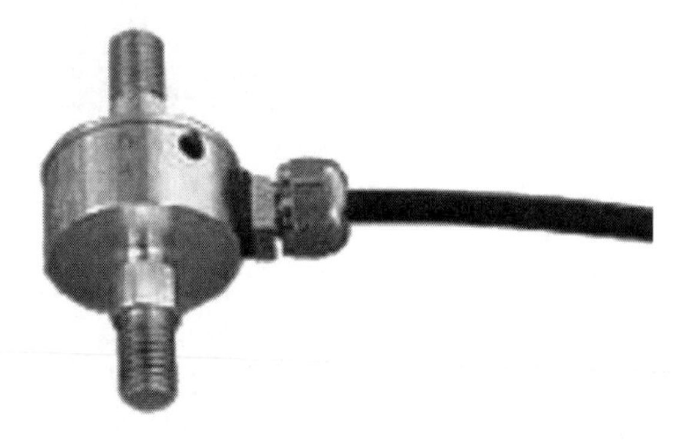

Figure 8.8 A view of a miniature load cell.

normally provided on either side of the load cell for easy attachment to the required parts. Miniature load cells can be obtained over a wide range from 50 N to 2 kN.

Load cells need to be calibrated before they are used in a centrifuge test. This is normally done by hanging known weights from the load cell and measuring the output. A calibration chart is then constructed and the calibration factor for the device is obtained in the units of N/V. It must be pointed out that many miniature load cells are quite sensitive to application of moments, that is, they can give erroneous readings if moments are applied to the load cell. Care must be taken to avoid transfer of any moment into the load cell body. The dynamic response of the load cells is very good and they can be used to measure time-varying forces quite reliably.

8.2.5 Bending moment and shear force measurement

In many centrifuge tests there will be a need to instrument the objects of inclusions such as a model pile foundation, retaining wall, or a tunnel to measure bending moments. Knowing the bending moments and any axial forces, we can work out the shear forces from equilibrium considerations in most instances.

In order to measure bending moments, for example, on a tunnel section, we can use miniature strain gauges. These strain gauges are stuck on the model tunnel wall at the desired locations. Miniature strain gauges can be obtained as wire or foil gauges and have a typical resistance of about 120 Ω. They are driven by a direct current excitation voltage V_{ex} in the range of 3 to 5 V. Depending on the wiring, the strain gauges can measure either axial strains or bending strains. In many instances the axial force measuring load cells discussed in Section 8.2.4 have strain gauged internal elements.

Figure 8.9 shows a view of a miniature strain gauge. Four such gauges are normally used to form one measurement point for the bending moment. These four gauges are wired as in a typical Wheatstone bridge arrangement,

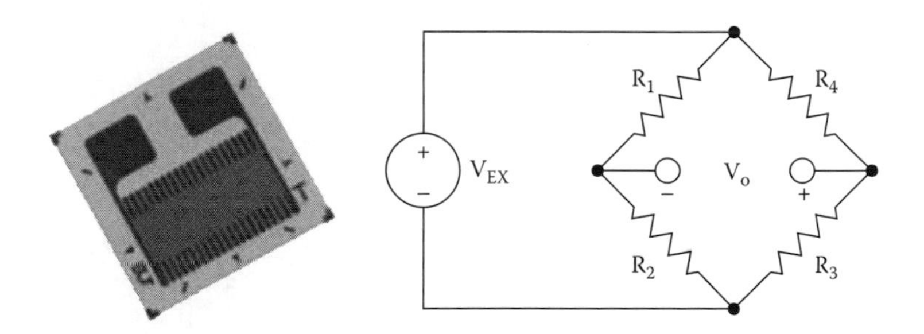

Figure 8.9 A view of the strain gauge and the Wheatstone bridge circuit.

shown in Figure 8.9. In order to measure bending strains, gauges R_1 and R_3 are wired together and stuck onto the side generating tensile strains, while gauges R_2 and R_4 form the second pair and are stuck onto the side generating the compressive strains. Such an arrangement is called a full-bridge configuration and gives four times the output of a single strain gauge. It also has the advantage of eliminating any axial effects on the bending strain and is fully temperature compensated; that is, the output from the bridge does not vary with temperature. This is a big advantage when long centrifuge tests are planned, as we do not want the output from our strain gauge bridge to vary with time. In some cases it may be necessary to use a half bridge either due to space limitations, lack of access to one side (either compressive or tension side) of the bending element, or for economic reasons. In this case, dummy resistors are used outside the centrifuge model to complete the Wheatstone bridge circuit. The output from such a bridge will obviously be only half of the full-bridge circuit and this arrangement may also be vulnerable to temperature changes and axial effects.

Strain gauging to measure bending moments is an intricate operation, particularly when there are a large number of measuring points present and the object to be strain gauged is small. Special tools are usually required to access and gauge the inside of a tunnel or a pile. An example of a model tunnel that is well-strain gauged is shown in Figure 8.10. This tunnel is gauged to measure both bending moments and tunnel lining forces at several locations.

Similar to other instrumentation discussed in this chapter, bending moment gauges need to be calibrated before use in a centrifuge test. This is normally done by subjecting the object to known loading conditions that will induce known bending moments at the locations where the gauges are present. Using these, calibration charts are prepared and the calibration factors for each measurement point are obtained in the units of N mm/V. Also the gauge factors of the strain gauges themselves are quite small, usually giving an output of few millivolts under applied bending moments. Therefore, the signal from the bending moment gauges need to be amplified by a factor of 100 or even a 1000. Electrical noise can become a problem

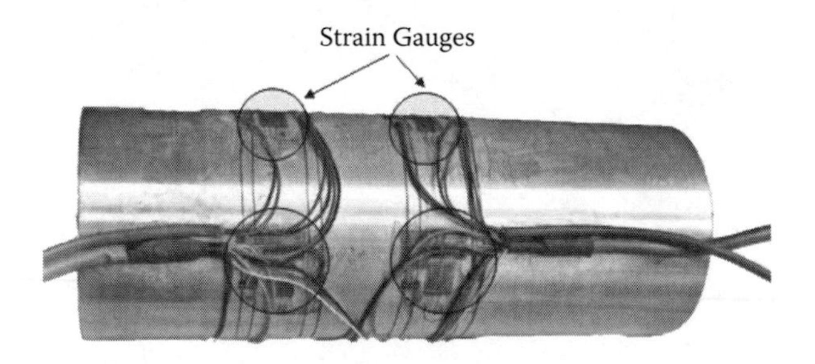

Figure 8.10 Strain gauged model tunnel.

with such high amplifications and due care must be taken to minimize the noise. The tunnel shown in Figure 8.10 has been used in an extensive study on seismic actions on tunnels; more details of this study can be found in Lanzano et al. (2012).

Apart from gauging individual objects such as model tunnels or retaining walls, individual cells that can measure axial compression, shear force, and bending moment have been developed. These are often referred to as Stroud cells (Stroud, 1971). These cells have an upper plate and a lower plate connected by four intricately machined thin legs that are strain gauged to measure the axial force, shear force, and bending moment.

8.2.6 Inclinometers

Measurement of inclination is often carried out in the field to establish angles of a slope or to determine the verticality of a structure. In a centrifuge test we are often required to establish the verticality of our structures or measure the angle they make with respect to the high gravity direction. With the advances in MEMS accelerometers, it is now possible to create miniature inclinometers quite easily. The object whose verticality needs to be known during different stages of a centrifuge test can be fitted with a MEMS accelerometer. The output from this device will change depending on the angle it makes with direction of the g field, thereby recording the inclination of the object to which it is attached.

Stringer, Heron, and Madabhushi (2010) made use of a series of MEMS accelerometers fixed onto a vertical, flexible strip passing through a sloping ground. The changes in the MEMS accelerometer measurements were used to determine the change in inclinations at different points of the strip as the slope suffered a failure during a centrifuge test. A similar technique was previously used to measure slope movements in the field (Abdoun and Bennett, 2008).

8.2.7 Earth pressure measurement

Normal PPTs are designed to measure the pore water pressure and not the total stress in the soil body. The porous stones in front of the diaphragm prevent the direct action of the soil on the active diaphragm allowing only fluid pressure to be measured. If we wish to measure earth pressures directly, then we need to use specially designed earth pressure cells (EPCs). One such device is shown in Figure 8.11.

These devices can measure earth pressure acting next to a solid inclusion in the soil model such as a retaining wall or a tunnel. The size of the device is about 5 mm in diameter. The active diaphragm has miniature strain gauges which register a change in voltage proportional to the applied earth pressure. Calibration of these EPCs can be carried out by subjecting them to water pressures of different magnitudes and obtaining a calibration factor in the units of kPa/V. However, this hydrostatic calibration factor may

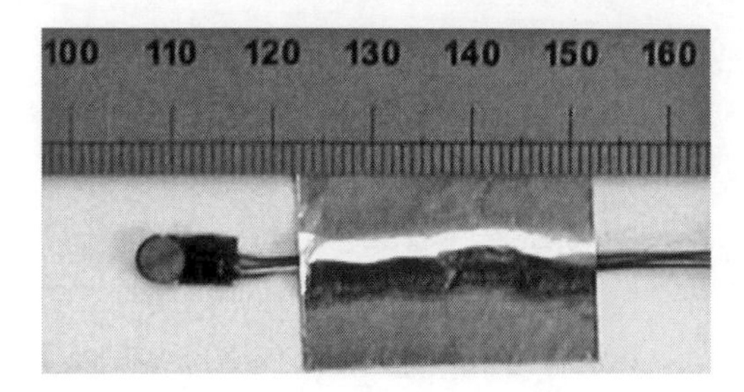

Figure 8.11 A view of a miniature earth pressure cell.

not be applicable directly when using the EPCs in soils. It may be necessary to subject the cells to vertical pressures in an oedometer placed in soils of different particle sizes to obtain a more realistic calibration factor.

Unlike measurement of pore water pressure, the measurement of earth pressure in a centrifuge test is a far more difficult proposition. In order to obtain good earth pressure measurement the stiffness of the pressure cell and that of the surrounding soil should match. This is to ensure that the diaphragm of the device deforms the same amount as the soil. If the diaphragm is too flexible relative to the soil, then soil arching can result and the earth pressure cell will register false readings. Similarly if the diaphragm is too stiff, the soil may not deform next to the diaphragm of the earth pressure cell and therefore may not mobilise the correct earth pressure. Although the earth pressure cells shown in Figure 8.11 were used in several centrifuge tests, they seem to work better in earthquake tests as the shaking of the soil tends to destroy any soil arching (Khoker and Madabhushi, 2010).

Another device that has become recently available to measure earth pressures in soil is the tactile pressure sensing system. This involves using a pressure sensing film that can be placed in a centrifuge model. The thickness of the film is only 0.1 mm and it is very flexible. Films of different sizes can be obtained. A view of one such film is shown in Figure 8.12.

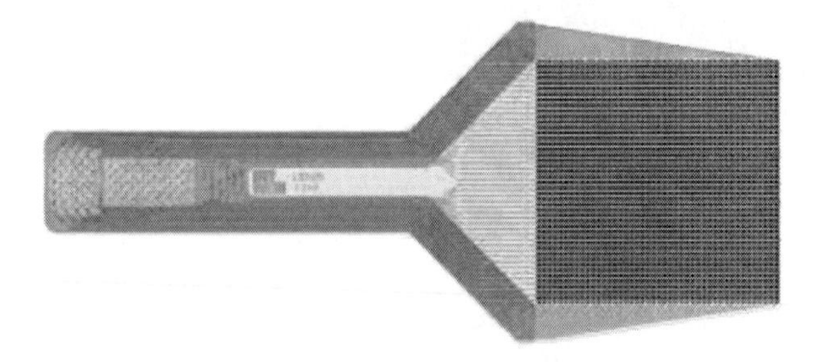

Figure 8.12 Tactile pressure sensing film.

The film contains a grid of horizontal and vertical lines. Pressure sensing occurs at each of the intersection points on the grid, typically at about 2000 points. Specialist software is used to obtain pressure distribution from the multiplexed signals obtained from the film. This system of pressure sensing has been used at various centrifuge centers with some success, including ETH, Zurich (Springman et al., 2002) and more recently at RPI in the United States and other centers.

8.3 DEFORMATION MEASUREMENT

Measuring soil deformations has been an important part of centrifuge modelling of various boundary value problems. Availability of views of the soil model cross-section through the Perspex window in a plane strain model makes it convenient to obtain images. Early attempts at obtaining soil deformations involved placing lead shot in the soil next to the windows and x-raying the soil models before and after the centrifuge test. This gave the overall soil deformations from which displacement fields and even strain fields could be determined. The pre- and post-test x-raying was complemented with in-flight imaging using film cameras. The plane strain centrifuge models were filmed in-flight using stationary camera systems outside the centrifuge that would take photographs as the centrifuge arm swept past at high speed. An electronic trigger was used to sync the shutter opening in the cameras and the lights at the precise moment when the centrifuge arm was directly underneath the camera. On-board cameras were not possible as moving shutters of a camera would not function in the high-gravity environment of a centrifuge. This meant that specialist 70-mm Hasselblad cameras were used to take pictures at specific points in the centrifuge test. This system was reasonably successful in the 1980s. The photographs obtained from centrifuge tests were analysed afterwards using a separate system. This consisted of manually pin-pointing markers in the photographs of the centrifuge model using a joy stick, which transferred the coordinates of that point into a computer program. By comparing the coordinates from successive images soil displacements could be obtained. Again the displacement fields determined from this procedure could be used to obtain soil strains.

8.3.1 Digital imaging

During the last decade instrumentation used in centrifuge modelling has seen huge advances. One of these advances took place in the area of digital imaging. With the advent of digital cameras it became possible to mount these cameras on the plane strain centrifuge packages, viewing the cross-section of the centrifuge model through the transparent side. Images could be taken at regular intervals up to one image every few seconds. More recently the resolution of digital cameras has increased and now 5-megapixel resolution has become common place, with advanced cameras having up to 20-megapixel resolution. Lighting on the centrifuge packages has improved with the use of LED lights. Polarizing plates are available to remove unwanted reflections from the Perspex sides of the centrifuge models appearing in the digital images. The cameras themselves need to be strengthened to be able to sustain the high *g* forces. Usually this involves such actions as supporting the lens system or switching off the zoom facility. Remote access cameras are preferable as they allow controlling and downloading of images during the flight. Otherwise, cameras that store images at regular intervals onto a memory chip can also be used, but the images are only available at the end of the centrifuge test.

Typically the digital cameras and lights are fixed directly in front of the Perspex window looking at the cross-section of the centrifuge model. An example of this is shown in Figure 8.13, which shows the digital camera setup for a soil model with a tunnel present in front of a retaining wall. A grid of circular markers is painted on the Perspex window as seen in this figure. This grid offers a reference frame against which the soil movements can be measured using the particle image velocimetry (PIV) techniques described in Section 8.3.2. Also in Figure 8.13 it can be seen that multiple digital cameras are present, which allows imaging of different regions of

Figure 8.13 Digital imaging setup for a centrifuge test with a tunnel in front of a retaining wall.

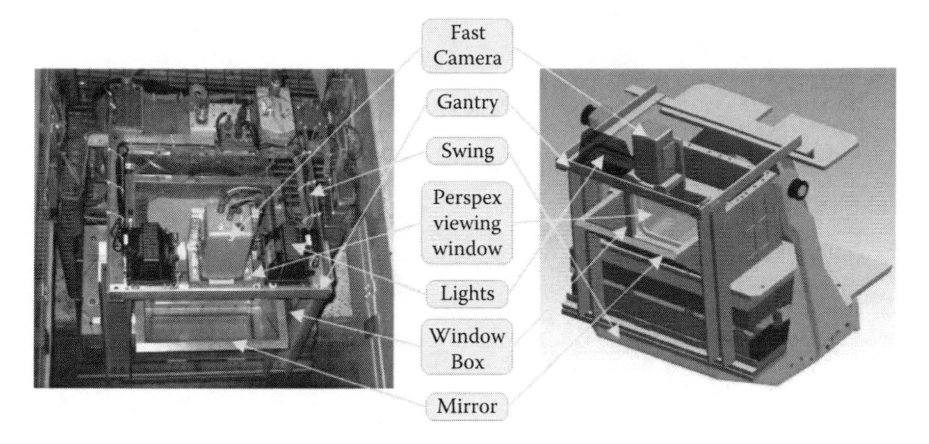

Figure 8.14 Special arrangement for a fast camera set up with a 45° mirror.

the centrifuge model at high resolution. If each 20-megapixel camera has a view covering 100 mm × 100 mm of the centrifuge model, then each pixel will have a size of 0.022 mm. This would allow accurate tracking of the movement of the soil relative to the grid of markers.

Another advancement that took place with digital imaging is the availability of high-speed cameras. These cameras allow imaging the centrifuge model at up to 1000 frames per second, while still taking images at 1-megapixel resolution. Initially these were black and white cameras but now even color cameras are available. Such high-speed cameras allow us to take rapid pictures during dynamic centrifuge tests where rapid dynamic loading such as earthquake loading is applied. Earlier versions of the fast cameras had a long body and required special arrangement, as shown in Figure 8.14, and the model container had to be fitted with a 45° mirror (Cilingir and Madabhushi, 2011). More recent versions of these fast cameras are quite small and can be used directly in front of the model container, as shown in Figure 8.13.

8.3.2 Particle image velocimetry

Availability of such high-resolution images that can be obtained with ease in any centrifuge model test demanded the development of automatic techniques that can process the images from a centrifuge test and give us the displacement vectors. Early attempts at comparing CCTV images from centrifuge tests were carried out by Allersma (1990). This system was further improved at the City University by Taylor et al. (1998). White, Take, and Bolton (2003) adapted the PIV as used in fluid mechanics to problems in the

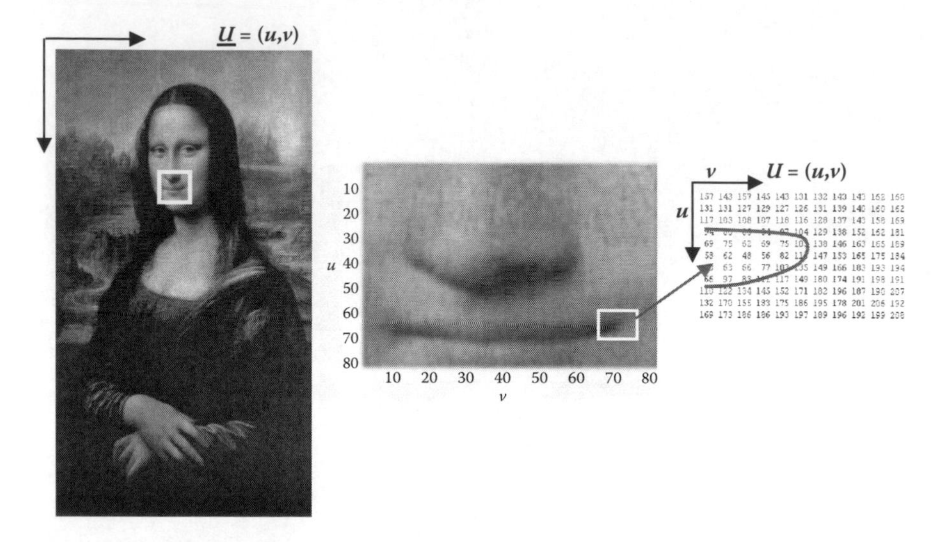

Figure 8.15 A digital image expressed as a two-dimensional matrix.

geotechnical area. The MATLAB®-based software called Geo-PIV developed by White and Take (2002) aims to:

1. Take advantage of high-resolution imagery and convert that into a high-precision measurement of ground deformations.
2. Remove the need to use artificial markers in the soil and instead use the natural texture of the soil to track its movements.
3. Automatically account for camera movement or lens distortion during a centrifuge test.

Any digital image can be viewed as a two-dimensional matrix of brightness at different pixel locations. For example, let us consider the photograph of the Mona Lisa in Figure 8.15. To make it simple, let us extract a portion of this image near her lips. This extracted region can be expressed

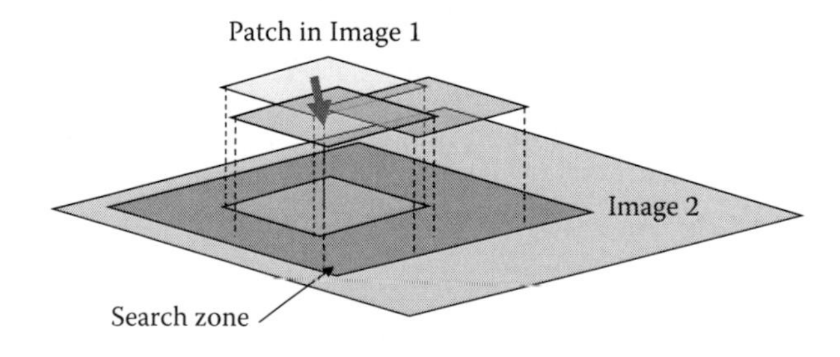

Figure 8.16 Illustration of searching for matching patches in two images.

as a two-dimensional matrix and the shape of her lips is reflected by the values of the brightness in the matrix as highlighted in Figure 8.15. Thus, facial features can be extracted as a series of numbers. For an 8-bit image the value of brightness can take a value between 0 and 255.

It must be noted that there is no physical scale associated with the image in Figure 8.15. When dealing with digital images where no quantification is required, lack of a physical scale is not a problem. The two-dimensional matrix of brightness is obtained in the image space. However, when real measurements need to be made from digital images we need to associate physical dimensions to the images. This is done by overlaying a rectangular grid of markers at precisely known locations on the Perspex windows in front of the soil model, as seen in Figure 8.13. By triangulating the grid of markers a reference plane is created in the object space. The movement of the soil model measured in the image space can be transformed into object space, as the location of the markers is known in both object and image spaces.

The main idea of the PIV analysis relies on comparing successive images from a centrifuge test. Each image is divided into a set number of patches. The movement of each patch between the images is tracked. This is done by obtaining an intensity map of the patch on the first image, which is then compared with the intensity maps of all the patches within a search zone in the second image, by calculating a cross-correlation function. This process is illustrated in Figures 8.16 and 8.17. When a good fit is obtained

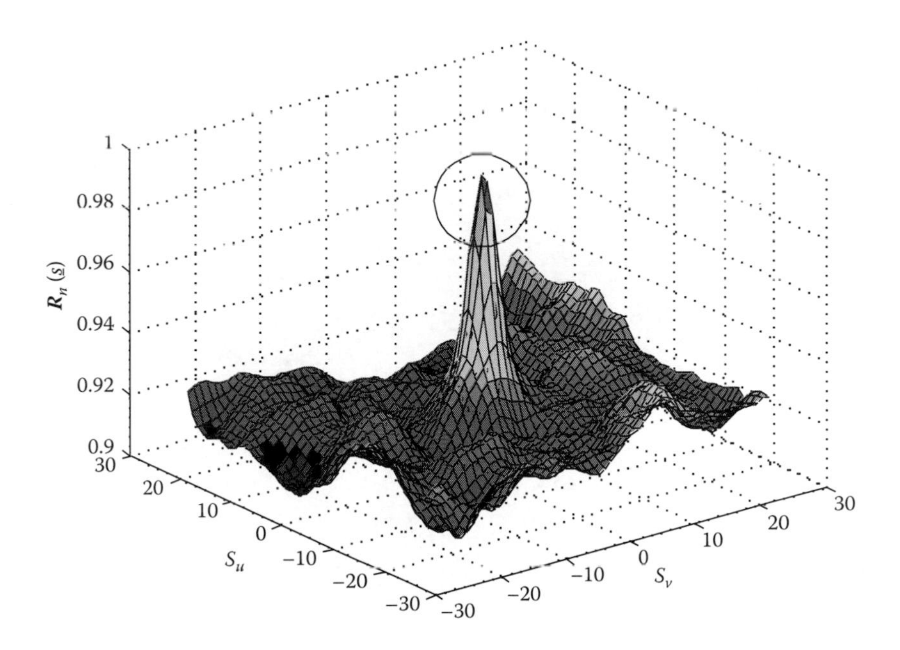

Figure 8.17 Cross-correlation of the intensity maps for patches from images 1 and 2.

it is determined that the patch from the first image has moved to the location where the best fit is obtained on the second image. The process is then repeated for each of the patches on the first image.

By the end of the PIV analysis the movement of each patch from image 1 to image 2 will be obtained. Further, knowing the locations of the grid of markers in the two images, we can convert each of the movements in the image space into quantifiable physical movements in the object space. Any changes in the locations of the grid markers can be interpreted as the distortion of the Perspex window, camera support system, and the lens system due to the *g* forces and the physical movements can be corrected for these. Once the displacement vectors are obtained, volumetric and shear strain fields in the soil model can also be obtained. Further, the size of the patches and the search zone are user defined in the Geo-PIV software. This allows the user to control the computing time it takes to carry out the PIV analysis on a set of images. The smaller the patch size and the larger the search zone, the higher will be the computational effort required to carry out the PIV analysis.

A few examples of the PIV analyses are considered next. Figure 8.18 presents a centrifuge test in which a suction caisson is loaded. The failure of this foundation under asymmetric loading is expected to be due to rotation. The PIV analysis of the problem from successive images is shown in Figure 8.19 that was obtained by White et al. (2008). The displacement field of the soil obtained below the foundation confirms the rotation mode of failure.

Another example of the PIV analysis is taken from seismic loading on a shallow foundation, as shown in Figure 8.20. A block foundation is subjected to horizontal shaking on a small shaking table. A high-speed camera was used to take images of the foundation during the application of the

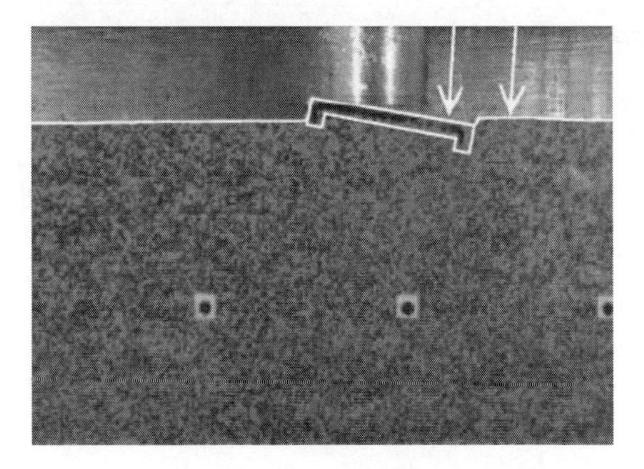

Figure 8.18 Asymmetric loading on a suction caisson (after White et al., 2008).

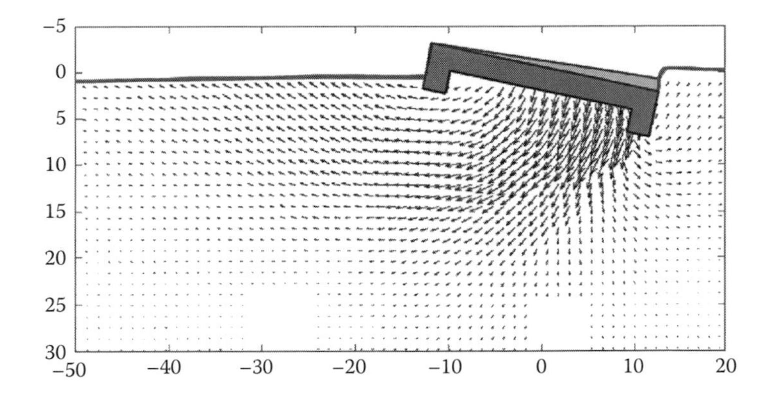

Figure 8.19 PIV analysis of the asymmetrically loaded suction caisson (after White et al., 2008).

earthquake loading. Figure 8.21 shows the shallow bearing capacity failure mechanism obtained from the results from a PIV analysis of the images is presented. A theoretical failure envelope proposed by Paolucci and Pecker (1997) is overlain on the PIV analysis. The results from the PIV analysis agree reasonably well with the theoretical solution. More details of this research are presented by Knappett, Haigh, and Madabhushi (2006).

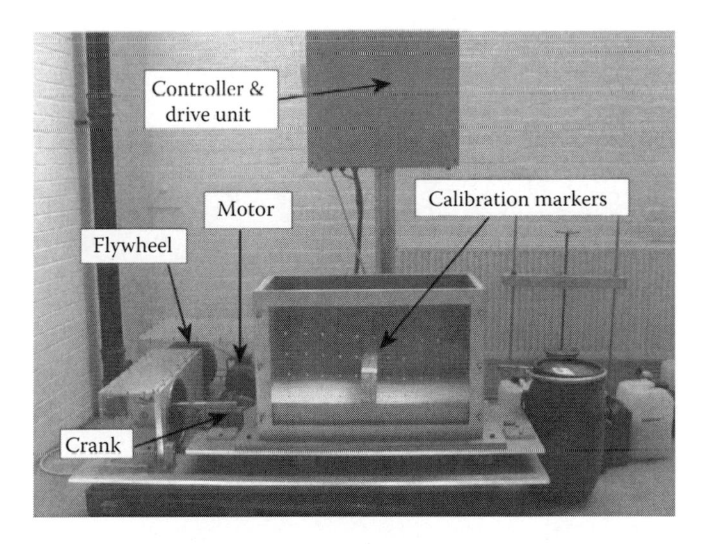

Figure 8.20 Seismic loading on a shallow foundation.

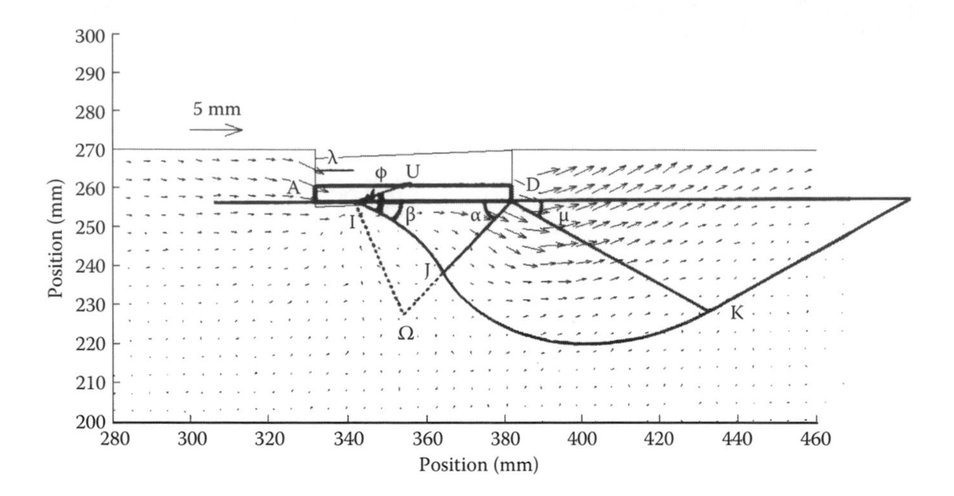

Figure 8.21 PIV analysis showing the shallow bearing capacity failure mechanism.

8.4 SUMMARY

In this chapter we considered the various types of instrumentation used in centrifuge modelling. It must be pointed out that the list of instrumentation considered herein is not exhaustive. Instead the most commonly required measurements such as displacements, pore pressures within the saturated soils, and accelerations in dynamic centrifuge tests are considered and the corresponding instruments used are explained. Detailed descriptions of the instruments themselves were also avoided; instead their usage was highlighted. Some of the new instruments that became available are also considered. MEMS accelerometers are proving to be both cheap and reliable instruments to measure accelerations and inclinations. However, care must be taken in choosing appropriate MEMS accelerometers as some of them have built-in filters. Similarly the basic strain gauging used to measure bending moments in various inclusions such as retaining walls and tunnels in soil models was also described. Again there is a large variety of strain gauges, the details of which were not discussed. The main objective was to show how they are used and what is possible rather than a comprehensive coverage of strain gauges themselves. Finally, developments in digital imaging and the technique of PIV analysis used to obtain soil deformations in the centrifuge model were discussed.

Chapter 9

Centrifuge data acquisition systems

9.1 INTRODUCTION

The instrumentation discussed in Chapter 8 predominantly produces analog signals. By this we mean that the output from the instrument comes as a continuous voltage, normally within the range of ±10 V. If the signal is too small, we may have to use an amplifier to bring it into this range. We would like to log the analog signal coming from the instrument using a computer. Of course computers can only log signals that are digital. In this chapter we will look into the systems that are used in centrifuge modelling to acquire data, normally called data acquisition systems. A schematic of a simple data acquisition system is shown in Figure 9.1. In this figure we can see that the origin of the signals starts from the instrumentation in the centrifuge model, which are transmitted to a set of filters and amplifiers by the transducer wires. Each instrument signal is carried separately to a designated filter and an amplifier, so that the user will have control over what amplification they want to apply to bring the signal into the desired range of ±10 V. After amplification the signal is passed onto an analog/digital converter normally called an A/D converter. This converts the signals into digital signals which are then logged by the software on a computer. The transmission between the A/D converter and the computer storage (either memory or hard disk) happens via the computer bus and is controlled by the software. The communication between the software and the A/D converter is usually two way, so that the software can control the A/D converter, for example, for setting the sampling rate, that is, the number of samples per second that are to be logged for each channel. The digitized data coming from the A/D converter for each channel can be stored in a file on the hard drive. Most software allow a variety of file formats for storing the data, so that it can be easily accessed after the centrifuge test for post-processing and plotting as desired.

In this chapter we look at the conversion of analog signals from our instrumentation into digital signals. We pay particular attention to the intricacies of signal processing such as filtering that are necessary to obtain

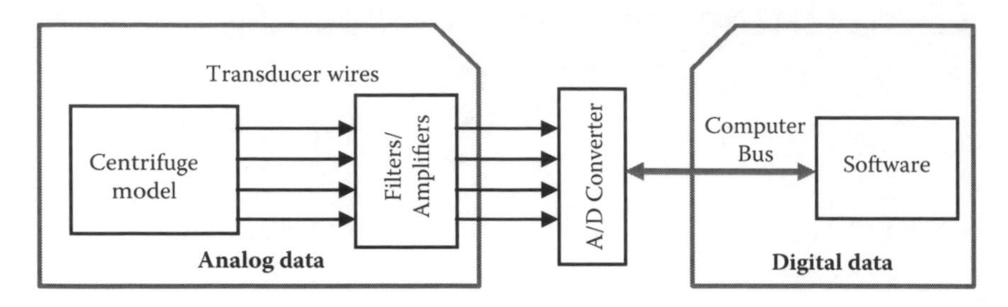

Figure 9.1 Schematic of a data acquisition system.

good data from a range of instruments. It is acknowledged that the field of signal processing is vast and we will limit the discussion only to the essential aspects that are commonly required while acquiring data in a centrifuge test.

Data acquisition systems have also developed significantly over the last decade. The primary driver for this development is the fast processors that are available in computers and the fast data acquisition cards. Another important innovation that took place is the location of the A/D converters and the computers. Historically all the signals from the transducers in the centrifuge model were carried through a set of electrical slip rings into the centrifuge control room, where the computers were present. This required a large number of slip rings as each instrument would require two slip rings to carry the signal in a double-ended format. Also the distance to the control room from the centrifuge meant long transmission distances and an increased chance of the signal being corrupted by ambient electrical noise. More recently, the A/D converters and the computers are being flown on the centrifuge, usually located closer to the center of the centrifuge axis. This minimizes the *g* forces seen by the computers. Further, the availability of solid-state hard drives enhanced the reliable functioning of computers in a high-gravity environment. Moving the data acquisition onto the centrifuge reduced the requirement for a large number of electrical slip rings. These days we only require a few slip rings to run an Ethernet connection onto the centrifuge and the on-board computers can be made available on a network.

9.2 ANALOG TO DIGITAL CONVERSION

A key stage in the data acquisition process as outlined in Figure 9.1 is the analog to digital conversion. As explained earlier, instruments produce an analog signal which is continuous with time. A digital signal by definition

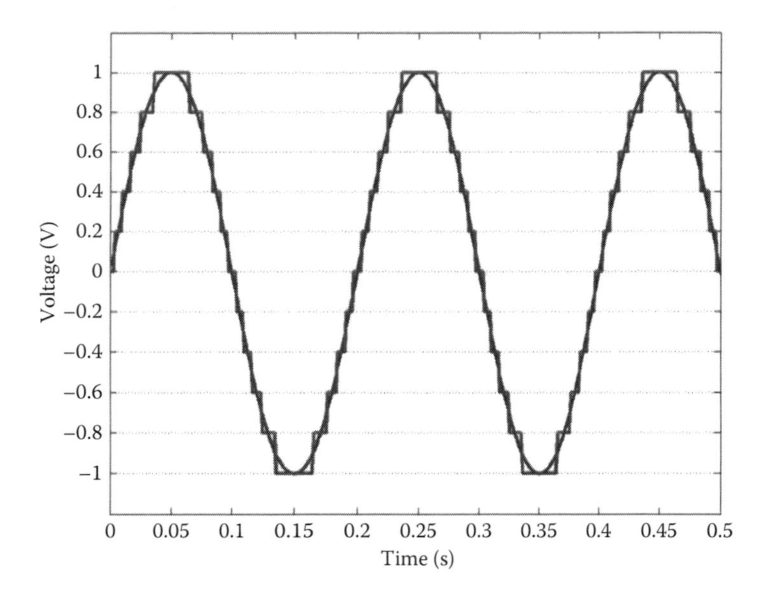

Figure 9.2 Analog and digital signals.

is a set of discrete values at each of the time increments. The range of the A/D card is divided into a number of steps. Thus, a continuous signal appears as a series of step functions. This is illustrated in Figure 9.2, which shows a continuous sinusoidal signal of ±1 V. This signal is digitized at 10 discrete levels, each a step of 0.2 V, and fits the continuous sine wave. At each time instant only a discrete value is logged. Clearly, the more discrete levels there are, the more closely the digital signal will match the analog signal.

The process of converting a continuous signal into a series of discrete values enables us to log the digital signal using computers.

9.3 FUNDAMENTALS OF DIGITAL DATA LOGGING

There have been rapid advances in digital data logging in the past decade or so, primarily due to advances in the computer industry and the rapid increase in processor speeds. Also the ability to log and store vast amounts of data has increased. In order to understand the data acquisition systems used in centrifuge testing, we need to know a few basic ideas regarding the digital data logging. In later sections of this chapter the post-processing of the digital data obtained in a centrifuge test will be considered.

9.3.1 Precision

The accuracy of a data acquisition system relies on the precision of the A/D converter. As explained in Section 9.2 the more levels of digitization there are, the better is the match between the analog and digital signals. In other words the digitization steps are so small that the digitized signal can match the changes in the analog signal with time closely.

The A/D converter changes the continuous analog signal into a number of discrete steps. For example a 12-bit converter will have 2^{12} or 4096 levels to cover the range of the analog signal. In order to demonstrate the increase in the precision of data logging with number of steps, let us consider the previous example in Figure 9.2. Here a signal of ±1 V is digitized by an A/D converter having 10 discrete levels, each with a step size of 0.2 V. So an instrument output giving a signal of +0.61 V will be recorded as +0.8 V, an error of nearly 31.2 percent. If the same range of ±1 V is digitized using a 12-bit A/D converter, then this range will be covered by +2048 to –2048 discrete steps with a step size of 2.4414×10^{-4} V. Thus, a signal of +0.61 V coming from an instrument will be recorded as 0.61010742 V, with an error of 0.0176 percent. This error is sometimes called the quantization error. Similarly if a 24-bit A/D converter is used the quantization error will be even smaller. Thus, it is straightforward to see that with increased levels of digitization the accuracy of the measurement is also increased.

9.3.2 Amplification

The analog signal produced by some of the instruments used in centrifuge modelling can be quite small, perhaps only a few millivolts. The signal therefore requires amplification. This can be done in two ways. Traditionally a power amplifier is used to increase the amplitude of the signal. Such amplifiers are applied for each of the instrument channels as outlined in Figure 9.1 and different gains can be used for different channels. It is also common to standardize the possible gains in units of ×10, ×100, or ×1000 depending on how small the signal from the instrument is. Another way to achieve the same effect of amplification of the signal is through the A/D converter. The range of the A/D converter can be made to match the largest value of the expected signal from the instrument. Normally this can be done by choosing the range of the A/D converter as ±1.25 V, ±2.5 V, ±5 V, or ±10 V.

It is also possible to combine these two ways of amplification as required, to obtain a good signal from the instrument. This is explained with the aid of an example. Let us suppose that a strain gauge produces a maximum signal of +40 mV and a minimum signal of –30 mV in response to the strains induced by an applied time-varying load in a centrifuge test. Let us suppose that an amplification of ×100 is applied on this channel to increase the signal to a maximum of +4 V and a minimum of −3.5 V. If we choose the range on the 12-bit A/D converter to be ±5 V, we can see that both the maximum and minimum values will still be in range. The value of each step of the digitized signal will be 10/4096 = 0.002441 V. Thus, the maximum signal of +4 V is expressed as 1638 steps and the minimum signal of −3.5 V is expressed as 1434 steps. If we choose the range on the 12-bit A/D converter to be ±10 V, then each step of the digitized signal will be 20/4096 = 0.004882 V. In this case the maximum and minimum values of the signal are expressed as 819 and 717 steps, respectively. Thus, by choosing the smallest possible range into which both the maximum and minimum values of the signal will fit, we can effectively amplify the signal. In this example, by choosing the range as ±5 V we are effectively using an amplification factor of ×200.

9.3.3 Sampling rate

The sampling rate can be defined simply as the number of data points logged per second. It is expressed in hertz and is sometimes called the sampling frequency f. For example, if we log 10 samples per second then the sampling frequency is 10 Hz. The separation between successive data points is often called data point spacing Δt and will have the unit in terms of time (seconds). The data point spacing is related to the sampling frequency as:

$$\Delta t = \frac{1}{f} \tag{9.1}$$

Clearly the higher the sampling frequency the more well-defined the time-varying signal we are trying to log will be. In other words at higher sampling frequencies, the number of data points will be high and the data point spacing will be small, allowing us to follow intricate changes in the signal with time. This is illustrated with the help of an example of a time-varying signal defined as:

$$S(t) = 1.0\sin(\omega t) + 1.0\sin(2\omega t) + 1.0\sin(3\omega t) \tag{9.2}$$

Let us consider a value of ω = 62.3 rad/s (i.e. 10 Hz) in Equation 9.2. This continuous signal is logged at 1000 Hz, 100 Hz, and finally at 50 Hz, as shown in Figure 9.3. Clearly the signal is fully captured when logged at a sampling frequency of 1 kHz. When the sampling frequency is cut to 100 Hz, the smoothness of the sine wave function in Equation 9.2 is not captured. Instead the digitized signal looks angular. When the sampling frequency is cut to 50 Hz, the digitized signal loses much of the higher

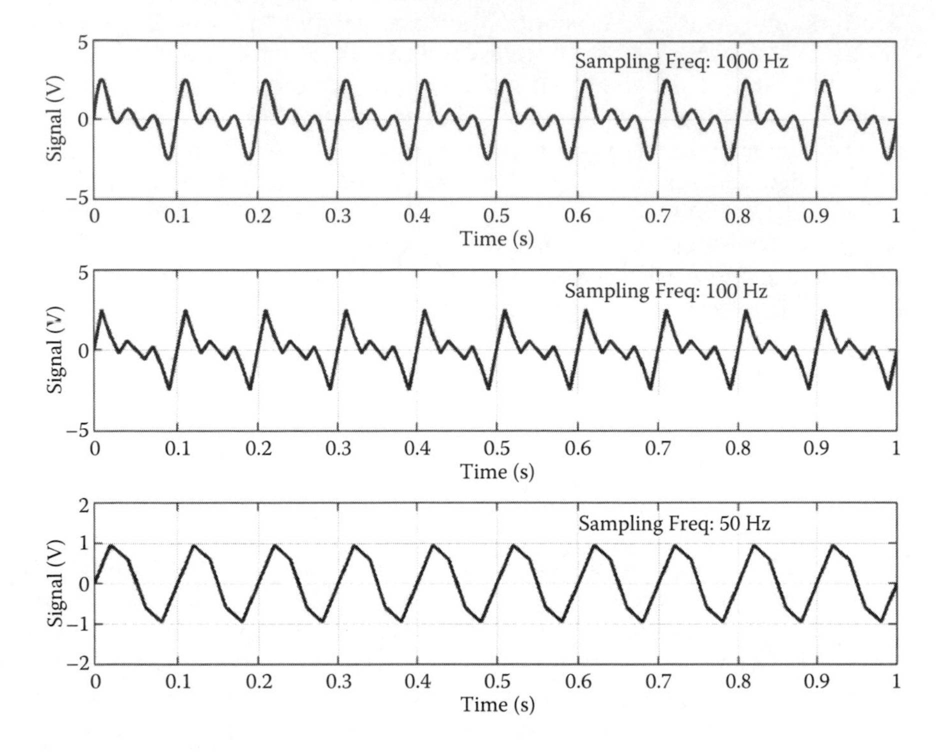

Figure 9.3 The effect of sampling frequency on logged data.

frequency components and the peak amplitude recorded also drops as seen in Figure 9.3. From this example it can be seen that it is very important to log signals at a sufficiently high sampling frequency to capture them accurately. Further, sampling at lower frequencies can result in the recording of quite spurious data from the instruments in the centrifuge tests.

9.4 ELECTRICAL NOISE

Electrical noise can impinge on analog signals coming from instruments in a centrifuge test. Many electrical devices, such as motors, transformers, and inverters, generate electrical noise. This electrical noise may be superimposed onto the signal. A common frequency at which the electrical noise is generated is the power supply frequency. In the United Kingdom the power supply frequency is 50 Hz, while in the United States it is 60 Hz.

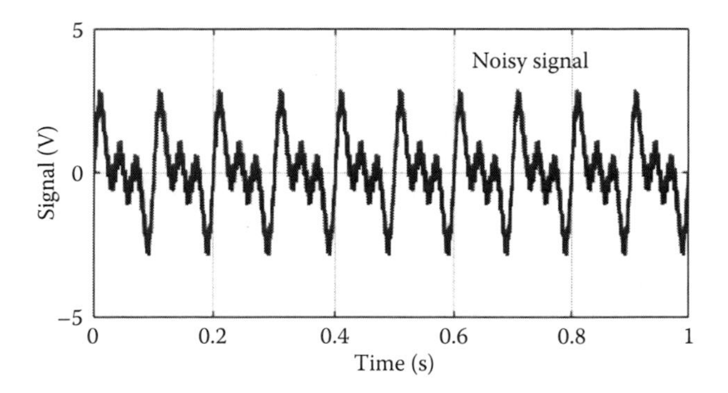

Figure 9.4 Noisy signal.

An example of noise being superimposed on the signal of the form given by Equation 9.2 is shown in Figure 9.4. In this example, the noise is superimposed at frequencies of 50 Hz, 250 Hz, and 500 Hz. The signal itself was logged at 2 kHz. Comparing this figure to Figure 9.3, although the basic form of the signal can still be seen, there are unwarranted and spurious higher frequencies superimposed on the signal.

It is possible to avoid electrical noise by having analog filters that remove the noise before the signal from the instrument gets digitized. In fact it is better to have some analog filtering to avoid aliasing of signals. This aspect is discussed in more detail in Section 9.7. Despite this it is possible and may be necessary to carry out digital filtering during the post-processing of the centrifuge data. In the next section, the theoretical basis for signal processing is presented. The techniques used in digital filtering are discussed in Section 9.8.

9.5 TIME AND FREQUENCY DOMAINS

The signals from instruments in a centrifuge test are usually time varying and therefore are logged with time. These are called time histories. Signals obtained in the time domain can be transformed into the frequency domain. This is done using a Fourier series. Any periodic, time-varying signal can be expressed as the summation of a series of cosine and sine functions (Fourier, 1822).

$$f(x) = a_o + \sum_{n=1}^{\infty} [a_n \cos(nx) + b_n \sin(nx)] \tag{9.3}$$

The Fourier coefficients for a periodic function with a periodicity of 2π can be evaluated using Euler formulae as:

$$a_o = \frac{1}{2\pi}\int_{-\pi}^{+\pi} f(x)\, dx \tag{9.4}$$

$$a_n = \frac{1}{\pi}\int_{-\pi}^{+\pi} f(x)\cos(nx)\, dx \tag{9.5}$$

$$b_n = \frac{1}{\pi}\int_{-\pi}^{+\pi} f(x)\sin(nx)\, dx \tag{9.6}$$

Another way of expressing the Fourier series in a more concise manner may be to utilize the relationships:

$$\cos(nx) = \frac{e^{inx} + e^{-inx}}{2} \tag{9.7}$$

$$\sin(nx) = \frac{e^{inx} - e^{-inx}}{2i} \tag{9.8}$$

and write:

$$\mathrm{f}(x) = \sum_{n=-\infty}^{+\infty} c_n\, e^{inx} \tag{9.9}$$

where the Fourier coefficients are given by:

$$c_n = \frac{1}{2\pi}\int_{-\pi}^{\pi} f(x)e^{-inx}\, dx \tag{9.10}$$

In Equation 9.3 inclusion of more terms increases the accuracy with which the signal is reproduced by the summation of the cosine and sine functions. Also, for periodic functions with different periodicities, the

Euler formulae in Equations 9.4 to 9.6 can be scaled simply by a factor of $2\pi t/T$ instead of x (Kreyszig, 1967). These can be written as:

$$a_o = \frac{1}{T}\int_{-T/2}^{+T/2} f(t)\, dt \tag{9.11}$$

$$a_n = \frac{2}{T}\int_{-T/2}^{+T/2} f(t)\cos\left(\frac{2\pi nt}{T}\right) dt \tag{9.12}$$

$$b_n = \frac{2}{T}\int_{-T/2}^{+T/2} f(t)\sin\left(\frac{2\pi nt}{T}\right) dt \tag{9.13}$$

Thus, if we have our signal *f(x)* as a periodic function with any periodicity, we can express it by the summation of cosine and sine functions and evaluate the Fourier coefficients. These Fourier coefficients define the components of the signal at different frequencies, which when added together will reproduce the original signal. In other words by working out the Fourier coefficients we are transforming the signal from the time domain into the frequency domain.

The same concept of moving a continuous signal from the time domain into the frequency domain can be applied to a digitized signal (Newland, 2005). Although the digital signal is now expressed as a set of discrete values as explained earlier, a discrete Fourier transform (DFT) can change this signal from the time domain into the frequency domain.

Let us suppose that our digital signal is expressed at N points as:

$$x_n = [x_o, x_1, x_2, \ldots x_{N-1}],$$

each separated equally by the data point spacing Δt. Following Equations 9.9 and 9.10, we can write:

$$X_k = \sum_{n=0}^{N-1} x_n\, e^{-i\, 2\pi kn/N} \tag{9.14}$$

X_k is a complex number that retains both the amplitude and phase information of the function x_n at specific frequencies. Numerically we can work out the value of X_k for any digitized signal using Equation 9.14. In fact, specialized algorithms have been developed to calculate efficiently the fast Fourier transform (FFT) of a digital signal. Newland (2005) discusses such algorithms and their relative merits. These days computational platforms like MATLAB® offer specific functions to calculate the FFT of a digital signal efficiently. It is also possible to revert back from the frequency domain into the time domain by carrying out an inverse transformation of the form:

$$x_n = \frac{1}{N}\sum_{k=0}^{N-1} X_k e^{i\,2\pi kn/N} \tag{9.15}$$

By employing this inverse discrete Fourier transform (IDFT) we can recover the original signal without any loss of information. Again there are MATLAB functions available that will recover the original signal from the Fourier components.

Let us reconsider the example of a digitized signal shown in Figure 9.4. We can apply the FFT to this signal and plot the amplitude of the signal with respect to frequency as shown in Figure 9.5. In this figure we can see that in addition to the signal components at 10, 20, and 30 Hz given by Equation 9.3, we have other components at 50, 250, and 500 Hz. It is these additional components that are making our signal look noisy.

As the FFT we worked out has components at each of the frequencies, we can simply delete all the unwanted components at higher frequencies. Of course this would involve removing both the real and complex parts of the Fourier components. For example in the FFT shown in Figure 9.5, we can simply delete all components at frequencies greater than 40 Hz. If we did that and then applied the inverse Fourier transform, we can obtain the "clean" signal.

This is illustrated in Figure 9.6, which shows the removal of the components at frequencies greater than 40 Hz and then obtaining the clean signal by performing an inverse fast Fourier transform (IFFT). This "cleaned" signal compares very well with the original signal seen in Figure 9.3 that was sampled at 1 kHz.

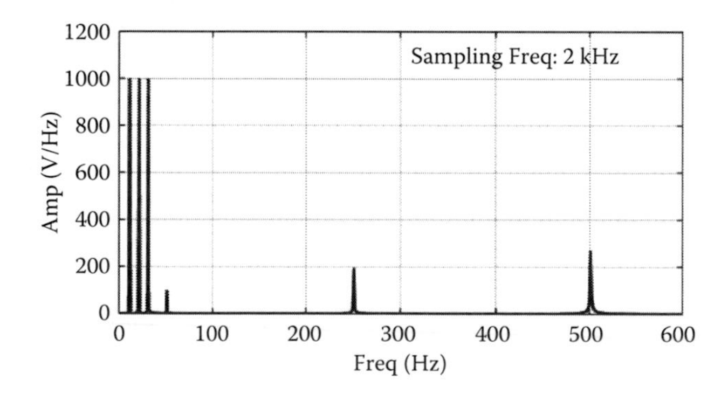

Figure 9.5 Fast Fourier transform (FFT).

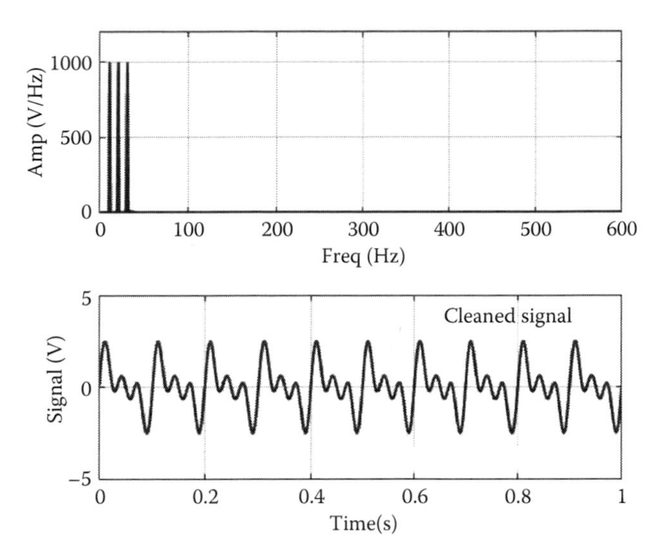

Figure 9.6 Cleaning of a noisy signal.

9.6 SIGNAL-TO-NOISE RATIO

In Section 9.4 we said that electrical noise can be superimposed on signals coming from the instruments in a centrifuge test and in Section 9.5 we looked at a way of cleaning up the signal using FFTs. However, we need to know if this is always possible. For example, what happens if the amplitude of the noise is very high and worse still, it is comparable to the amplitude of the signal generated by the instrument in a centrifuge test. This can often be a problem for instruments such as strain gauges which produce relatively small amplitude signals. If the frequency of the noise is well known, such as the power supply frequency, and it is very different from the frequency of the signal we can still use the FFT technique outlined in Section 9.5 to clean up the signal. However, if the noise occurs at the same frequency as the signal from the instrument, it will not be possible to use the FFT technique to remove the noise.

In signal processing the term signal-to-noise (S/N) ratio is used to define a good quality signal or how well the signal stands out from background noise. This ratio typically measures the ratio of power of the signal from the instrument to the power of the noise. As the power of the signal is proportional to the square of the amplitude, we can write:

$$\frac{S}{N} = \frac{P_{signal}}{P_{noise}} = \frac{A_{signal}^2}{A_{noise}^2} \tag{9.16}$$

It is common to express the S/N ratio in log scale as many signals have a wide frequency range.

$$\frac{S}{N} = 10\ log_{10}\left(\frac{A_{signal}^2}{A_{noise}^2}\right) = 20\ \log_{10}\frac{A_{signal}}{A_{noise}} \tag{9.17}$$

Typically we would require an S/N ratio of about 5, so the amplitude of the signal from the instrument must be at least 1.78 times larger than the amplitude of the noise. If the noise level in a data acquisition system on a centrifuge is about 20 mV, then the smallest signal from the instrument must be at least 35.6 mV to give us a good S/N ratio. The same idea of expressing the S/N ratio in log scale is behind the concept of the 3 dB point. This will be considered later in Section 9.8.2.2.

9.7 ALIASING AND NYQUIST FREQUENCY

When logging signals from instruments in a centrifuge test we need to have a good S/N ratio, as explained in the previous section. In addition we need to be aware of certain fundamentals of signal processing.

9.7.1 Aliasing

Aliasing of a signal from an instrument in the centrifuge test can happen if the sampling frequency is too low. We have seen earlier in Section 9.3.3 that the signal must be sampled at a sufficiently high sampling frequency to capture it properly. In this section we shall investigate what happens if the sampling frequency is low. This is best explained by considering a simple example.

Let us consider that the signal from an instrument is a simple sine wave with an amplitude of ±1 V and a frequency of 9 Hz, as shown in Figure 9.7. If we were to sample this signal at a frequency of 1 Hz, that is, one sample per second, then we can see in Figure 9.7 the values match the 9-Hz signal very closely. Thus, if we sampled at this low frequency we would think that the actual signal is the curve that passes through the **x** marks and therefore the dashed curve line is the actual signal. This discrepancy of two signals becoming indistinguishable when sampled at different frequencies is called temporal aliasing.

In fact this type of aliasing can happen with signals at multiple frequencies. As a result the actual signal from the instrument in a centrifuge test can be aliased with multiple spurious frequencies just by virtue of sampling frequency. In other words, any other sine wave frequencies that fit the same data points marked by **x** in Figure 9.6 can be aliased onto

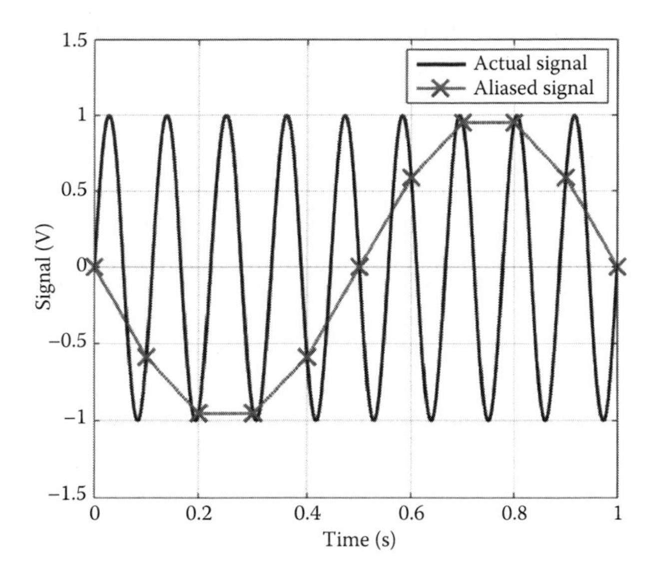

Figure 9.7 Aliasing of a signal.

the signal. One way of avoiding aliasing in data logging is to use anti-aliasing filters before the analog signals are sent to the A/D converter (see Figure 9.1). Another way to reduce the effects of aliasing is to over sample the signal, that is, to use a high sampling frequency. This way we can avoid aliasing of high frequency signals. The theoretical background for aliasing is explained next.

9.7.2 Nyquist frequency

In information theory, the Nyquist theorem of sampling states that if f_h is the highest frequency present in a signal, the sampling frequency in order to capture the signal perfectly must be greater than or equal to 2 f_h.

$$f_{sampling} \geq 2\ f_h \tag{9.18}$$

The Nyquist frequency can be related to the sampling frequency as:

$$f_{Nyquist} = \frac{1}{2}\ f_{sampling} = \frac{1}{2\ \Delta t} \tag{9.19}$$

In Equation 9.19 Δt is the data point spacing. In other words if we sample a signal at 2 kHz, the highest frequency in the original signal must be less than 1 kHz. This theorem requires high sampling frequencies in order for us to capture the high frequency components in a signal coming from the instruments in a centrifuge test. In most centrifuge tests, the frequencies in a signal are quite low and therefore the sampling frequencies can also be low. A notable exception to this is the dynamic centrifuge tests such as earthquake tests, where the scaling laws require the rate of loading to be in tens of hertz (e.g., 50 Hz to 250 Hz). In these tests the sampling frequency required may be a few kilohertz per channel logged (e.g., 2 to 20 kHz per channel).

The reason for the requirement that the sampling frequency must be twice the highest frequency present in the signal is explained with the help of Figure 9.8. In this figure we reproduce the plot shown in Figure 9.5 of the FFT of a noisy signal logged at 2 kHz. This time the frequency axis is extended to the full 2 kHz as opposed to plotting only the range of interest as we did in Figure 9.5. In this example we have logged the signal at 2 kHz, which means the Nyquist frequency is 1 kHz as marked by the vertical line in Figure 9.8. We can see in this figure that the noisy signal shown has the main frequencies at 10, 20, and 30 Hz, with noise present at 50, 250, and 500 Hz. All these frequency components get reflected about the Nyquist frequency. Thus, we get additional components at 1990, 1980, and 1970 Hz for the main signal and the noise components at 1950, 1750, and 1500 Hz.

Another way to look at the Nyquist frequency is that only half the Fourier spectrum has useful information while the other half simply has reflected

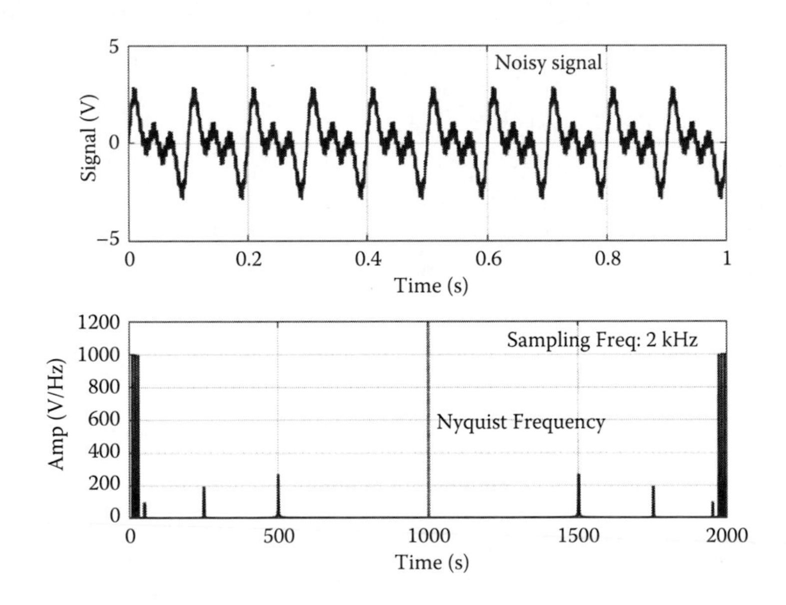

Figure 9.8 Nyquist figure showing the replication of Fourier components.

components. So if we want to log the signal accurately with all the higher frequencies intact, we need to make the Nyquist frequency to be large; that is, we need to increase the sampling frequency to be high according to Equation 9.19. The highest sampling rate possible in a centrifuge data acquisition system is usually determined by the speed of the A/D converter. We can get A/D converter speeds to be of the order of 1 MHz throughput. Thus, if there are 64 channels in the A/D converter then the maximum frequency with which we can sample would be about 15 kHz per channel. This gives us a Nyquist frequency of 7.5 kHz and this would limit the highest frequency signal that can be logged.

9.8 FILTERING

In previous sections the need for filtering of signals produced by the centrifuge instrumentation was introduced. Electrical noise from power supply frequencies, electrical inverters, transformers, and other electrical devices can all produce noise that can interfere with the signals from the instruments. Removing the noise from the signals can be done in many ways. These are considered next.

9.8.1 Analog filtering

One way to remove noise from signals is the use of analog filters. These filters are made from electrical components or chips and can filter analog signals. These filters are inserted between the instruments producing the signal and the A/D converter as shown in Figure 9.1. If amplifiers are used to boost the signal from some of the centrifuge instruments, then filtering is done before the amplifiers. This prevents the unnecessary amplification of the noise as it is removed before reaching the amplifier. Analog filters are also used for anti-aliasing purposes to prevent high frequency noise components aliasing onto lower frequency signals as explained in Section 9.7.

A simple analog filter can be constructed using resistors and capacitors (RC). An example of such a filter is shown in Figure 9.9. It can be used when the signal (V_{in}) from an instrument is measured relative to ground.

In the signal coming from an instrument, this circuit will allow low frequencies below a cut-off frequency to pass but will stop the high frequencies from going through. The cut-off frequency for this type of RC circuit can be calculated as:

$$f_c = \frac{1}{2\pi RC} \tag{9.20}$$

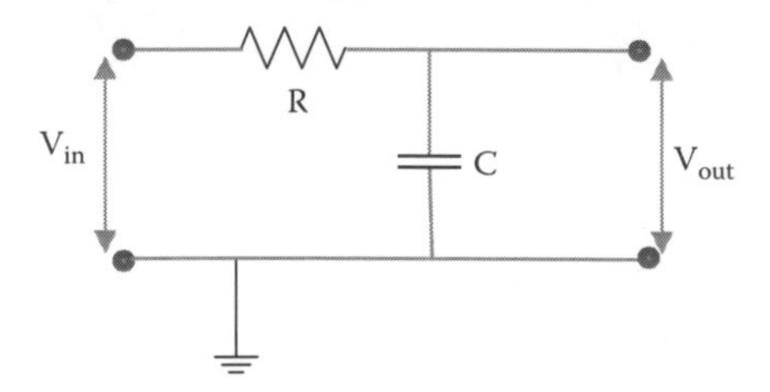

Figure 9.9 A simple RC filter circuit.

The effect of the RC circuit on the signal from an instrument can be calculated using a transfer function. For a simple RC shown in Figure 9.9 the transfer function can be calculated as:

$$\frac{V_{out}}{V_{in}} = \frac{f_c}{\sqrt{R^2 + f_c^2}} \tag{9.21}$$

One of the side effects of using an RC filter is that it introduces a phase difference between the input signal (V_{in}) and the output (V_{out}). This phase difference may be calculated as:

$$\varnothing = \tan^{-1}(2\pi f RC) \tag{9.22}$$

We can design the circuit to have the desired cut-off frequency by using appropriate R and C values. This type of filter is called a first-order filter and reduces the power in the signal by half whenever the frequency doubles, that is, increases by an octave. Let us choose a resistor with 5 kΩ resistance and a capacitor with a 50 nF capacitance. With these components the cut-off frequency can be determined using Equation 9.20 as 636.62 Hz. With these particular components, we can plot the filter characteristics as shown in Figure 9.10. The transfer function in this figure remains at 1 up to about 100 Hz and starts to roll down after that. At 10 kHz it falls to about 0.1. The cut-off frequency for which we designed the RC filter is also marked on Figure 9.10. It can be seen that instead of an ideal "step function"-like filter, we obtained a more gently rolling down filter. So using this filter will ensure that all frequencies below 100 Hz are unaffected, and frequencies above 100 Hz will start to attenuate albeit gently. Filter characteristics similar to the idealised "step function" can be obtained using higher-order filters and this will be discussed later in Section 9.8.2.3. As mentioned before, using RC filters introduces a phase lag between the input signal and the output. For the particular components we have chosen, the phase angle is plotted in Figure 9.10. In this figure it can be seen that at lower frequencies the

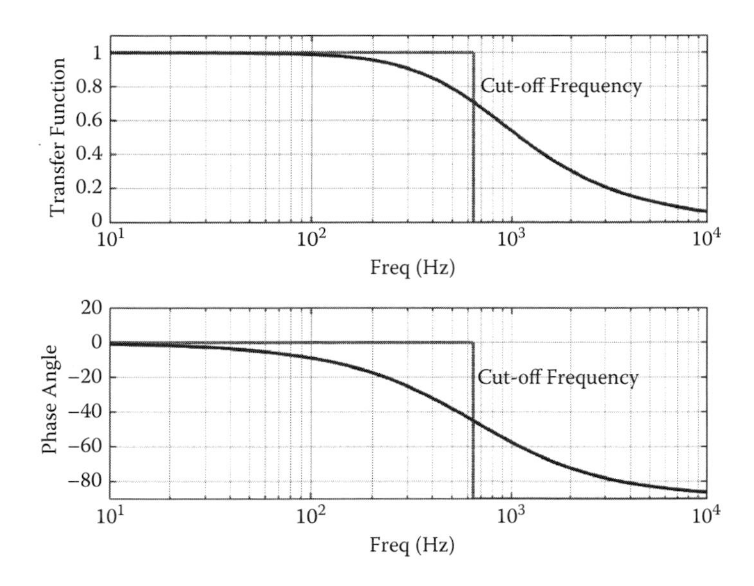

Figure 9.10 Characteristics of a first-order RC filter.

phase angle is close to zero. However, at higher frequencies it increases. At 100 Hz we get a phase difference of about 10° which can be considered to be small. By the time we reach the designed cut-off frequency of 636.62 Hz the phase difference increases to nearly 45°. At very high frequencies the phase lag approaches 90°. In a way the phase difference is the price we pay for using RC filters. As centrifuge modellers we need to be aware of this. If all the signals coming from a centrifuge test are put through the same type of RC filters, then they are all affected in the same way and we will still be able to compare them with each other, without worrying about the phase lags. However, if the instruments produce signals at different frequencies, then we must understand that the phase lags due to RC filters will affect different frequency signals differently as seen in Figure 9.10.

It is normal to use some form of RC filter in centrifuge data logging for anti-aliasing purposes. Normally we will organize the cut-off frequency to be about 3 kHz or higher, so that the phase lags at lower frequencies will be quite small.

9.8.2 Digital filtering

Digital filtering is used during post-processing of the data obtained from the instruments in a centrifuge test. It is carried out either by specific software or numerical tools developed to carry out the filtering. The main aim of digital filtering is to improve the quality of the signal. Digital filtering can be done on the stored data from a centrifuge test, after the testing has been completed or during the centrifuge test as the digitized data starts to flow from the A/D converter channels. It is good practice to store the raw data and carry out any signal processing after the test. However, in some circumstances digital filtering may be carried out before displaying the signals during a test or the filtered data is used to control an actuator through a D/A converter in a control loop. The latter is often used for example in a force-controlled load application, where the signal from a load cell is digitized and filtered before passing through a control loop back into the actuator via a D/A converter.

There are many ways in which digital filtering can be carried out. Some of these are discussed below.

9.8.2.1 Smoothing functions

The simplest form of filtering to reduce the noise in a signal may take the form of using a smoothing function. This involves taking the data at a given time instant in each signal and averaging it with its neighboring data points. Then we move to the next data point and repeat the operation. This process is also called a moving average. If our signal consists of N data points in an array $[x_1, x_2, x_3 \ldots, x_N]$, then we can perform a three-point moving average using:

$$[x_{new}] = \sum_{i=2}^{N-1} \frac{x_{i-1} + x_i + x_{i+1}}{3} \tag{9.23}$$

If we take the noisy signal from Figure 9.4, we can apply the three-point moving average from Equation 9.23 and see the effect of smoothing. This is presented in Figure 9.11, marked “Single pass smoothing.” In this figure the time axis is shortened to 0.2 seconds and the smoothed signal is overlain on the original signal, so that the effect of smoothing is clear. In this figure we can see that despite using Equation 9.23, the smoothed signal still retains the higher frequency ripples on the main signal although their amplitude is reduced due to averaging. Of course, we can apply the same process of moving averages several times, with each pass using the smoothed array from previous pass. An example of this is also presented in Figure 9.1 where the noisy signal is subjected to eight passes of moving point average. In this figure we can clearly see that all the high frequency components are removed from the noisy signal.

Instead of using a three-point moving average, we could use a five- (or more) point moving average. It is possible to show theoretically that the moving point average is similar to application of a low pass filter and changing the number of passes used is equivalent to changing the filter characteristics.

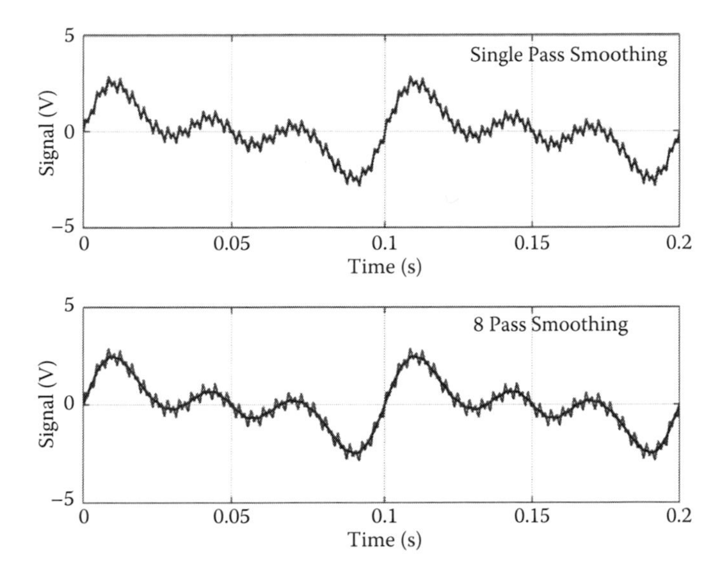

Figure 9.11 Single and multi-pass smoothing.

9.8.2.2 Turning frequency and 3-dB point

While considering analog filters we have seen that a first-order filter results in reduction of the amplitude of the output signal that starts before the cut-off frequency and continues to roll down gently above the cut-off frequency. We would like to know how the power in the input signal changes with the application of the filter. As we did with the S/N ratio given by Equation 9.17 in Section 9.6, we can express the power in the signal as the square of the amplitude. Thus, "gain" is defined as the ratio of amplitudes of the signal before and after filtering and is normally expressed in a log scale as:

$$Gain = 10\ \log_{10}\left(\frac{A_{output}^2}{A_{input}^2}\right) = 20\ \log_{10}\frac{A_{output}}{A_{input}} \tag{9.24}$$

The units for gain are decibels (dB) as the origin for this concept is in sound and noise measurement.

Let us suppose that we would like to know when the power of an input signal drops by half by application of a digital filter. For power to be halved, the amplitude of the output relative to input must drop by $1/\sqrt{2}$, that is, by 0.707. Using this value in Equation 9.24 gives as a gain of −3.01 dB. Thus, we can define a 3-dB point of a filter as the point where the power of the

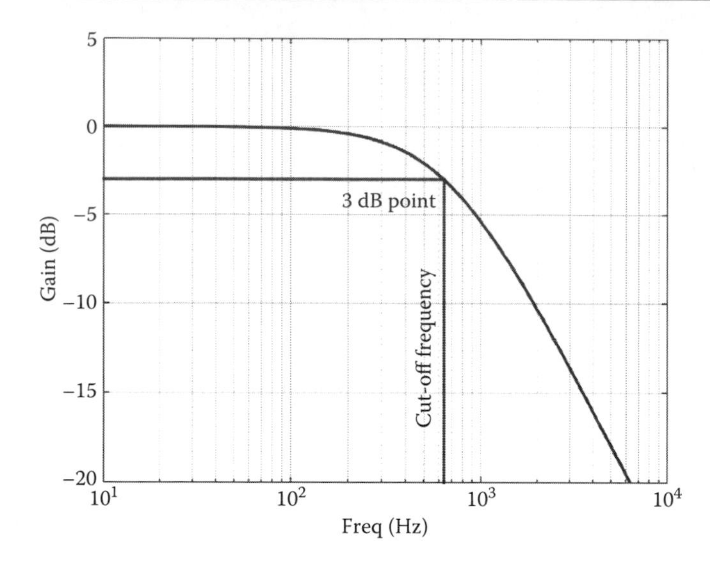

Figure 9.12 Turning frequency and 3-dB point.

signal is halved. The corresponding frequency at the 3-dB point is defined as the cut-off frequency or turning frequency of the filter.

Let us reconsider the first-order filter with a cut-off frequency of 636.62 Hz shown in Figure 9.10. This is now plotted in terms of gain defined by Equation 9.24, as shown in Figure 9.12 on which the 3-dB point and the cut-off frequency are marked. The slope of the filter characteristic beyond the cut-off frequency is about 20 dB per decade and is linear. The first-order filters have quite a gentle roll-off and therefore affect quite a wide range of frequencies on either side of the cut-off frequency. As the output signal drops in amplitude by a factor of 10 every decade, it takes quite a large frequency range before the output signal drops off significantly. The other aspect of the filter that needs to be kept in mind is the phase difference that is introduced in the output signal. At the 3-dB point the phase difference between the output and the input signal is as much as 45° (see also Figure 9.10). We need to use higher-order filters to achieve a steeper roll-off that would be closer to the ideal filter, which would attenuate the output signal significantly beyond the cut-off frequency.

9.8.2.3 Higher-order digital filters

Higher-order filters are used to have a much sharper cut-off beyond a desired frequency and avoid a gentle roll-off. As discussed before for a first-order filter the slope of the gain function shown in Figure 9.12 beyond the cut-off frequency is about –20 dB per decade. For a second-order filter

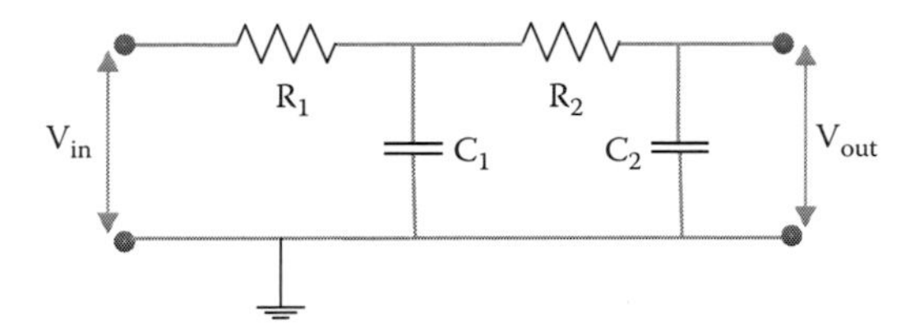

Figure 9.13 Second-order RC filter.

this increases to −40 dB per decade, for a third-order filter to −60 dB per decade, and so on. Butterworth worked with analog filters although the same theory can be applied to digital filters. He showed that by adding successive filter elements, and choosing resistors and capacitors of appropriate values, filter characteristics close to an ideal filter can be obtained. For example a second-order filter can be constructed as shown in Figure 9.13 (Williams and Taylors, 1988).

In fact we can repeat this process and add more and more R and C components to get closer to an ideal filter. Butterworth filters of up to the eighth order are commonly used in processing the test data from many centrifuge tests. The transfer functions for the higher-order Butterworth filters can be obtained by using:

$$\frac{V_{out}}{V_{in}} = \frac{1}{\sqrt{1+\left(\frac{f}{f_c}\right)^{2n}}} \tag{9.25}$$

where n is the order of the filter and f_c is the cut-off frequency. The resulting transfer functions are plotted in Figure 9.14, for the same cut-off frequency of 636.62 Hz used before. In this figure we can clearly see that the filter characteristics get closer and closer to the "ideal" filter as the order of the filter is increased. Of course an ideal filter will have a perfect vertical line at the cut-off frequency.

Although Butterworth filters are extensively used, there are other types of filters available as well. Notable among these are Chebyshev type I and type II filters and elliptic filters, which are popular. In Figure 9.15 the transfer functions for eighth-order Butterworth, Chebyshev type I and type II filters, and the elliptical filter are presented for the same cut-off frequency we used earlier, that is, 636.62 Hz. In this figure we can see that the eighth-order Butterworth filter is quite stable although the roll-off is not as steep as some of the other filters. For the Chebyshev type I filter, the transfer function oscillates before the cut-off frequency but is stable afterward. This means that the output signal components lower than the cut-off frequency can get affected by this filter, which is not ideal. On the other hand, its

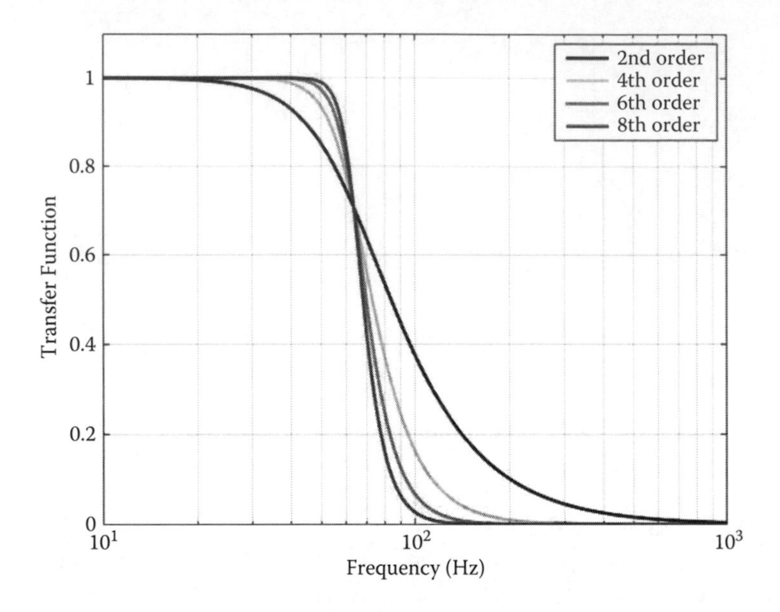

Figure 9.14 Transfer functions for higher-order Butterworth filters.

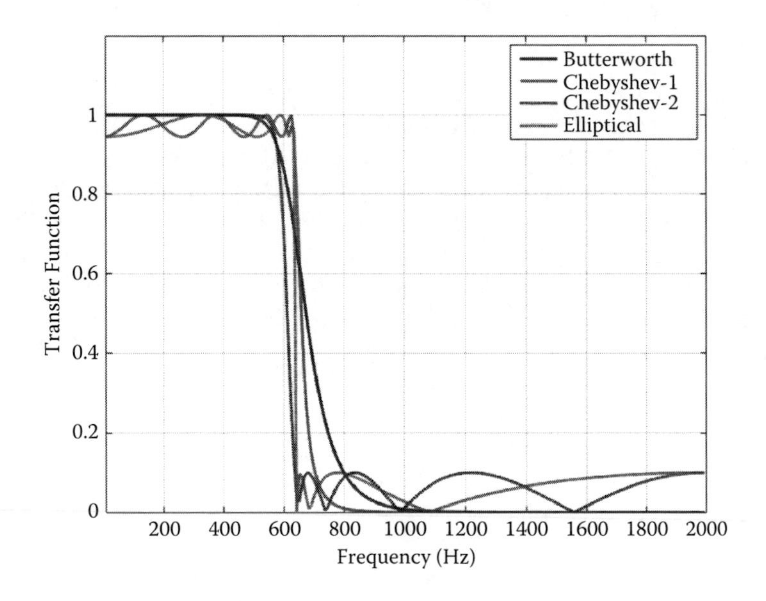

Figure 9.15 Comparison of transfer functions for eighth-order filters.

roll-off is much better than the Butterworth filter. For the Chebyshev type II filter, there are no oscillations before the cut-off frequency but large oscillations ensue post–cut-off frequency. This would mean that output signal components higher than the cut-off frequency can get affected adversely. The elliptical filter has a transfer function with oscillations both before and after the cut-off frequency, although it has the steepest roll-off which is quite close to the cut-off frequency. In practice we will choose the order and type of filter depending on the type of noise present in the data coming from the instrumentation in a centrifuge test.

We will now see how these digital filters are used to remove the noise from a signal. For this, let us revert back to the noisy signal that we have considered in Figure 9.4. This signal had high frequency noise components. By applying the transfer functions from the digital filters shown in Figure 9.15 we can remove the noise from the signal. This is demonstrated in Figure 9.16. In this figure the noisy signal is overlain by the "cleaned up" signal using two different filters, that is, the Butterworth eighth-order filter and the Chebyshev type I filter. We can see in this figure that in both cases the high frequency noise components are removed in the cleaned up signal.

We could have used other types of filters such as the Chebyshev type II filter or the elliptical filter to achieve the same result. However, as discussed before when there are many high frequency components in the signal, these types of higher-order filters can allow some noise to pass through due to the

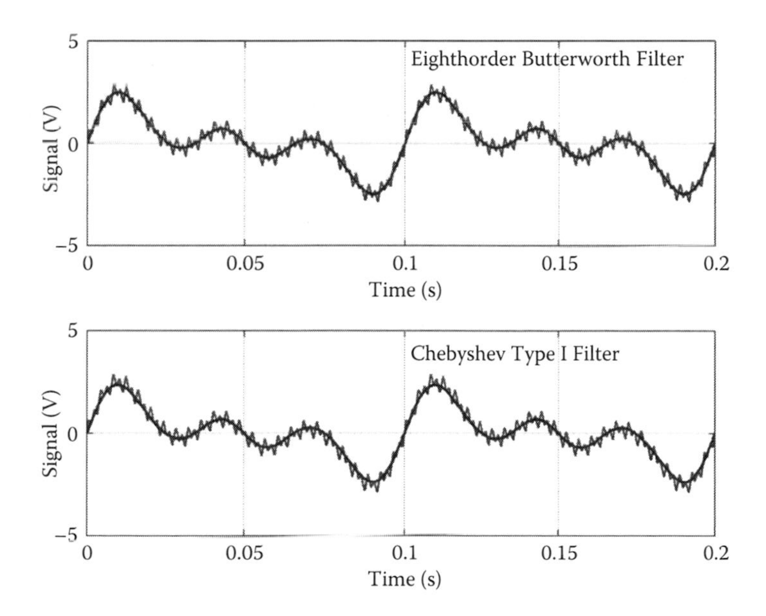

Figure 9.16 Filtering of a noisy signal using different filters.

ripples in the transfer function. We should therefore be careful in the type of filter that is used in any given case.

Another aspect that warrants careful attention is the phase difference that is introduced into the signal during filtering. We can minimize this by passing the input signal through the filter in the forward direction and then taking the filtered signal and passing it through the same filter in the reverse direction. This removes any accrued phase difference after the forward pass. In MATLAB this is achieved by using the "filtfilt" function.

9.9 SUMMARY

In this chapter we have considered the data acquisition systems used in a centrifuge to acquire data from the instruments. We were required to understand some basic concepts of A/D conversion that is used to convert analog signals from the instruments into digital signals. Then we considered the fundamentals of digital data logging such as precision, need for amplification of signals, and sampling frequency. In addition we had to recognize that electrical noise can be superimposed onto the signals coming from the instruments in a centrifuge test. We have looked at the necessity of logging signals at a high sampling rate and what happens if we undersample a signal that has high frequency components. We learnt about aliasing of a signal with another signal at a different frequency. Anti-aliasing filters are required in a data acquisition system to prevent this.

In order to understand the ways in which we can remove this electrical noise, we had to look at the theory of Fourier transforms that convert a signal in the time domain to the frequency domain. Fast Fourier transforms (FFTs) are used to achieve this operation in digital signals. Once the signal has been expressed in a frequency domain, we can remove the noise components and perform an inverse Fourier transformation (IFFT) to recover the cleaned up signal. We then looked at different ways in which filtering can be performed, from simple analog filters built from resistors and capacitors to digital filtering that can be carried out during post-processing of centrifuge test data.

The techniques presented in this chapter are routinely used during acquisition of data and processing of the data coming from various instruments in a centrifuge test. The whole data acquisition and consequent signal processing must be viewed as a single operation with the view to obtain the best data possible in every single centrifuge test.

Applications

Chapter 10

Shallow foundations

10.1 INTRODUCTION

A shallow foundation is defined as the foundation where the width is greater than or equal to the depth at which it is placed below the ground level. This definition was originally proposed by Terazaghi and Peck (1967) and is appropriate for most foundations except very narrow or very wide ones. The term shallow foundation encompasses strip foundations below walls, individual pad foundations below columns, and raft foundations.

Shallow foundations are the most common form of foundation for a vast number of civil engineering structures. These types of foundations have been historically used worldwide. For many soil conditions that commonly occur in the field such as dense sand layers or stiff clays, the shallow foundations offer a simple solution for supporting the superstructure loads coming from columns or walls. Typically shallow foundations are used for single-story buildings as well as for up to five- or six-story buildings. Sometimes they are used to support bridge pier foundations, particularly if the span of the bridge is relatively small. Shallow foundations are typically used when the expected bearing pressures on the normally consolidated clayey soils below are typically 50 or 60 kPa but can be used for bearing pressures of up to 100 kPa. For sandy soils the allowable bearing pressures can be quite high, for example, 200 to 300 kPa, but the design is often governed by the settlement of the foundation rather than the bearing capacity. More details on allowable bearing pressures can be found in Tomlinson (1986).

In simple cases, we only expect vertical loading to be transferred onto the shallow foundation, as shown in Figure 10.1. The vertical load V coming from the superstructure is transferred onto the foundation. The soil below the foundation reacts to this by generating a reacting pressure or a bearing pressure q_f as shown in Figure 10.1, in order to maintain vertical equilibrium. Clearly for a given shallow foundation as the applied vertical load V increases the bearing pressure q_f also increases, until the shear strength of the soil is exceeded. At this point the foundation will fail due to formation of shear planes. Knowing the mechanism of failure, that is, the

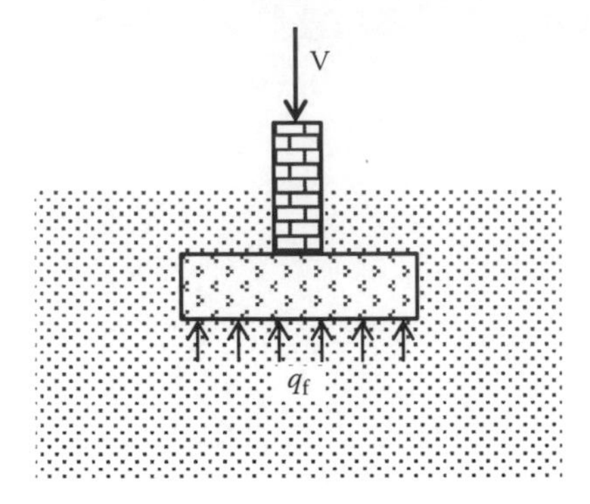

Figure 10.1 A typical shallow foundation supporting a wall.

shear zones that develop below the foundation, it is possible to estimate the largest bearing pressure that can be sustained by a given foundation. This limiting bearing pressure is also commonly termed the bearing capacity of the foundation.

For simplified cases of soil conditions, the bearing capacity of a shallow foundation can be estimated quite easily using plasticity theorems. This is considered in more detail in Section 10.2. For such cases we do not generally resort to centrifuge modelling. However, we may want to engage in research to investigate whether estimation of the bearing capacity that relies on a certain assumed failure mechanism is actually valid. For this we have to see if the actual failure mechanism that is mobilised below a shallow foundation actually matches the one we assumed in our plasticity-based calculations. Of course in a similar vein, we may engage in research to see if the correction factors recommended by researchers or codes of practice such as Eurocode 7 for the shape of the foundation, ground inclination, etc., are valid or not. Similarly, the bearing capacity of shallow foundations in layered soils, especially with weak clay layers interspliced with stiff clay layers, can be investigated using centrifuge modelling and compared with analytical or numerical predictions available in the literature.

10.2 BEARING CAPACITY OF SHALLOW FOUNDATIONS

The bearing capacity of shallow foundations has been extensively researched since the inception of soil mechanics as an independent branch of civil engineering. The most common approach is the use of bearing capacity factors originally proposed by Terazaghi as shown in Equation 10.1. These factors can be determined by knowing the soil properties such as its unit weight, friction angle, cohesion (or interlocking between soil grains), and the depth at which the foundation is placed.

$$q_f = cN_c + qN_q + 0.5B\gamma N_\gamma \tag{10.1}$$

where N_c, N_q, and N_γ are the bearing capacity factors, c and γ are the cohesion/interlocking between soil grains and the unit weight of soil, q is the overburden pressure at the foundation level, and B is the width of the foundation.

The bearing capacity factors in the above equation are usually determined from design charts or tables for any given soil properties; for example, see Tomlinson (1986). The bearing capacity determined in this fashion is usually reduced by a factor of safety to determine the allowable or safe bearing pressure. There are many correction factors that have been developed for each of the bearing capacity factors to include various effects, such as the shape of the foundation, ground inclination, and location of the water table relative to the shallow foundation.

Eurocode 7 suggests the use of a shape function for undrained conditions, given as:

$$s_c = 1 + 0.2\frac{B}{L} \tag{10.2}$$

where L and B are the footing dimensions. This factor is used as a multiplier to the N_c term in Equation 10.1. Other factors will not be considered here, but can be readily found in any standard textbook on the subject.

The concept of bearing capacity is well established and most textbooks on soil mechanics, such as Knappett and Craig (2012), include the estimation of bearing capacity for a variety of soil and loading conditions. For simplified soil conditions such as clayey soils under undrained conditions (i.e., subjected to rapid loading), the bearing capacity of the foundation can be obtained quite easily, if the undrained shear strength of the material is known. Typically this is done by assuming that the soil behaves like a perfectly plastic material and that the theory of plasticity is applicable. Following Prandtl's theory of indentation into metals, failure mechanisms with active, passive, and mixed shear zones as shown in Figure 10.2 can be assumed and the bearing capacity can be estimated.

Using the upper bound theorem of plasticity and equating the work done in displacing the footing to the energy dissipated in the active, mixed shear, and passive zones, we can show that the bearing capacity for soil with undrained shear strength of c_u is given by:

$$q_f = (\pi + 2)c_u + q_s \tag{10.3}$$

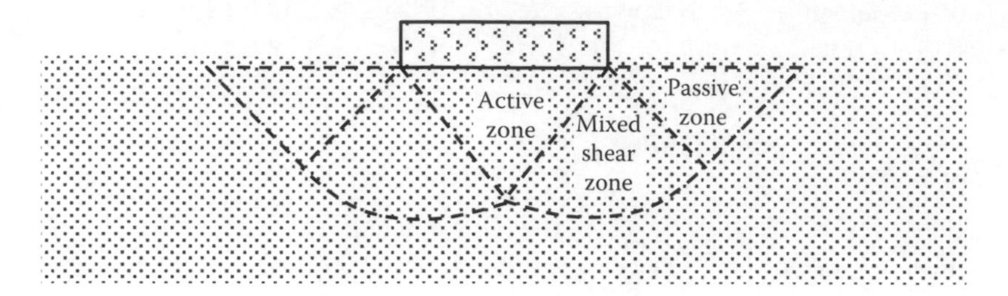

Figure 10.2 Classical bearing capacity failure mechanism.

where q_f is the bearing capacity and q_s is the surcharge pressure present at the foundation level. The surcharge pressure could be due to the overburden stress at the foundation level plus any pressure loading applied at the ground surface. Similarly by using the lower bound theorem of plasticity and considering the equilibrium and principal stress rotation from below the foundation to the free field, we can show that the bearing capacity is exactly same as the one given by Equation 10.3. Based on the two theorems of plasticity, as we get exactly the same upper bound and lower bound values for the bearing capacity, Equation 10.3 must represent the exact solution. Similar expressions can also be derived for drained materials.

Although this type of simple plasticity-based analysis gives good estimates of the bearing capacity of shallow foundations, its validity relies on the actual failure mechanism that develops below the foundation. Further, it does raise many questions.

- In Figure 10.2 do the active and passive zones deform as "whole rigid blocks" with only the mixed zone showing continuous shearing?
- Does the soil outside these zones participate in the foundation failure?
- How deep does the failure mechanism extend as a proportion of the footing size?
- What role do the soil properties have in terms of the failure mechanism's size and shape?

These questions are really research questions although their results have an obvious and direct practical importance in the design of shallow foundations.

10.3 MODELLING OF A SHALLOW FOUNDATION IN A LABORATORY

Attempts have been made by many researchers to determine the bearing capacity of shallow foundations experimentally in a laboratory. While it was possible to measure the load-settlement curves for shallow foundations, very few researchers were able to actually visualize or determine the failure mechanism that was mobilised below the foundation. Furthermore, carrying out tests on small-scale model foundations in a laboratory raises the issue of low effective stresses within the soil body.

Some of the early research into visualizing the failure mechanism below a shallow foundation was carried out at the University of Cambridge by Chan (1975). In order to avoid the problem of low effective stresses at the model scale, Chan developed a system using a pneumatic airbag that could be used to pressurize the soil surface to create the required surcharge pressure q_s. A cross-section of the experimental setup that was developed by Chan is shown in Figure 10.3. The cross-section was visible through a transparent Perspex window and was photographed during the loading of the foundation.

In this experimental research kaolin clay samples were tested under a wide variety of soil conditions from normally consolidated to overconsolidated soils. Loading was applied both rapidly and slowly to simulate undrained and drained conditions. Some tests were conducted with different load increments. The settlement of the foundation was monitored during all tests. An example of the settlement curve is shown in Figure 10.4. The time axis in this figure is plotted on a log scale. Prior to application of footing load, the kaolin clay was normally consolidated under a surcharge pressure of about 200 kPa. In this test the applied bearing pressure on the footing was 72 percent of the ultimate bearing capacity q_f.

It was possible to observe the failure mechanism that developed below the foundation by placing lead shot in the clay body and radiographing the clay

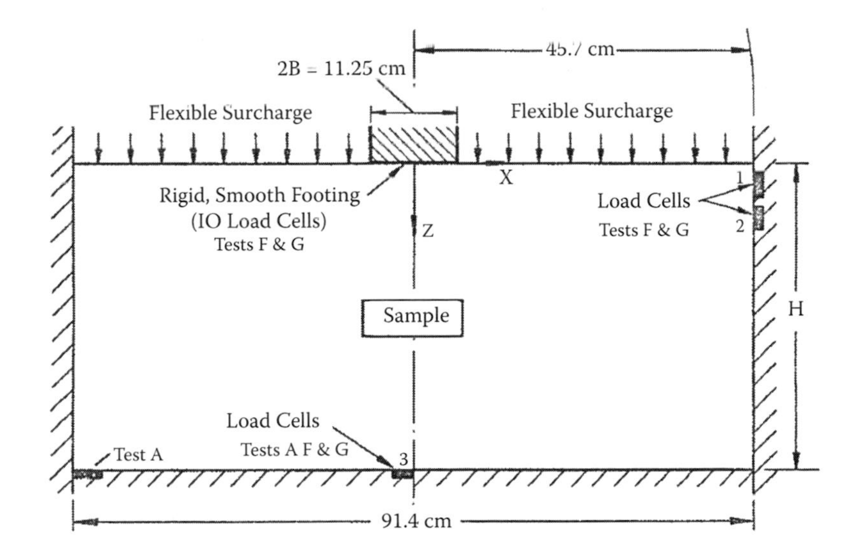

Figure 10.3 An experimental setup to investigate the bearing capacity of shallow foundations (after Chan, 1975).

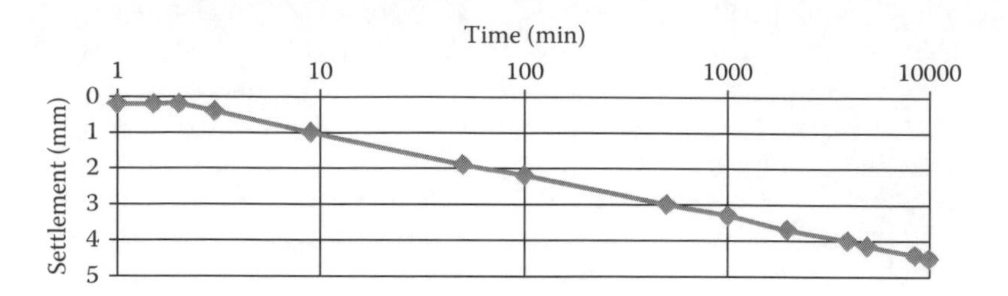

Figure 10.4 Typical settlement curve of a shallow foundation on a normally consolidated kaolin clay (data from Chan, 1975).

at different stages of loading. An example of this is shown in Figure 10.5, which shows the deformation in soil body as the footing settles in the first 3000 minutes after the application of the load. The observed displacement vectors clearly show the vertical settlements below the footing which start to rotate on either side. Comparing Figure 10.5 with Figure 10.2 we can see that the idealised failure mechanism assumed in the bearing capacity calculations may not be completely accurate.

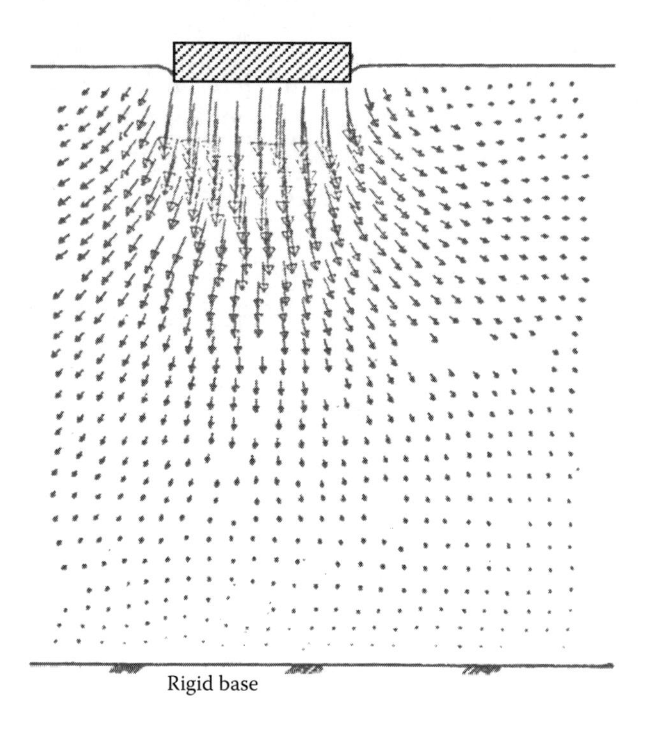

Figure 10.5 Failure mechanism below a shallow foundation under undrained conditions in a normally consolidated clay (after Chan, 1975).

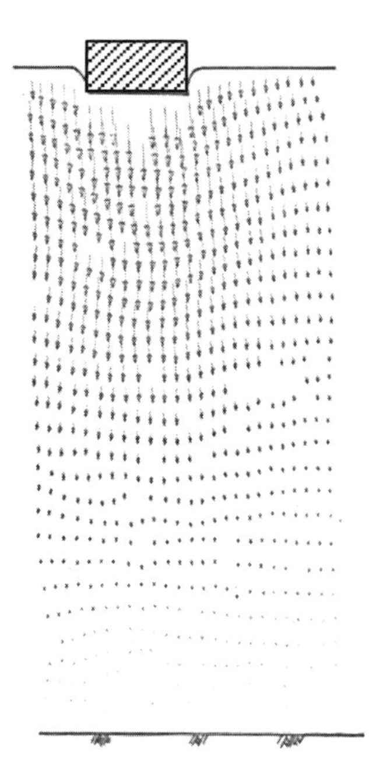

Figure 10.6 Consolidation settlement below a shallow foundation (after Chan, 1975).

The footing settlements shown in Figure 10.5 continue beyond the 3000 minutes as seen in Figure 10.4. This stage is predominantly due to the consolidation of the clay. The displacement vectors from 3000 to 10,000 minutes are shown in Figure 10.6. Unlike the previous stage, all the displacement vectors are nearly vertical throughout the soil body. However, the differential settlements between the free field and the region below the footing does result in some shear being mobilised in the soil body.

Although this type of research shows the importance of investigating the failure mechanism below shallow foundations, its questions remain due to the laboratory-based testing that was undertaken. Although the stresses within the soil body were relatively high because of the use of the pneumatic air bags, this increases the vertical stress everywhere in the soil body. In contrast, in the real problems the stress level increases with the depth. Therefore, the stress gradients are not reproduced in this type of testing.

10.4 CENTRIFUGE MODELLING OF SHALLOW FOUNDATIONS

Centrifuge modelling of shallow foundations was conducted by several researchers, including Ovesen (1975) and Kimura, Kusakabe, and Saitoh (1985), although in the early days it was not possible to obtain images

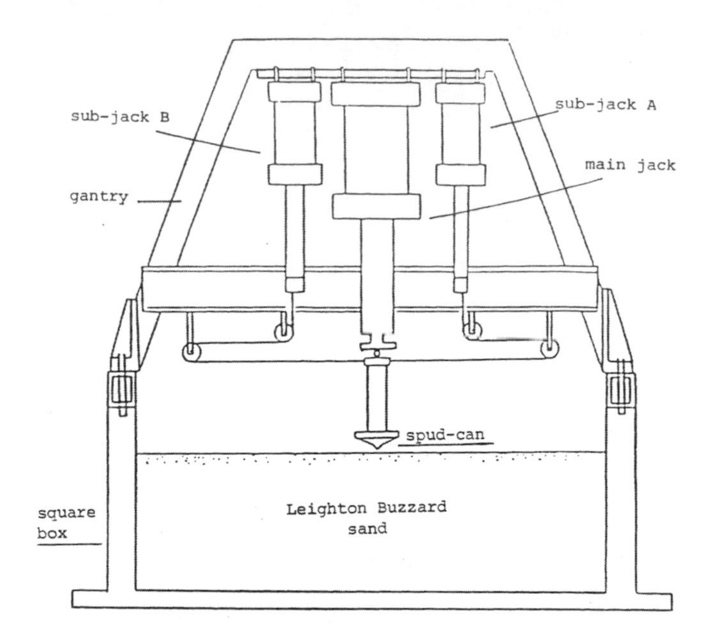

Figure 10.7 A centrifuge testing rig for application of combined loading on shallow foundations (after Shi, 1988).

of sufficient quality and resolution to determine the failure mechanisms. However, more complex loading problems could be investigated. For example, Shi (1988) investigated the bearing capacity of shallow foundations under the combined action of axial, lateral, and moment loading. His experimental setup used for the centrifuge testing is shown in Figure 10.7. It allowed loading of the foundation in vertical and horizontal directions. The main jack allowed loading the footing in the vertical directions while the sub-jacks allowed controlled lateral loading of the foundation. The moment loading is generated due to the height of the loading point above the base of the footing. In this setup the moment loading is not independent of the horizontal loading. The shallow foundations investigated were spud-can foundations on sandy soils used by the offshore oil industry at that time.

Shi carried out a series of centrifuge tests at 60 *g*. Combinations of a wide variety of loading conditions were considered. Examples of the data that were obtained during this centrifuge tests are shown in Figures 10.8 and 10.9. The data is shown in these figures at model scale. We can use the scaling laws discussed in Chapter 4 to change these to prototype scale. Figure 10.8 shows the vertical settlement of the footing when it is subjected to cyclic variation in the horizontal load. The vertical load was maintained constant during this process, but the experiment was repeated under several levels of vertical load. In Figure 10.8 it can be seen that the vertical

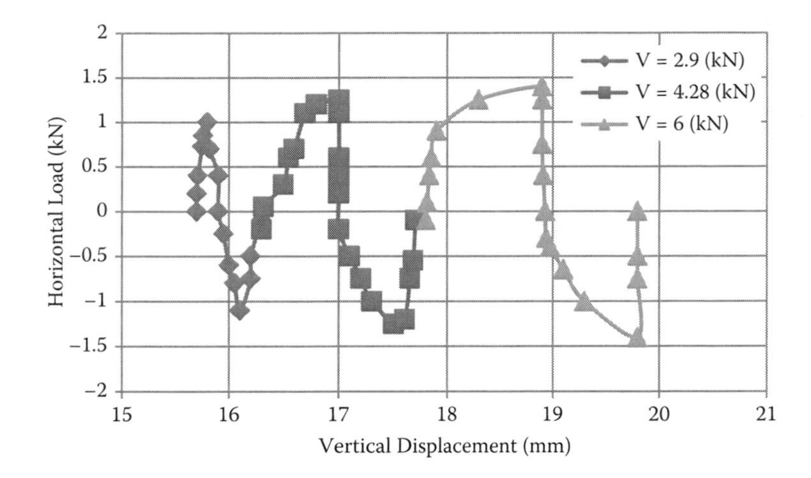

Figure 10.8 Settlement of the foundation under applied horizontal load cycles (data from Shi, 1988).

settlement continues to accrue with the cycles of horizontal loading and the magnitude of this settlement increases with the increase in the vertical load applied to the footing. This research showed that the bearing capacity of the shallow foundation is rapidly reduced on application of the horizontal load.

The experimental data can also be plotted to investigate the stiffness of the soil. In Figure 10.9 the horizontal load is plotted against the horizontal displacement of the footing. Again the data is presented at the model scale. In this figure it can be seen that the stiffness of the soil mobilised is smaller when the applied vertical load is smaller. Also with the increase in the vertical load, the size of the loop and the enclosed area within the loop are reduced, which indicates that less energy is being dissipated at this loading level. The data shown here are just examples of the quality of information that can be obtained from centrifuge test series like this.

More recently McMahon (2012) has investigated the failure mechanisms below shallow foundations exploiting the recent advances in high-resolution digital imaging and the development of PIV analysis described in Chapter 8. He conducted a series of centrifuge tests at 100 *g* and tested multiple footings placed on the surface of clay and sandy soils. In this section we will only consider examples of the clay tests conducted on normally consolidated and over-consolidated clay layers. The experimental setup used in this test series is shown in Figure 10.10. The footings were made of aluminum alloy and had a relatively smooth interface with the foundation clay. The footings were loaded either by dead weight or by using pneumatic cylinders after the clay was reconsolidated under the applied *g*

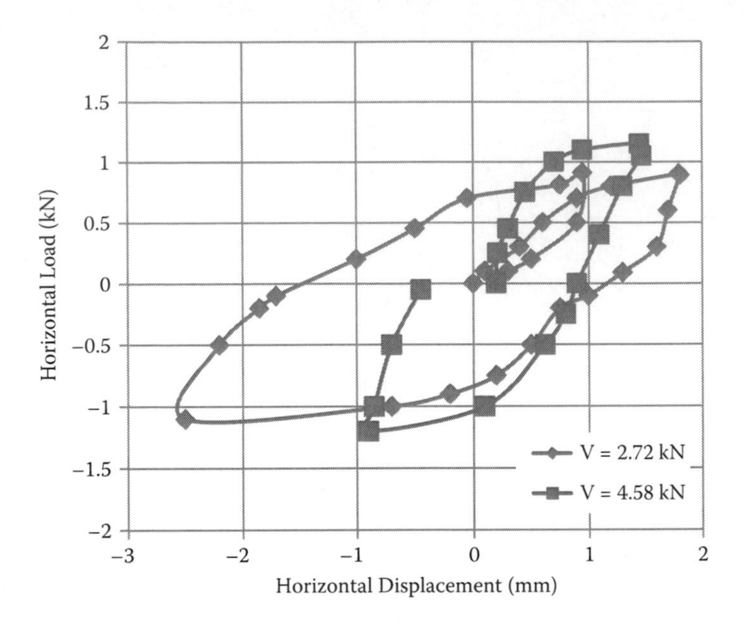

Figure 10.9 Horizontal load-displacement loops (data from Shi, 1988).

field. The footings are placed next to the Perspex window as indicated in Figure 10.10. A digital camera system is used to take images of the foundations through the transparent Perspex window at rapid intervals which are later processed to reveal the soil deformations.

The first series of tests were carried out on soft clay which is nominally taken to be normally consolidated. The footings were loaded using deadweight and simulated undrained conditions in the clay layer. The soil deformations obtained for this case are shown in Figure 10.11. The vertical scale indicates the soil deformations and the horizontal scale indicates the geometric scale. It must be pointed out that these are at model scales and we need to use the centrifuge scaling law for displacement to convert them into prototype scale. In this figure it can be seen that the soil deformations extend to a depth less than the width of the footing. In addition the displacement vectors change in size with depth. Overall the deformations observed in this test are different from the classical failure mechanism shown in Figure 10.2. This can also be compared to the failure mechanism obtained in laboratory-based 1 *g* testing carried out by Chan (1975) and shown in Figure 10.5. The failure mechanism in Figure 10.11 is even shallower than the one in Figure 10.5.

The second series of tests were conducted on a nominally over-consolidated clay layer. In these tests the loading of the footings was carried out using the pneumatic cylinders shown in Figure 10.10 to ensure rapid loading

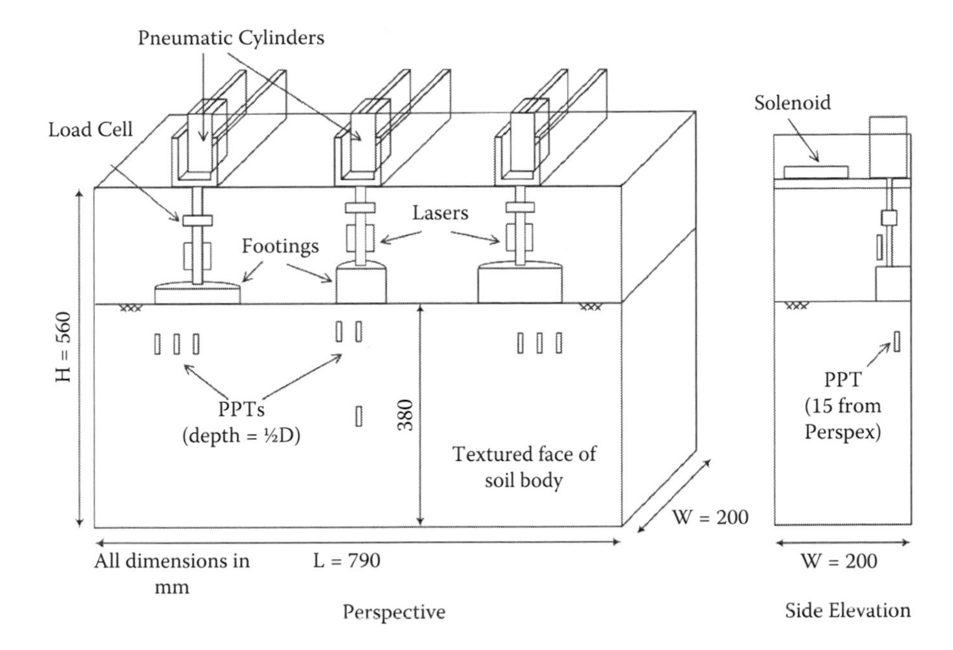

Figure 10.10 Centrifuge experiment setup for loading of footings (after McMahon, 2012).

and simulate undrained conditions. A typical result from this series of tests is shown in Figure 10.12. In this figure it can be seen that the soil deformations are much smaller compared to the case of the normally consolidated clay shown in Figure 10.11. The deformations extend to a much shallower depth relative to the footing width. This is to be expected as the over-consolidated clay is much stiffer than normally consolidated clay. This results in much smaller soil deformations below the footing and the mechanism does not

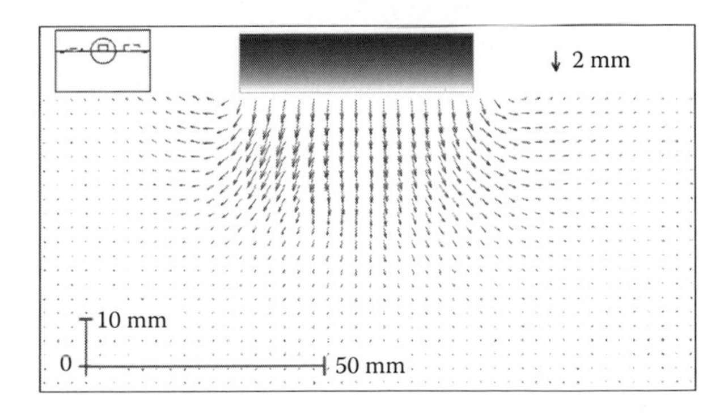

Figure 10.11 Soil deformations in normally consolidated clay below a shallow foundation (after McMahon, 2012).

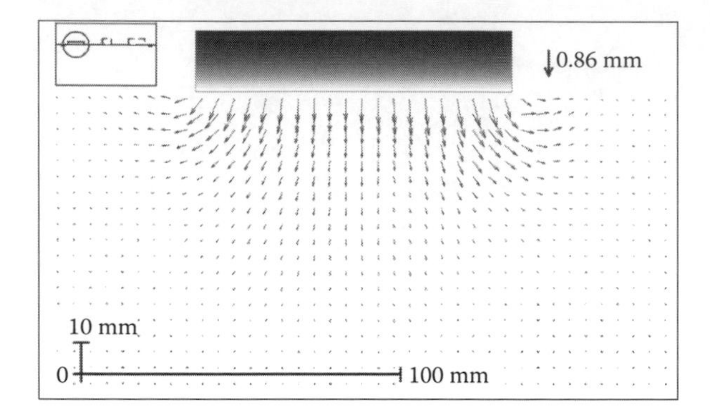

Figure 10.12 Soil deformations in over-consolidated clay below a shallow foundation (after McMahon, 2012).

penetrate that deep below the footing. It is easier for the soil to squeeze out laterally rather than mobilise a deep failure mechanism.

10.5 MODELLING OF MODELS

Some examples of modelling the shallow foundations in a centrifuge were discussed in the previous sections. While these examples deal with advanced centrifuge modelling, simple cases of shallow foundations can be modeled in a straightforward fashion. Graduate students can attempt to use centrifuge modelling to simulate shallow foundations in a simple manner. In Cambridge, the graduate module that deals with advanced geotechnical modelling involves such an exercise. As part of this the students test circular shallow foundations of different diameters at different *g* levels. One of the aims of this exercise is to determine the load-settlement curves for various circular footings and understand the concept of bearing capacity and settlements in granular soils. The data used in this section is available for free download from www.tc2teaching.org and can be used by anyone for further processing and analyses. A typical cross-section of the centrifuge model used in these tests is shown in Figure 10.13. The soil deposit that is commonly used in these tests is prepared from uniformly graded Hostun sand. This sand is deposited at the required relative density using the automatic sand-pouring equipment described in Chapter 7 into a standard 850-mm-diameter tub. Students normally are divided into two groups and each group prepares a soil model with a different relative density. It is usual that one group prepares a soil deposit at a relatively low relative density of about 40 percent while the other group prepares a soil deposit at a high relative density of about 80 percent. The students then test the circular foundations at different *g* levels. This allows data from footings of different sizes to

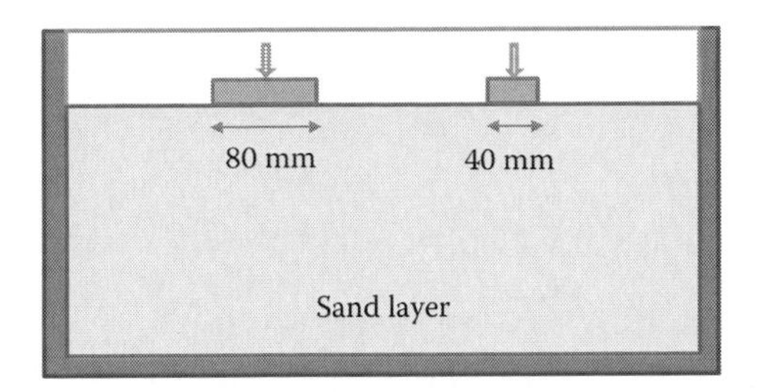

Figure 10.13 Cross-section of the centrifuge model of shallow foundation tests.

be analysed using the principle of modelling of models. In Figure 10.13 the circular footings shown have diameters of 40 and 80 mm. These footings are tested at 80 *g* and 40 *g*, respectively. The equivalent prototype diameter in each case is 3.2 m. Note that the centrifuge models in this case represent a truly axisymmetric problem. Application of a lateral load in any horizontal direction will make the problem a three-dimensional one.

In these experiments the one-dimensional actuator presented in Chapter 7 is employed. This actuator is used to push the footings into the sand layer under displacement control and the rate of penetration is kept constant. Also the footing is first pushed in to the required depth and then withdrawn back to the surface. A load cell between the footing and the actuator measures the load being applied on to the footing. Let us first consider the 40 *g* test in which the 80-mm-diameter footing is tested. The displacement-time history and load-time history are presented in Figure 10.14 at model scale. Note that the footing starts some 12 mm above the ground surface, as the load starts to pick up only after this displacement has occurred and the footing comes into contact with the ground. In this case the footing is pushed in by about 18 mm and in this phase it mobilises a peak resisting force of 21 kN. The maximum bearing pressure mobilised is therefore 4.2 MPa. In Figure 10.13 we can also see that as soon as the footing starts to withdraw towards the surface, the resisting load drops to a zero load.

In the second centrifuge test a footing of 40 mm is pushed into the sand layer at 80 *g*. The displacement-time history and load-time history are presented in Figure 10.15 at model scale. This footing comes into contact with the ground surface after it has travelled about 15 mm. The footing is pushed a further 13 mm into the sand layer as shown in Figure 10.15 and it mobilises a peak load of 10 kN at this stage. This corresponds to a bearing pressure of 7.96 MPa. Again the load on the footing drops as soon as the footing starts to reverse and move towards the ground surface.

Of course, you could download the data and replot it to obtain load versus settlement graphs in each of the centrifuge tests. One question that

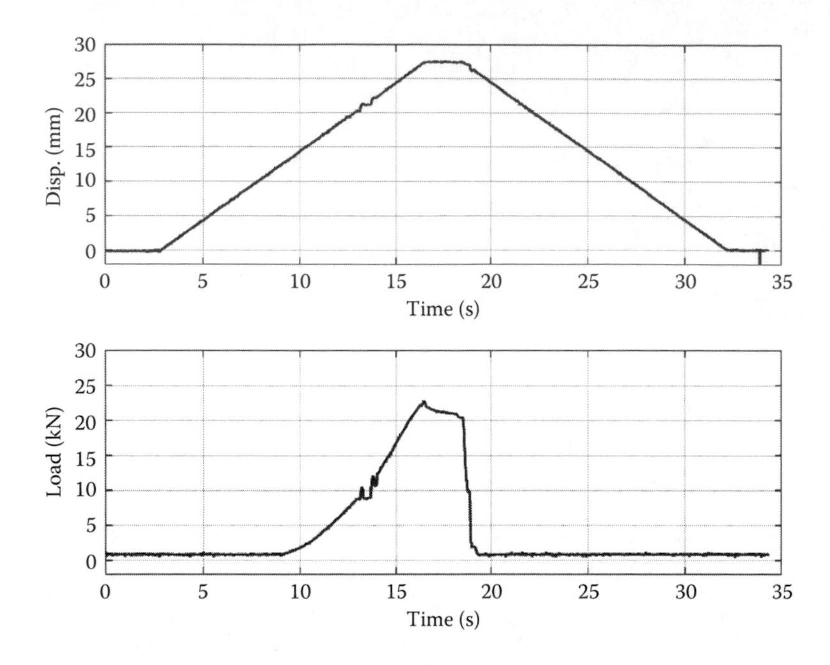

Figure 10.14 A large-diameter foundation tested at 40 g.

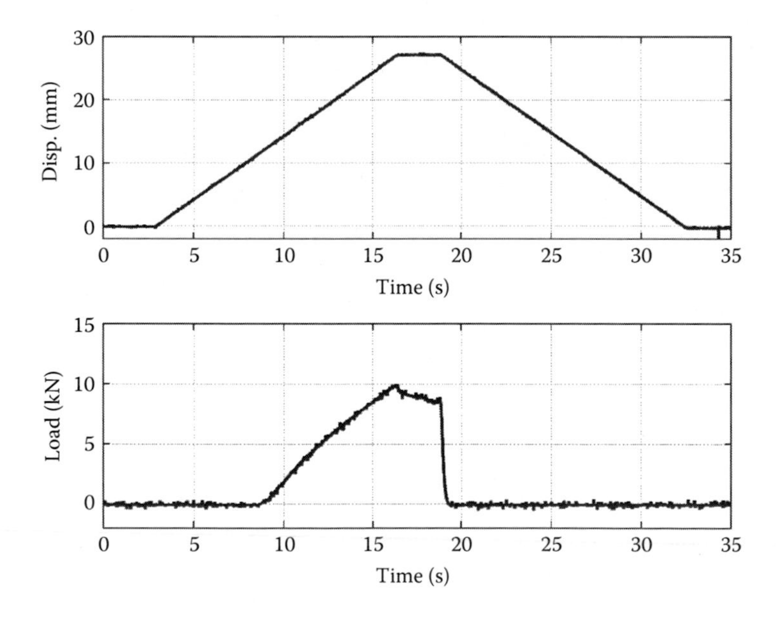

Figure 10.15 A small-diameter foundation tested at 80 g.

arises straightaway is why is the bearing pressure in a 40 *g* test so different from that obtained in an 80 *g* test? It is important to realize that the bearing pressure mobilised in granular soils is a function of the settlement of the footing into the ground. Unlike in the case of plastic clays where a failure mechanism is mobilised and the ultimate bearing capacity is reached, in granular soils you can mobilise higher and higher bearing pressures as you drive the footing settlements higher. While designing the footings in granular soils we are normally limited by the amount of settlement our structure can tolerate.

Revisiting the centrifuge data presented above, it is better to normalise the settlement of the footing with its width to obtain a nominal strain. If we do that and replot our data as bearing pressure versus nominal strain, then we can directly compare our data from the two centrifuge tests. This is shown in Figure 10.16 for both the 40 *g* and 80 *g* centrifuge tests and we can see that the two tests produce quite comparable data.

Overall, we can say that the modelling of models has worked well as a technique and we used this to show that a 3.2-m-diameter circular footing on sand produces satisfactory comparisons whether we test a 40-mm-diameter model footing at 80 *g* or an 80-mm-diameter model at 40 *g*.

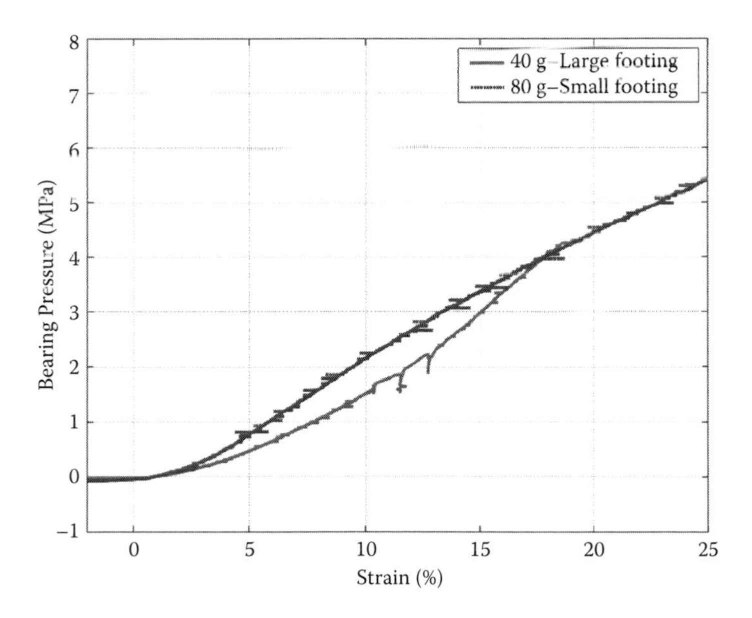

Figure 10.16 Modelling of models: comparison of a 40 *g* and 80 *g* test.

10.6 SUMMARY

In this chapter we have considered the centrifuge modelling of shallow foundations. The basic concepts of bearing capacity for shallow foundations based on assumed failure mechanisms such as the Prandtl's mechanisms can be tested against centrifuge test data. Early work carried out at laboratory scale on footings placed on normally consolidated clay and using radiography to determine the failure mechanism was considered and compared to more recent centrifuge experimental data on footings placed on normally consolidated and over-consolidated clays. Combined loading on shallow foundations placed on granular soils was also considered and the centrifuge-based experimental work was discussed. This shows the importance of considering combined loading as the vertical displacements of the footings can accrue due to horizontal load cycles, even if the vertical load is held constant. This type of loading can be important while dealing with offshore wind farm foundations, which can be sensitive to settlements and rotations. Finally we considered a simple example of a circular foundation and used the principle of modelling of models to interpret the data from a 40 *g* and a 80 *g* centrifuge test. Good comparisons were obtained from the data of these two centrifuge tests and were plotted as bearing pressure versus nominal strain.

Chapter 11

Retaining walls

11.1 INTRODUCTION

Retaining walls are commonly used to hold a soil mass back, next to an excavation or earth fill. These can be used to form an artificial embankment that supports a railway or a roadway or to create space below ground level. There are different types of retaining walls, such as:

- Gravity retaining walls
- Flexible, cantilever retaining walls
- Anchored retaining walls
- L-shaped walls
- Propped or braced retaining walls

Gravity retaining walls rely on their mass to resist the movement of the soil on the retained side by generating adequate frictional resistance at the base. Similarly, cantilever retaining walls rely on their flexural stiffness to hold the soil in place. Where the penetration depth of the wall into the soil is limited, they may require anchoring or propping. L-shaped walls are a variation on cantilever retaining walls where the weight of the retained soil contributes to the stability of the wall. These types of walls are illustrated in Figure 11.1.

The selection of any type of retaining wall from the above list depends on several factors, such as the height of the soil that needs to be supported, properties of the soil behind and in front of the retaining wall, wall sections available, and so on. The design of the retaining walls is usually dictated by the estimation of the soil pressure on the retained side, which is the driving force, and the passive pressure on the excavation side that provides the resisting force. The design is carried out to achieve the equilibrium of the wall when “limiting” active and passive pressures are mobilised on either side of the wall. Structurally, the walls are checked to ensure they can carry the induced shear forces and bending moments with adequate factors of safety. When the walls are anchored or propped as shown in Figure 11.1,

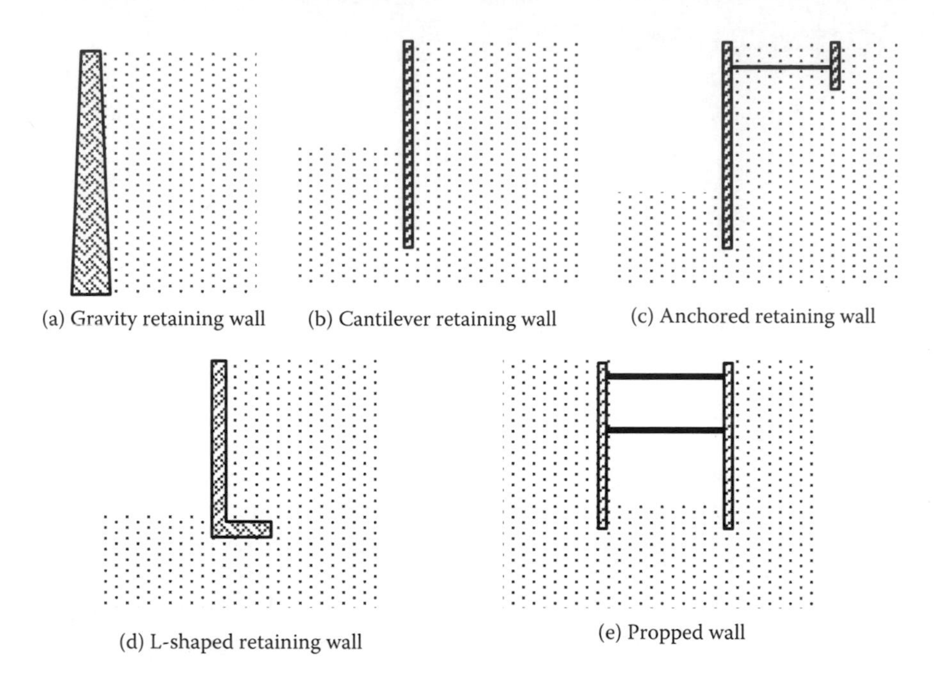

Figure 11.1 Schematic representation of various retaining walls.

the forces generated in the tie rods or props must also be calculated and checked for adequate design of these components.

Much of the research on retaining walls historically pertained to the stability of different types of walls, generation of active and passive pressures, wall friction effects, surcharge loading behind retaining walls, and staged excavation effects in front of the wall. Normally the retaining walls are not designed to reach plasticity; that is, formation of plastic hinges in a retaining wall is avoided. However, recent research has investigated the formation of such plastic hinges in cantilever retaining walls and the subsequent stability of the walls (for example, Viswanadham et al., 2009, based on centrifuge testing of cantilever retaining walls, and Bourne-Webb et al., 2011, based on finite element analysis). Bourne-Webb et al. looks at the provisions in the Eurocode 7 for the design of earth retaining structures.

11.2 RETAINING WALL MODELS AT LABORATORY SCALE

Model retaining wall tests have been carried out in soil mechanics from a very early time. Terazaghi (1934a, 1934b) conducted model tests on relatively large retaining walls at Harvard in order to investigate the pressure exerted by the soil on the retaining walls. Rowe (1972) describes the large-scale tests on retaining walls carried out at Manchester. At Cambridge, model retaining walls were tested by Milligan (1974), who took advantage of the radiographic techniques of the day to investigate the soil deformations behind the retaining walls. The experiments were carried out by excavating sand in front of the retaining wall using an electric vacuum cleaner. Lead shot was placed in the sand and x-ray images of the model were taken before and after soil excavation. A large number of tests were carried out on flexible and rigid retaining walls retaining both loose and dense sands. In some of the later tests he considered the rotation of the wall about a fixed point above the base, but these tests will not be considered here.

Milligan (1974) measured the deflection of the model retaining walls in each of the tests with loose sand backfill and by excavating the sand in front of the wall gradually. The wall deflections (Δ) and the excavated depth (D) can be normalised by the total height (H) of the model wall. These are replotted in Figure 11.2 for all of the loose sand tests. In this figure it can be seen that the wall deflection increases almost exponentially as the excavation depth increases in front of the wall. This is to be expected as the retaining wall will become unstable as the excavation depths increase beyond 65 percent of the total height of the wall.

Unlike previous researchers Milligan (1974) was able to obtain the displacement vectors for the backfill soil by using radiographic techniques. An example of this is presented in Figure 11.3. In this figure the displacement vector field is shown when the sand in front of the wall has been excavated to 142 mm (i.e., D/H of 0.473) and for the case when the excavation reached a depth of 192 mm (i.e., D/H of 0.64). Clearly the magnitude of the soil displacement vectors increases with the increase in D/H ratio. The

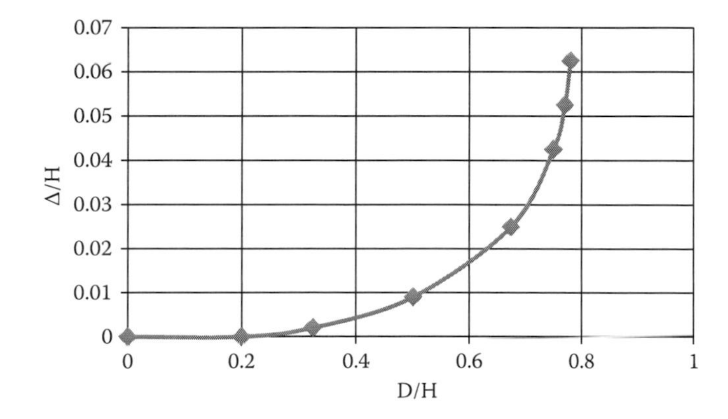

Figure 11.2 Normalised deflection of a retaining wall (after Milligan, 1974).

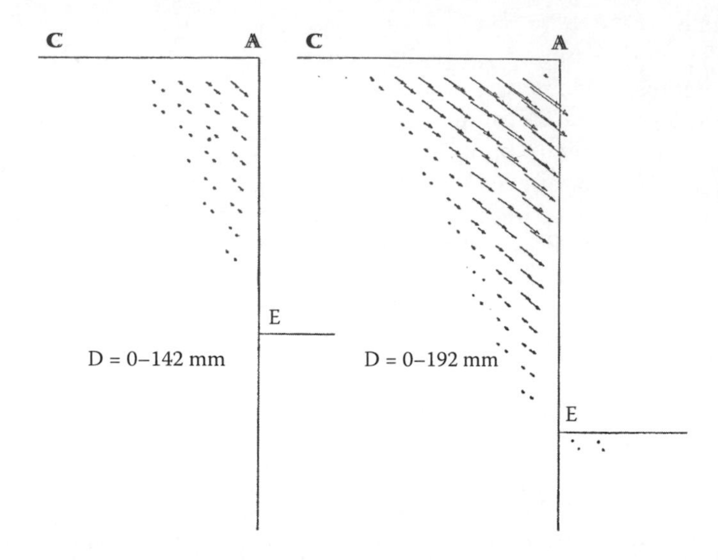

Figure 11.3 Displacement vectors behind a cantilever retaining wall (after Milligan, 1974).

development of the active wedge behind the wall can be visualised and becomes quite clear in Figure 11.3.

Based on this research, Bransby and Milligan (1975) were able to relate the mobilisation of shear strain (γ_s) in the soil behind the wall to the wall deflections as:

$$\gamma_s = 2\ sec\psi \frac{\Delta}{H} \tag{11.1}$$

where ψ is the dilatancy angle of the sand.

The relationship shown in Equation 11.1 was utilised by Madabhushi and Chandrasekaran (2005) in determining the point of rotation of a sheet pile wall by using the moment minimization technique.

11.3 SIMULATING RETAINING WALLS IN A CENTRIFUGE

Retaining walls can be modeled in a centrifuge quite easily as the models are constructed on the laboratory floor at 1 *g* conditions and then subjected to high gravity. It is also possible to carry out staged construction sequences such as excavation and deployment of multilevel props in a high-gravity environment such as the propped diaphragm wall shown in Figure 11.1(e). These are discussed in more detail in Chapter 13. Single-level props positioned at the top of the retaining walls will be considered in Section 11.7.

In order to model a retaining wall with a clay or sand backfill, the first step would be to create a model wall that has the correct flexural stiffness (EI) as the prototype. Similarly if the wall is anchored or propped then the anchor rods or props must have the correct axial stiffness (EA). In order to achieve this we need to think about the scaling laws for these parameters.

Using the principles we considered in Chapter 4, the scaling law for flexural stiffness can be obtained as:

$$\frac{(EI)_{model}}{(EI)_{prototype}} = \frac{1}{N^4} \tag{11.2}$$

This is based on the assumption of using the same material to make the model wall as the prototype wall and the second moment of area (I) simply scaling as a factor of $1/N^4$. However in centrifuge modelling, we often use aluminum alloys such as Dural to make the model walls due to the ease of machining and fabrication, and the ease with which such model walls can be strain gauged to measure the bending moment. The retaining walls in the field are often made from steel as in the case of sheet pile walls. If this is the case then the ratio of Young's modulus of the two materials must be considered as in Equation 11.2. Further, in a plane strain problem the length of the retaining wall is normalised and flexural stiffness per meter length is considered. In such cases the scaling law for flexural stiffness per meter will be:

$$\frac{(EI)_{model/m}}{(EI)_{prototype/m}} = \frac{1}{N^3} \tag{11.3}$$

Using Equation 11.3 it is possible to estimate the model wall thickness for a centrifuge test conducted at a known g level. For example, if the centrifuge test is to be conducted at 100 g then the equivalent model wall thicknesses that represent standard wall sections can be calculated as shown in Table 11.1. The Young's modulus of the steel is taken as 210 GPa and that of Dural is taken as 70 GPa, for the calculations shown in Table 11.1. Also

Table 11.1 Typical sheet pile wall sections and equivalent flat plate model walls

Prototype section	*I (cm⁴)*	*EI (MNm²)*	*Model wall thickness* (mm)*
Larssen LX8	12863	27.01	1.67
Larssen LX38	87511	183.77	3.16
Frodingham 1BXN	4947	10.39	1.21
Frodingham 5	49329	103.59	2.61
Giken 350WL-14	112000	235.20	3.43
Giken 500WX-16	293000	615.30	4.72

* For a 100 g centrifuge test.

Table 11.2 Typical tie rods sections

Prototype section	*A (mm²)*	*EA (MN)*	*Model tie rod* ϕ^* *(mm)*
GEWI-20	314	65.94	0.35
GEWI-63.5	3167	665.07	1.10
A36/32	804.25	168.89	0.55
A36/102	8171.28	1715.97	1.77
Williams #8	490.87	103.08	0.43
Williams #28	6221.14	1306.44	1.54

* For a 100 *g* centrifuge test.

the wall thickness is calculated by considering that the model wall is made from a flat plate of Dural and the complex shapes of the prototype sections are not reproduced at model scale.

Similar estimations of model wall thicknesses can be made for other prototype sections and for centrifuge tests that are planned at different *g* levels.

In order to model anchored retaining walls the axial stiffness (EA) of the anchor rods in the prototype must be scaled accurately. The scaling Law for axial stiffness (EA) can be written as:

$$\frac{(EA)_{model}}{(EA)_{prototype}} = \frac{1}{N^2} \tag{11.4}$$

As with the flexural stiffness, care must be taken about the material used in making the model anchor rods. If Dural is used then the appropriate Young's modulus for this material must be used. In Table 11.2 the model tie rod diameters are shown for some of the standard prototype anchor rods used in the field. These model rod diameters were calculated assuming that they are made from Dural and that the centrifuge test will be conducted at 100 *g* and using Equation 11.4. While modelling the anchored retaining walls care must also be taken with respect to the spacing of the tie rods in the prototype. The spacing distance must be scaled down in the centrifuge model using the usual scaling law for length. For example, if the tie rods are placed at 2 m spacing in the prototype, then the spacing in a centrifuge model test carried out at 100 *g* will be 20 mm. Similarly the anchoring detail of the rods to either a whaling beam or a dead-man anchor must be accurately reproduced in the centrifuge model.

11.4 CENTRIFUGE TESTING OF CANTILEVER RETAINING WALLS

Cantilever retaining walls are commonly used to retain moderate heights of soil. Sheet piling is used to create such retaining walls using available standard sections shown in Table 11.1. Centrifuge modelling of such

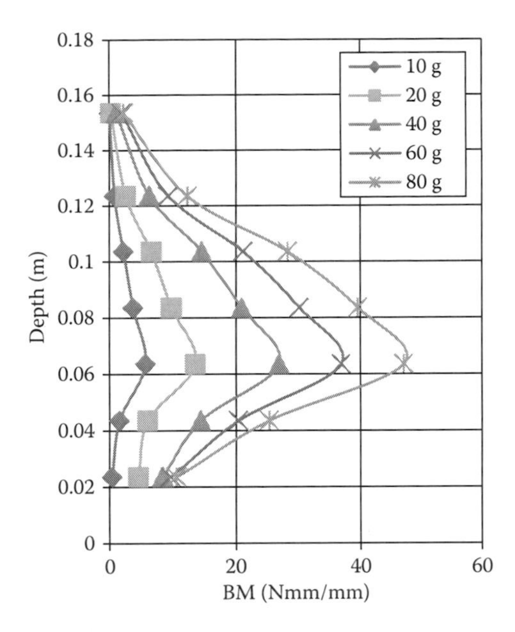

Figure 11.4 Variation of bending moment with g level.

retaining walls is straightforward. In this section we will consider a flexible, cantilever retaining wall tested at 80 *g* as reported by Madabhushi and Zeng (2006). The cross-section of the model is similar to the one shown in Figure 11.1(b). The model was created by air pluviating the sand. A model retaining wall made from a 3.3-mm-thick Dural plate and having a height of 180 mm. The wall was placed into the sand such that it penetrated to a depth of 90 mm below excavation level and retained the sand to a height of 90 mm. The retained height of the soil will be equivalent to 7.2 m at prototype scale. The model wall was strain gauged at several levels and the bending moments in the wall were recorded as the centrifugal acceleration was increased. These are shown at model scale in Figure 11.4. Clearly the bending moment increases as the *g* level rises. At 80 *g* the peak bending moment was recorded to be 48 Nmm/mm at a height of 65 mm above the base of the wall (115 mm below the top of the wall). We can use the scaling law given in Equation 11.2 to relate this to the equivalent prototype. Thus, at the prototype scale this corresponds to a peak bending moment of about 24.6 MNm/m at a depth of 9.2 m below the top of the wall.

No images of this plane strain model were obtained during centrifuge testing and therefore it is not possible to see the soil deformations behind the wall. However, Madabhushi and Zeng (2006) report the centrifuge test data and compare it with results from the finite element analysis of cantilever sheet pile walls. Soil deformations were obtained from the deformed mesh in the finite element analysis. Similar work was carried out for saturated backfill as reported by Madabhushi and Zeng (2007).

When designing retaining walls in the field adequate factors of safety are assumed and the walls never reach failure; that is, they are prevented from suffering excessive lateral displacements or rotations, or forming plastic hinges. However, in the event of unexpected loading coming onto these walls and if a plastic hinge forms in the sheet pile wall, it would be interesting to see how the earth pressure distribution behind and in front of the wall and consequent bending moments in the wall change. Viswanadham et al. (2009) investigated this problem in a series of centrifuge tests carried out at the Indian Intitute of Technology, Bombay. They tested a wall 400 mm high with a sand backfill. These model walls were nearly twice as high as the ones considered by Madabhushi and Zeng (2007). The model wall was made from 3-mm-thick Dural plate and had 24 levels of strain gauges, which measured bending moments in the wall. The test was conducted by increasing the *g* level. Figure 11.5 shows the bending moment distribution with the depth of the wall, plotted at model scale. At lower *g* levels of 20 *g* and 30 *g*, the bending

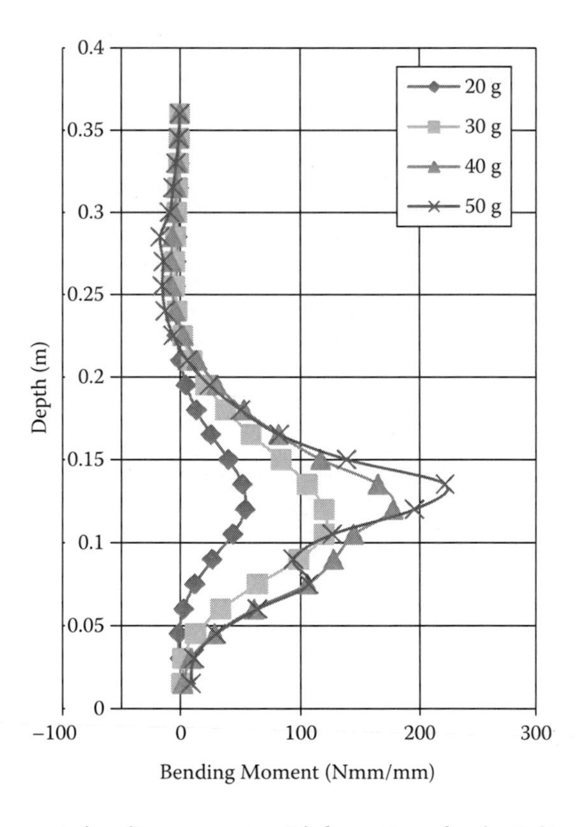

Figure 11.5 Changes in bending moments with formation of a plastic hinge.

moment distribution is smooth. As the centrifugal acceleration reaches about 40 *g* the bending moment distribution starts to change and at 50 *g* there is a clear movement of the location of the peak bending moment upwards and the distribution starts to show a kink. Viswanadham et al. (2009) show that beyond 40 *g* the plastic moment capacity of the wall is exceeded and this was confirmed with the observation of a plastic hinge in the wall.

11.5 ANCHORED RETAINING WALLS

Anchored retaining walls with typical cross-sections as shown in Figure 11.1(c) are used commonly to form quay walls and other seafront structures. Centrifuge modelling can be an effective way to investigate the stability of such structures and even determine the forces generated in the anchor rods. Cilingir et al. (2011) reanalysed some of the earlier centrifuge test data on anchored retaining walls in dry and saturated sands. The main thrust of this work was to look at the stability of these structures under earthquake loading, but here we will consider only the static behaviour. By having anchors holding back the retaining wall the bending moment distribution in the wall changes considerably. The centrifuge tests were carried out at 80 *g*. The model wall thickness was 3.3 mm and made from Dural. Anchor rods were connected 30 mm below the model ground surface. Load cells were used to measure the force in the anchors. The bending moment distribution obtained in the centrifuge tests with dry and saturated backfill is shown in Figure 11.6 at prototype scale. Clearly large negative bending moments are generated in the wall compared to the cantilever sheet pile wall.

In these centrifuge tests, at 80 *g* anchor forces of 20.4 N and 23.1 N were measured in the dry and saturated tests, respectively. These anchor forces correspond to prototype forces of 130.7 and 147.8 kN in the anchor rods. Suitable anchor rod sections from Table 11.2 can be chosen to safely carry these forces.

11.6 CENTRIFUGE TESTING OF L-SHAPED WALLS

There has been considerable research on L-shaped retaining walls of the kind shown in Figure 11.1(d). Wind (1976) reports a series of 15 centrifuge tests carried out on L-shaped retaining walls with sand backfill. Bending moments (M) in the wall were recorded at a few locations. These measurements were expressed as equivalent earth pressure coefficients (K_e) using the equation:

$$K_e = \frac{M}{\left(\frac{1}{6}\gamma\ Ng\ H^3\right)} \tag{11.5}$$

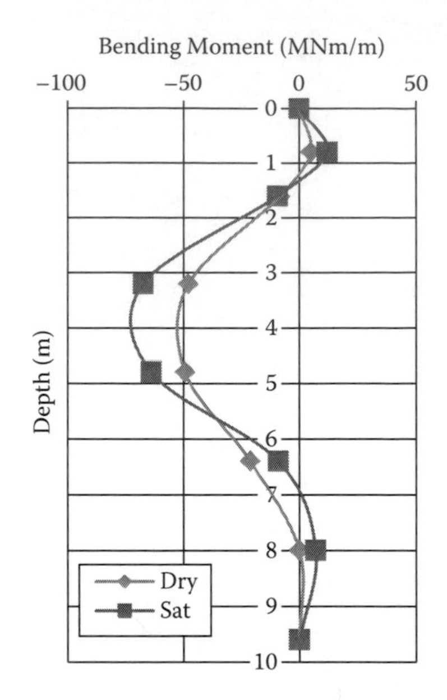

Figure 11.6 Distribution of bending moment in an anchored retaining wall.

where H is the height of the retained soil, N is the geometric scaling factor for the centrifuge test, and γ is the unit weight of the soil. The variation of the equivalent earth pressure coefficient (calculated from Equation 11.5) with the depth of the wall in a 40 *g* centrifuge test is shown in Figure 11.7. The depth in this figure is plotted at model scale.

Mak (1983) investigated rigid L-shaped walls with sand backfill in centrifuge tests carried out at 60 *g*. He considered the presence of a shallow foundation behind the wall as would be the case say for a quay wall with dockside structures. A typical cross-section of his model is shown in Figure 11.8.

Mak measured the settlement of the strip footing as he increased the bearing pressure on the footing as shown in Figure 11.9. This is quite an interesting soil-structure interaction problem. Unlike the shallow foundations seen in Chapter 10, these footing settlements are controlled by both the usual bearing capacity mechanism and the movement of the wall. Mak showed that as the bearing pressure increases the footing settlement increases up to a point and beyond that the footing settlement increases even when the bearing pressure is reduced. This is due to the translation and rotation of the L-shaped wall.

An example of the wall's lateral translation and rotation is plotted in Figure 11.10. The wall's translation increases with increasing bearing pressure on the strip footing and just like the settlement of the footing, beyond

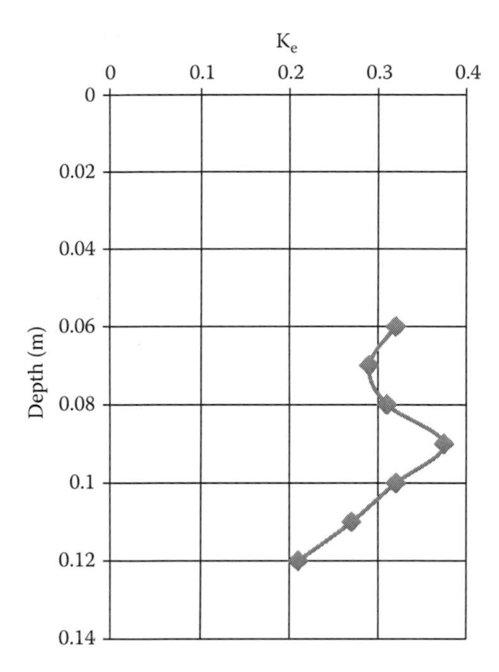

Figure 11.7 Variation of earth pressure coefficient with wall depth (data from Wind, 1976).

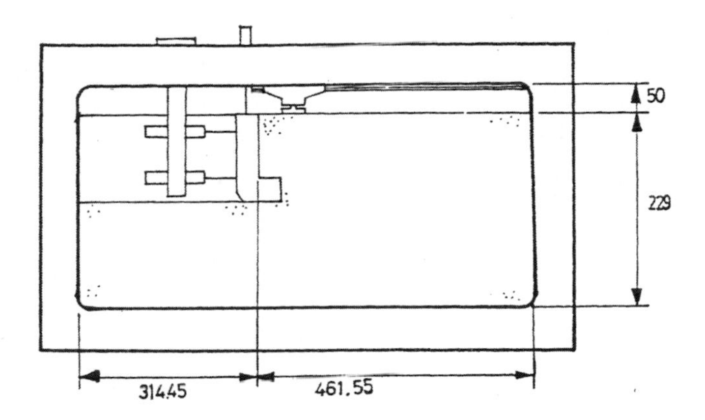

Figure 11.8 Cross-section of the centrifuge model showing an L-shaped wall (after Mak, 1983).

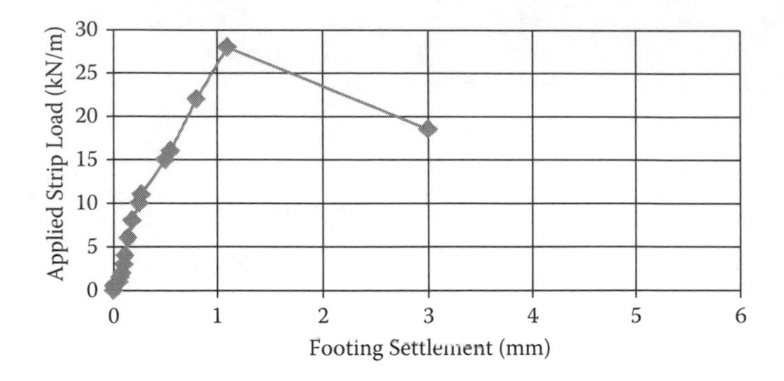

Figure 11.9 Settlement of the footing (after Mak, 1983).

a threshold value it starts to increase even with decrease bearing pressure. The rotations are quite interesting. For low bearing pressure the wall rotates counterclockwise and then changes direction and starts to rotate in a clockwise fashion. Again a threshold value is reached beyond which the rotation increases even when the bearing pressure from the foundation decreases.

Mak (1983) was able to use a Hasselblad camera with a 70-mm film to take images of the retaining wall through the Perspex window on the side of the container. This camera could not be mounted on the centrifuge. Instead the camera was positioned above the roof of the centrifuge chamber and took images through a glass window exactly at the moment the

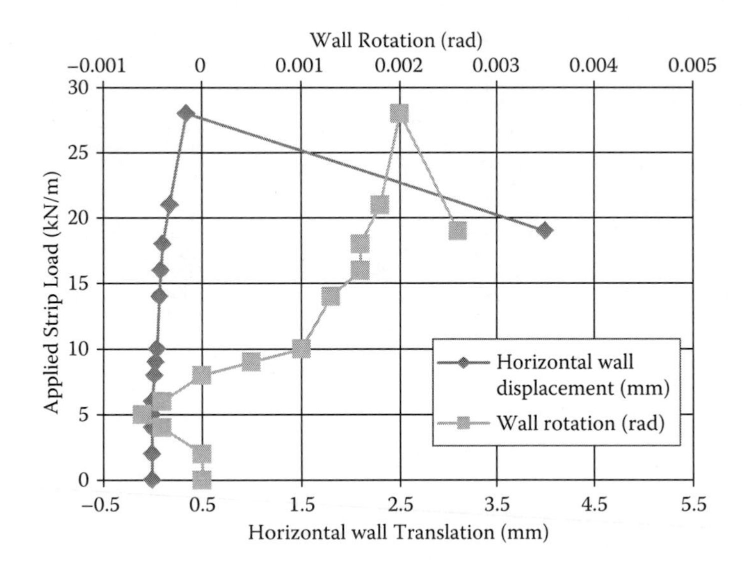

Figure 11.10 Lateral translation and rotation of the L-shaped wall (after Mak, 1983).

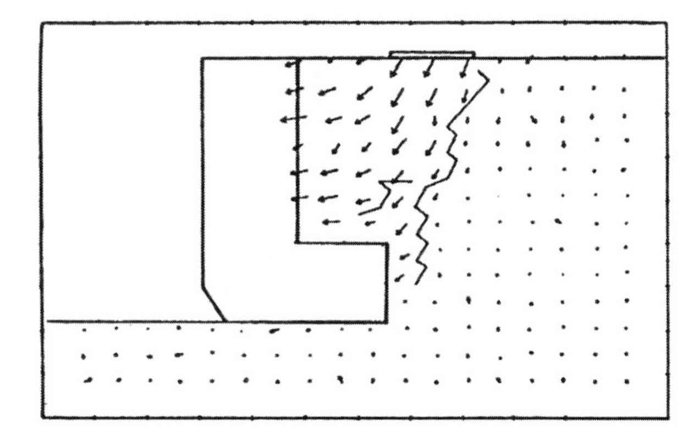

Figure 11.11 Soil deformations behind the L-shaped wall (after Mak, 1983).

centrifuge passed below. This required accurate syncing of the camera shutter and the centrifuge, which was achieved using an electronic trigger. The quality of images obtained from the Hasselblad camera was sufficient to obtain displacement vectors in the soil model at different *g* levels. This system had a distinct advantage compared to the radiographic techniques used earlier, as images could be obtained during the centrifuge flight and were not limited to before and after test images. Also no lead shot was required as simple reflective markers could be placed just on the surface of the soil next to the soil Perspex window interface. Using photogrammetric techniques, the soil deformations could be calculated by comparing images taken at different times in the centrifuge test. An example of the soil deformations behind the L-shaped wall is shown in Figure 11.11.

11.7 CENTRIFUGE MODELLING OF PROPPED WALLS

When we need to create underground space for new foundations, for example, a set of retaining walls are driven in and props are placed between them for support. Such walls are commonly used where the depth of penetration of the wall into the soil is limited or the wall encounters soft soils at depth. The props are normally deployed at different levels while excavation progresses between the two walls. This process of gradual excavation and placement of multistage props is considered in Chapter 13 as explained before. In this section let us consider centrifuge modelling of a set of retaining walls that are propped just at the top of the wall as shown in Figure 11.12. The walls are 6 mm thick and 200 mm high. They are embedded into sand to a depth of 60 mm. The single prop is at the top of the walls. A load cell is placed within the prop to measure the prop force. The centrifuge test was conducted at 40 *g*.

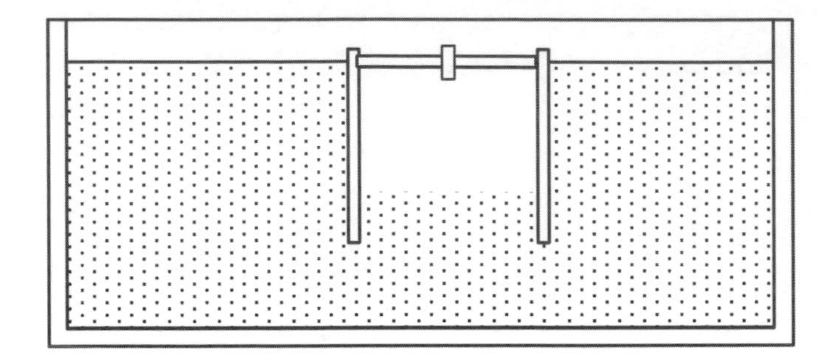

Figure 11.12 Cross-section of a centrifuge model of propped retaining walls.

The main purpose of these tests was to see the effect of earthquake loading on propped walls as part of the RELUIS project (Tricarico et al., 2013) but only the static part of the test will be considered here. The RELUIS project was carried out as a collaborative project between the University of Cambridge and the University of Parthenope, Naples.

The bending moments generated in the left and right walls at various g levels are presented in Figure 11.13. Note these data are presented at

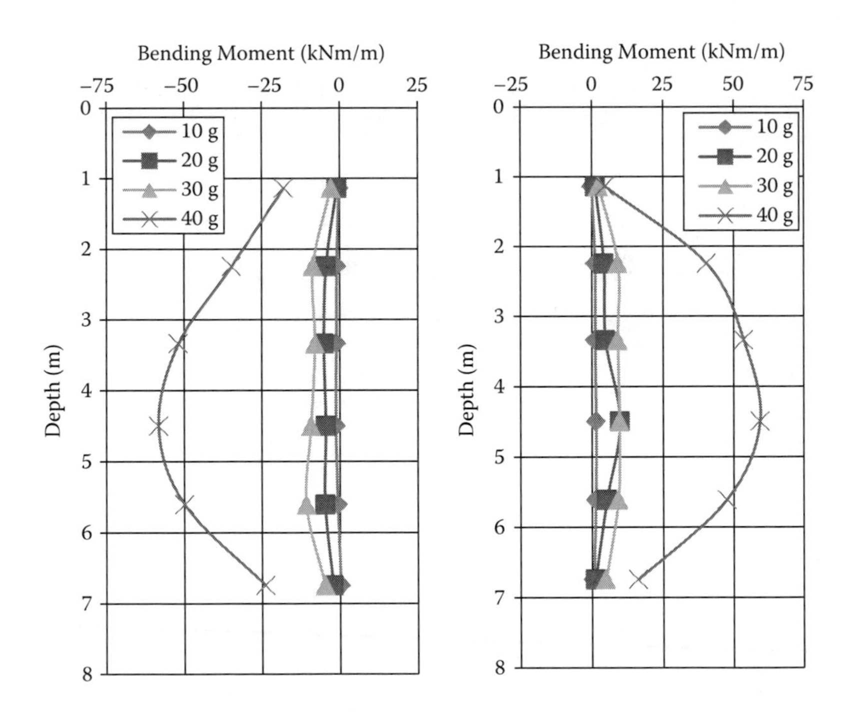

Figure 11.13 Bending moment distribution in left and right walls.

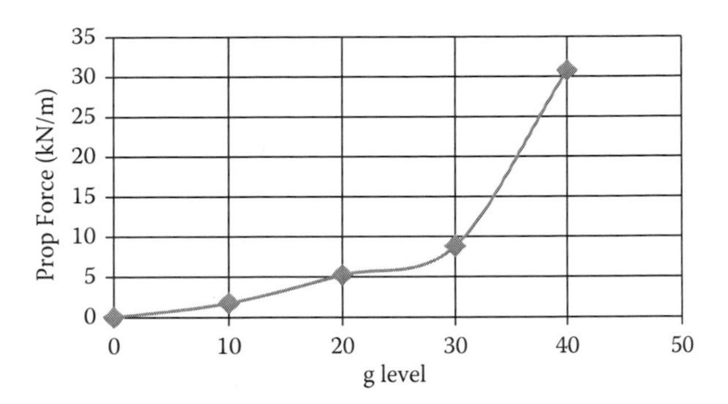

Figure 11.14 Increase in prop force with g level.

prototype scale. In this centrifuge test the bending moments do not pick up during the early phase of the test. There is a large jump in the bending moments between 30 and 40 *g*. However, the bending moment distributions in the left and right walls compare very well with each other both in terms of magnitude and shape. The location of the peak bending moment also agrees well between the two walls and occurs at a depth of 4.5 m below the soil surface.

The load cells in the props at the top of the wall recorded the prop forces. The change of prop force with increasing *g* level is plotted in Figure 11.14. Again the data is plotted at prototype scale and has been converted to kN/m using the spacing between the props. In this figure it can be seen that a peak prop force of 30 kN/m is recorded at 40 *g*. However, the increase in prop force is rather sudden between 30 and 40 *g*. This is consistent with the changes in the bending moment between these *g* levels presented in Figure 11.13.

11.8 SUMMARY

In this chapter we have considered centrifuge modelling of a wide variety of earth retaining structures. Research on retaining walls has been going on for a long time, initially using laboratory models, some of which are fairly large sized. In this chapter some of these were considered particularly when they were used to measure the soil deformations behind the retaining walls. With the developments in centrifuge modelling, retaining walls of different types have been tested using this technique. Bending moments generated in the retaining walls were measured using strain gauging. Initially radiographic techniques were used to take x-rays before and after centrifuge tests to identify the soil deformations behind the walls. Ingenious ways of taking images of the centrifuge packages in flight, with stationary cameras outside the centrifuge but synced with its rotation, were discussed. This technique was widely used to obtain soil deformations until recently when modern digital cameras were able to fly on the centrifuge and attached directly to the centrifuge packages.

One of the main topics with retaining wall research is the mobilisation of earth pressures on the active and passive sides of the walls. The measurement of earth pressures is a difficult task in geotechnical engineering. This was not considered in this chapter although earth pressures can be estimated based on the measurement of bending moments in the model walls in a centrifuge test. Some attempts were made to use miniature earth pressure measuring transducers; these are mostly successful only in earthquake tests where the shaking helps destroy the arching of sand around the transducers and allows more reliable measurement of earth pressure. Novel techniques like use of tactile measuring strips are coming online which should allow us to make more accurate measurements of earth pressures behind retaining walls in the near future.

Chapter 12

Pile foundations

12.1 INTRODUCTION

Pile foundations are the most commonly used form of deep foundations where the surface soil layers are either soft or weak. They are used worldwide to transfer axial structural loads into competent soil strata below. The functionality of pile foundations can differ from being structural load-carrying members. Sometimes they are used to stabilize soil slopes, such as the side slopes of a dam or an embankment, in which case they attract lateral loads. Piles can also be used to densify loose soils, in which case they are called compaction piles or displacement piles. The use of piles goes back many centuries, with the earliest recorded cases utilising timber piles. Modern piles are made from either steel or reinforced concrete. Further, most piles are used to carry compressive loads, although on occasions they may be required to carry tensile loads, particularly in offshore applications. In such instances they are called tension piles.

Pile foundations carry the axial loads coming onto them by generating two resistive components, shaft friction and end bearing. These are identified for a single pile in Figure 12.1. The pile needs to be in vertical equilibrium with the applied loading such that:

$$F = Q_{shaft} + Q_{end} \tag{12.1}$$

where F is the applied structural load, Q_{shaft} is the shaft friction, and Q_{end} is the end-bearing resistance generated at the tip of the pile.

The base resistance Q_b can be calculated as:

$$Q_b = A_b \sigma_b (N_q - 1) \tag{12.2}$$

where A_b is the base area of the pile, σ_b is the effective overburden pressure at the pile tip level, and N_q is the bearing capacity factor.

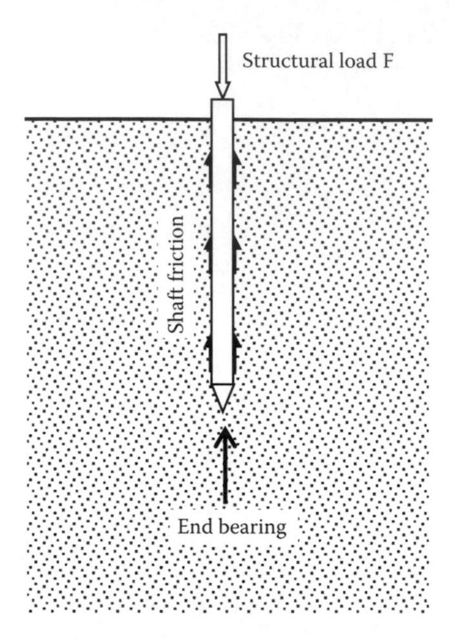

Figure 12.1 Load-carrying mechanism of a pile foundation.

The shaft capacity may be obtained by estimating the shear stress generated along the shaft, which can be calculated as:

$$\tau_s = K_s \, \sigma'_v \tan \delta_{cv} \tag{12.3}$$

where K_s is an earth pressure coefficient, σ'_v is effective vertical stress at a given elevation, and δ_{cv} is the friction angle between the pile material and the soil.

Broms (1966) related the values of K_s and δ_{cv} to the angle of shearing resistance of the soil ϕ' as shown in Table 12.1.

Table 12.1 Earth pressure coefficient K_s and pile-soil friction angle δ

		K_s	
Pile material	δ_{cv}	*Low relative density soil*	*High relative density soil*
Steel	20°	0.5	1.0
Concrete	$\frac{3}{4}\phi'$	1.0	2.0
Wood	$\frac{2}{3}\phi'$	1.5	4.0

Source: Broms (1966).

In order to obtain the shaft capacity due to skin friction, the shear stress must be integrated over the surface area of the pile using the following equation:

$$Q_s = 2\pi\ r \times \int_0^L \tau_s \qquad (12.4)$$

where r is the pile radius and L is the length of the pile.

On the first application of the load F the pile may have to undergo some vertical settlements to generate the two restive components in Equation 12.1. However, this is a function of the way the pile foundation is installed into the ground. Piles can also be classified based on the method of installation, such as: driven piles and bored or cast in situ piles.

The earth pressure coefficient K_s in Equation 12.3 depends both on the type of pile and the installation method (driven or cast in situ piles). The load-carrying mechanisms can be very different for these two types of piles. Driven piles can generate large horizontal stresses during the driving process, which leads to a large shaft friction. Also large stresses are generated at the tip of the pile during the driving process, particularly if the pile end is solid or it forms a soil plug. These stresses can remain high, even after the pile driving has stopped and the pile has been driven to the required depth. This gives these piles large end-bearing capacity. Bored piles, on the other hand, are created by auguring a vertical hole in the ground and placing the reinforcing steel cage and concrete. This construction process means that the horizontal stresses around the pile shaft are not large and therefore the initial shaft friction is not high. Similarly the end bearing can be small soon after casting the pile in situ. However, on application of the first load the pile can start to settle and mobilise both shaft friction and end bearing.

12.2 LABORATORY TESTING OF PILE FOUNDATIONS

Physical modelling of pile foundations has been carried out for a long time, with small-scale physical models being tested on a lab floor. Of course, the stresses and strains generated on such small scales may not represent the field conditions but some interesting and useful observations can be made. For example, it is possible to compare driving loads of piles in loose and dense sand samples. As small-scale models are physical realities in their own right, it is possible to made direct comparisons from tests. Difficulties arise only when trying to extrapolate the results to field-sized piles, for which centrifuge modelling is a much more appropriate technique.

Early work on testing of model scale piles at Cambridge was carried out by Swain (1979), who looked at an offshore jacket structure

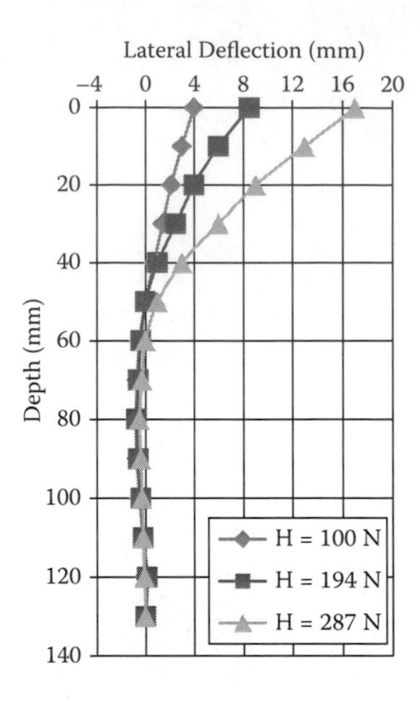

Figure 12.2 Lateral deflection of the pile (data from Swain, 1979).

supported on piled foundations. The model piles had a diameter of 7.94 mm and were made from Dural. The experimental program extended to testing of a 1:100 scale model of the offshore jacket structure with pile foundations extending into either dense sands or stiff clays with an over-consolidation ratio (OCR) of 8. Swain was also able to measure horizontal deflections in the pile using x-rays. He radiographed the model cross-sections at different stages of application of the horizontal loads. In Figure 12.2 the deflected shapes of the piles for various horizontal loads are plotted. For all these piles, it can be seen that beyond a depth of about 60 mm the pile suffers very small lateral deflections. This depth is normally termed the effective length of the pile and determines the depth of fixity of the pile.

Swain was able to measure the bending moments generated in the pile directly using foil strain gauges. Figure 12.3 shows the variation of bending moments with depth of the pile for model piles in dense sand and. Clearly for a given horizontal load, the curvature of the pile is going to be larger for piles in the clay compared to that in a sand layer. Therefore, the measured bending moments that depend on the curvature of the pile are larger in the case of the pile in a clay layer compared to that in a sand layer. In fact the pile in sand is subjected to nearly 2.5 times the lateral load compared to that in clay, and yet the bending moment generated is only 0.6 Nm compared to 1.9 Nm for the pile in clay.

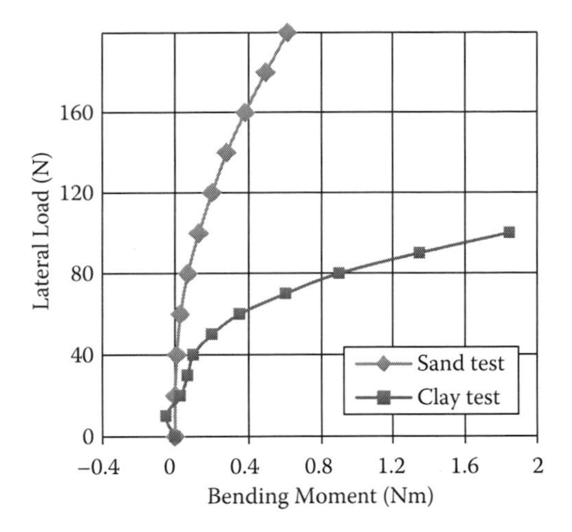

Figure 12.3 Bending moment distribution with pile depth (data from Swain, 1979).

In parallel with Swain's work, Williams (1979) carried out extensive testing of model piles in sands subjected to cyclic lateral loading. He tested piles of different diameters but we will consider the tests carried out at the same pile diameter and material as Swain (1979). Williams subjected his piles to cycles of horizontal loads and measured the horizontal deflection of the piles. He tested piles that extended 200 mm into the dense sand. Figure 12.4 shows an example of load-unload cycles applied to the piles.

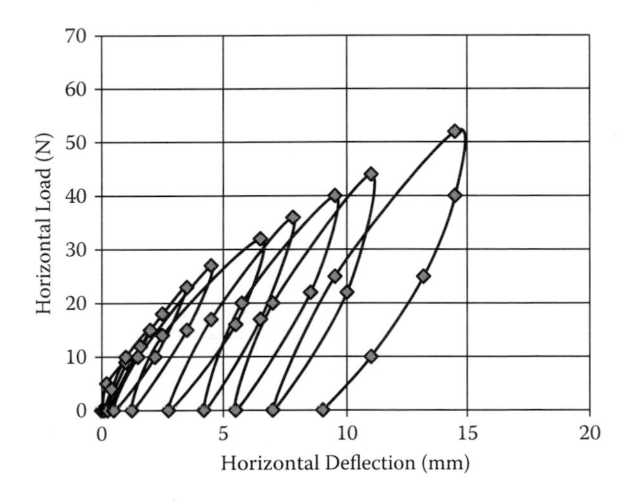

Figure 12.4 Cyclic loading of piles (data from Williams, 1979).

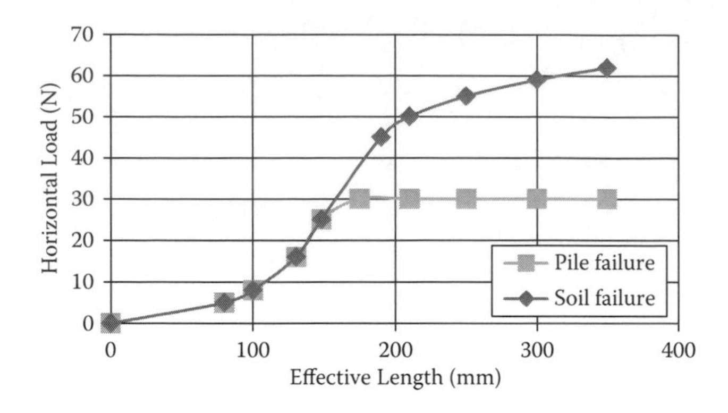

Figure 12.5 Soil versus pile failure (data from Williams, 1979).

In this figure it is interesting to note the accumulation of the irrecoverable horizontal displacement the pile suffers after each cycle. Further, these irrecoverable displacements increase in magnitude as the pile is subjected to cycles of large amplitude of lateral load.

Another interesting aspect of Williams's research was the observation of the formation of plastic hinges in piles subjected to large lateral loads. In Figure 12.5 the horizontal load is plotted against the effective length of the pile. At small effective lengths the pile will behave like a rigid member and cause the soil to fail. Similarly if the pile is flexible relative to the soil then it will bend rather than causing the soil to fail, when the effective length of the pile is large. However, if the pile is rigid and is embedded to large effective lengths, then the pile will fail by forming a plastic hinge, subject to the lateral loads becoming sufficiently large. Thus, it is possible to demarcate the failure criterion of a pile by considering its effective length, relative stiffness with respect to soil, and its plastic moment capacity.

These early experiments were run as a precursor to centrifuge testing and yet provide some interesting insights into the behaviour of laterally loaded piles in sands and in clays. Many of the model-making techniques developed during the course of this research were later used while attempting centrifuge modelling of laterally loaded piles.

12.3 CENTRIFUGE MODELLING OF PILE FOUNDATIONS

Research into pile foundations started utilising centrifuge modelling for over two decades. Various researchers have utilised centrifuge modelling to investigate the mechanisms of load transfer; mobilisation of the shaft friction and end bearing of piles with increasing pile settlements in different types of soil strata; different types of loading on pile foundations, such as horizontal or vertical loads under monotonic or cyclic conditions; and pile group effects; as well as to estimate the shaft friction and end-bearing capacities of piles given by Equations 12.2 to 12.4. When using centrifuge modelling for pile foundation problems, we need to ensure that the particle size effects described in Section 6.4 are taken into consideration while designing the model piles and in choosing the appropriate *g* level to conduct the centrifuge test. Typically, the model pile diameters need to be at least 25 times the mean particle size of the soil they are placed in. Similarly the model piles must have the same axial stiffness (*EA*) and bending stiffness (*EI*). The scaling laws for these parameters can be similar to those considered in Chapter 11 on retaining walls.

Model piles in centrifuge tests are often made from Dural and utilize hollow sections. The wall thickness is adjusted to achieve the required *EA* and *EI* values that simulate the right pile section in the prototype. This allows for the axial and bending behaviour of the pile to be captured accurately but not the failure of the pile. This is because Dural sections often have quite different yield stress compared to the actual steel or concrete pile sections utilised in the prototypes. To address this problem, recent research started utilising miniature plaster of Paris sections that are reinforced by steel cables. Recently Knappett et al. (2011) described the development of such model pile sections so that failure of the piles can be accurately reproduced in a centrifuge test.

12.3.1 Wished-in-place piles

In centrifuge modelling the small-scale models are built under laboratory conditions, under the so-called 1 *g* conditions. They are then subjected to increasing levels of gravity until the model reaches the desired *g* level. In case of centrifuge modelling of pile foundations, this presents an interesting choice. We can start by making our physical models of soil at 1 *g* and insert the model piles into this model ground. The stresses in the soil are quite small at 1 *g*. This means that the horizontal stresses next to the pile shaft and the bearing stresses below the pile tip will be small soon after the installation of the pile. However, as the model is subjected to increasing levels of *g*, the vertical stresses in the soil and consequently the horizontal stresses around the pile start to increase. Piles installed using this method are called "wished-in" piles. Although no great stress buildup occurs during pile installation, some amount of densification of the soil around the pile occurs as the pile is installed into the soil and displaces it. It can be argued that this method of installation of piles is akin to the construction of bored piles in the field, which have very little horizontal stresses in the soil surrounding the pile. End bearing is mobilised when these bored piles are subjected to first loading usually by suffering some amount of vertical settlement.

Wished-in piles are commonly used in centrifuge modelling as it offers a straightforward means of constructing the models and there is no requirement to have actuators to drive the piles in-flight. Under certain circumstances this would be a perfectly satisfactory way of conducting centrifuge testing of pile foundations. For example, in establishing the axial response of pile foundations in liquefiable soils subjected to earthquake loading, Stringer and Madabhushi (2011) show that the wished-in piles had a very similar dynamic response to model piles driven in-flight.

12.3.2 Piles driven in-flight

It is well known that the driven piles have larger load-carrying capacity due to the generation of large horizontal stresses during the driving process. Also, the driven piles suffer smaller settlement on application of the first load, as the end bearing and shaft friction are already fully mobilised. If we wish to simulate the driven pile behaviour in a centrifuge test it is necessary to install the piles in-flight. Clearly driving the piles at high *g* levels will generate the requisite large horizontal stresses around the pile shaft and mobilise the full end bearing when the pile is at its required depth. Also the same level of densification of the soil in the vicinity of the pile should occur in the centrifuge model as it would in the field.

However, installation of the piles in-flight does require additional actuators to drive the model piles. Many researchers use hydraulic jacks or electric actuators to push the piles into the soil. It must be noted that it is not necessary to have identical pile-driving equipment used in the field, such as single- or double-acting pile drivers, at a miniature scale in a centrifuge test. Jacking of model piles into the soil at high *g* will generate the requisite large horizontal stresses and pile tip stresses. For this reason, model piles can be installed into the soil using one- or two-dimensional actuators described in Section 7.3.

Care must be taken while installing piles at high gravity. A common problem is that the model piles made from thin-walled tubes made of Dural can suffer buckling during the jacking-in process. Careful calculations need to be made at the model design stage to prevent this and pile wall sections of suitable size must be chosen. Also pile tip wander can occur during installation. The connection between the model pile and the actuator must be carefully designed so that adequate pile head fixity is provided and eccentricity of load during the driving in phase is a minimum. This reduces the P-Δ effects that may cause pile failure during driving.

12.4 CENTRIFUGE MODELLING OF PILE INSTALLATION

As described in Section 12.3.2, in-flight installation of piles is desirable to create the correct stresses and strains in the soil surrounding the pile foundation. Centrifuge modelling of pile installation during flight started

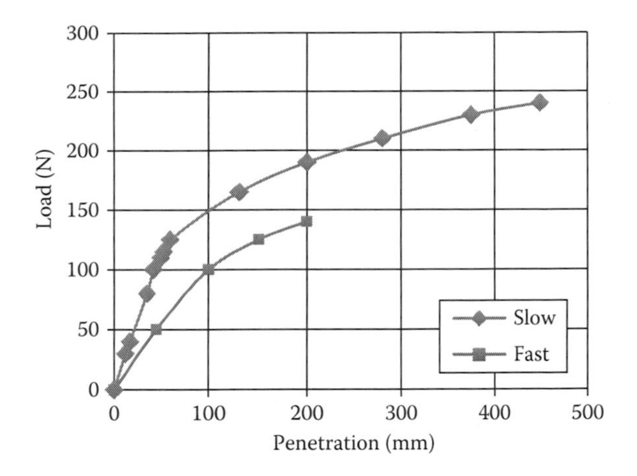

Figure 12.6 Pile load during installation (data from Clegg, 1981).

in Cambridge as early as 1981. In the early days this was achieved by using either pneumatic or hydraulic jacks, which pushed the model piles into the soil.

Clegg (1981) carried out testing of relatively small-sized piles some 3 m long and 0.5 m in diameter in the field. He followed this work with the newly available centrifuge facilities and carried out testing of 11 small-sized piles in stiff clays with an average OCR of 6. The model piles were made from Dural and had a diameter of 19.05 mm. The piles were closed ended and were tested at 40 *g*. He attempted to jack the model piles in at different rates of penetration. Figure 12.6 shows an example of the experimental data obtained during model pile penetration, with the pile head loads applied and the penetrations of the pile for two different rates of penetration. Clearly the faster penetration rate resulted in lower pile head load for any given penetration. This is due to the generation of the excess pore pressures in the clay in the vicinity of the pile shaft. During a slower penetration rate, the excess pore pressures have more time to dissipate giving rise to larger horizontal effective stresses acting on the pile shaft. This results in the resistance to pile driving being higher.

In more recent tests at Cambridge, piles are jacked in-flight using either the one- or two-dimensional actuators referred to in Section 7.3. Other centrifuge centers such as IFSTTAR in France use hydraulic jacks. The control of the actuator movement is excellent and the model piles can be driven under very controlled rates of penetration.

12.5 CENTRIFUGE MODELLING OF LATERALLY LOADED PILES

One of the early problems in pile foundations that was tackled using centrifuge modelling was that of laterally loaded piles. A theoretical framework for analyzing the laterally loaded piles called the *p-y* method was developed based on some field testing on 0.61-m-diameter piles by Matlock and Reese (1960). This method has become quite popular over the years and is used with some variations to the current day. As field testing of piles subjected to lateral loading is difficult and can be expensive, many researchers use centrifuge modelling to investigate this problem.

Barton (1982) describes the centrifuge modelling of lateral loading on piles used in the offshore oil industry. She carried out centrifuge testing of piles in dense sands at a relative density of about 80 percent. These were wished-in-place piles as explained in Section 12.3.1. The model piles of 12.7 mm diameter were subjected to lateral loading and the bending moments were measured at different depths along the pile. Figure 12.7 shows typical data obtained on the bending moment variation with the normalised depth of the pile measured during a 120-*g* centrifuge test. The peak bending moment occurs some 4 pile diameters below the soil surface.

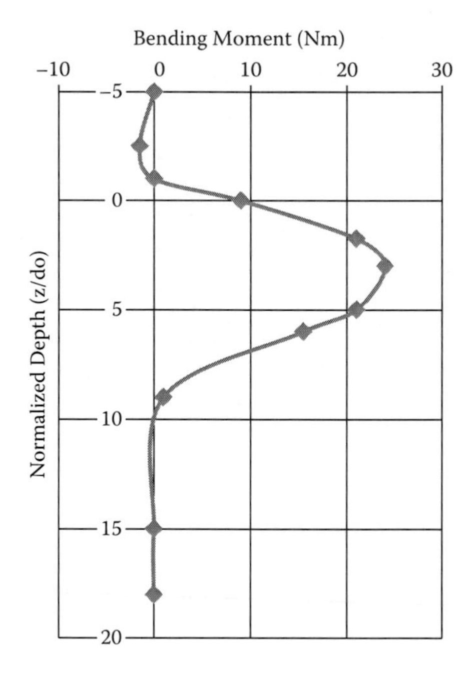

Figure 12.7 Variation of bending moment with depth (data from Barton, 1982).

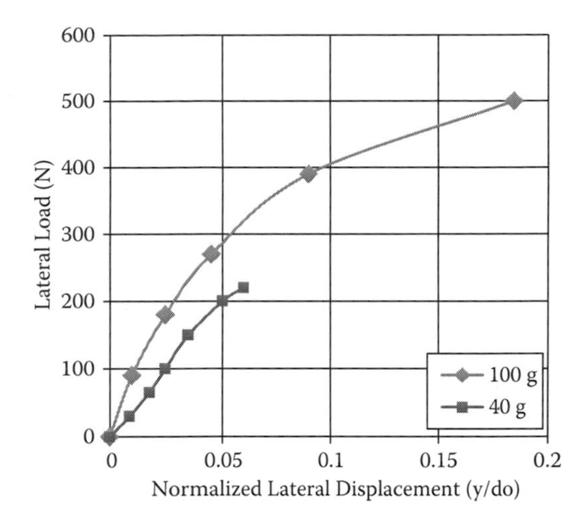

Figure 12.8 Lateral response of the pile at different *g* levels (data from Barton, 1982).

The model piles were tested under lateral loading at different *g* levels. The lateral load was applied to the model pile through an ingenious method. This involved having two identical blocks of concrete hanging over pulleys on either side of the model pile and maintaining horizontal equilibrium. The blocks were enclosed in watertight containers. When the lateral load was needed, water was pumped into one of the containers and the buoyancy force reduced its weight and caused a lateral load on the model pile. In Figure 12.8 the lateral response of the pile at two different *g* levels is plotted. Clearly in the 100-*g* test the pile shows a stiffer response and the peak lateral load is much larger, that is, 500 N (equivalent to a prototype lateral force of 5 MN). In a 40-*g* test the response of the pile is softer and the peak lateral load is only 210 N (equivalent to a prototype force of 0.34 MN).

Of course the main purpose of this research was to develop *p*-*y* curves for laterally loaded piles and compare them to those used in design. It is possible to obtain the *p*-*y* curves once the bending moment in the pile is measured. Double integrating the curve that fits through the bending moment data gives the lateral deflection of the pile (*y*) while double differentiating the same curve gives lateral soil pressures acting on the pile (*p*). An example set of *p*-*y* curves are plotted in Figure 12.9 at different depths along the pile in the 40-*g* centrifuge test. At shallow depths the *p*-*y* curve is softer and it gets progressively stiffer with increasing depths.

This research not only provided very valuable insights into the lateral response of piles but also provided experimental data against which numerical procedures could be validated.

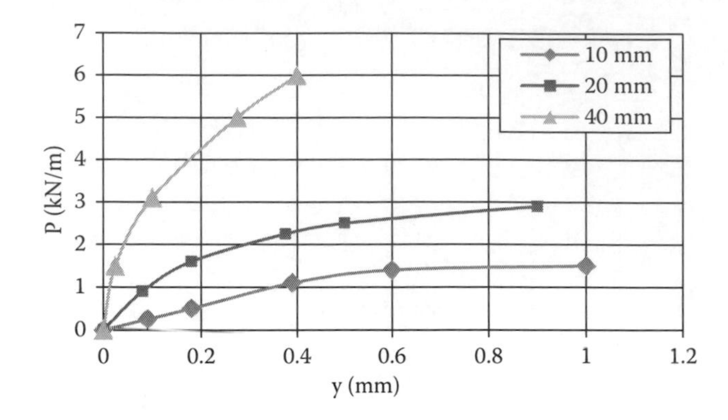

Figure 12.9 *p-y* curves obtained from a 40-*g* centrifuge test (data from Barton, 1982).

12.6 CENTRIFUGE MODELLING OF TENSION PILES

Tension piles are used where the foundations are subjected to pull-out forces. For example, in the offshore oil industry they are typically used to anchor tethers holding a floating structure to the seabed. Unlike normal piles, these piles see very small axial compressions but are called upon to resist large tensions in the form of pull-out forces.

Nunez (1989) carried out centrifuge modelling of tension piles in clay which simulated very large prototype piles with a length of 50 m and a pile diameter of 2.1 m. The main research question was the generation of skin friction around the pile shaft during the pull-out phase of the loading. An example of the data obtained during a 140-*g* centrifuge test is presented in Figure 12.10. In this figure the average skin friction mobilised as the

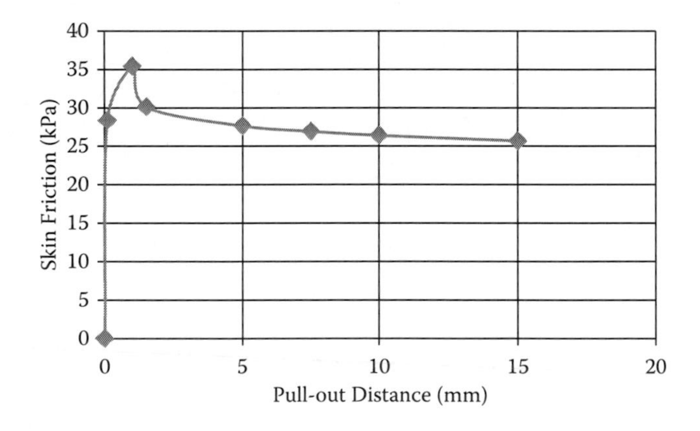

Figure 12.10 Mobilisation of skin friction in a tension pile (data from Nunez, 1989).

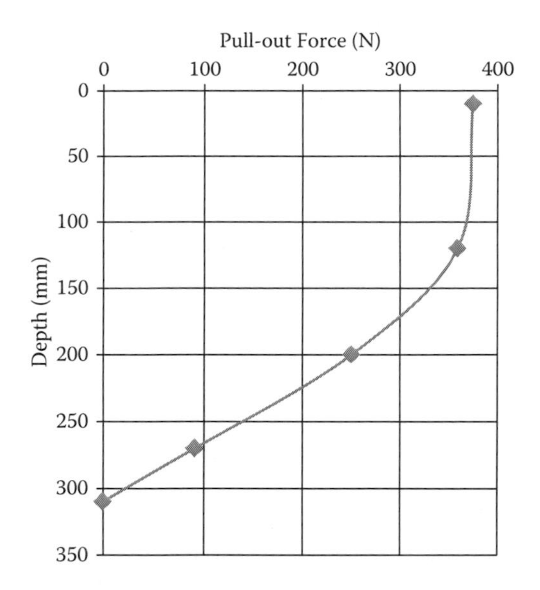

Figure 12.11 Load distribution with depth (data from Nunez, 1989).

pile is pulled out is presented. It can be seen that the peak skin friction is mobilised rapidly with very little pull-out distance. However, the peak skin friction does suffer some drop with increasing pull-out distances.

The model pile was segmented with load cells between segments that allowed measurement of tensile forces in each segment. This allows us to visualize the variation of tensile forces in the pile with depth. This is shown in Figure 12.11. In this figure we can see that the maximum pull-out force is recorded close to the surface. This is to be expected as the entire pull-out force of 380 N (equivalent prototype force of about 7.5 MN) must pass through these shallow segments. As we go deeper, the tensile force in the pile reduces as the skin friction mobilised along the pile shaft increases with depth. At a depth of about 320 mm the tensile force in the pile reduces to zero. This corresponds to a prototype depth of 44.8 m, thereby justifying the use of 50-m-long piles in the field.

This research contributed to the understanding of tension pile behaviour. The mobilisation of skin friction and generation of excess pore pressures in the vicinity of the pile were better understood and this helped in developing suitable guidance for the geotechnical practice.

12.7 NEGATIVE SKIN FRICTION IN PILES

Pile foundations are normally used to transfer structural loads through weak or soft soil layers to more competent soil strata below. When they pass through soft clay layers, they can attract additional axial loads due to the consolidation of the clay layers. Such a consolidating clay layer can apply a down drag as it settles and this is commonly called the negative skin friction. The presence of negative skin friction may result in the pile's load-carrying capacity being reduced as the pile has to carry the down-drag forces. Negative skin friction occurs when the settlement of the soil is larger than the pile settlement. As the clay layer at shallow depths settles more during consolidation compared to regions of the clay deep below, the negative skin friction is larger at shallow depths and is expected to decrease with depth. In the lower regions, the pile will settle more than the clay thereby generating positive skin friction. Therefore, there must be a horizontal plane that passes normal to the pile above which negative skin friction acts on the pile while below this plane positive skin friction is generated. Such a plane is called the "neutral plane" as it demarcates the compressive and tensile forces acting on the pile.

Centrifuge modelling offers a way to understand and quantify the negative skin friction acting on piles passing through soft clay layers. Lee (2001) has investigated the problem of negative skin friction in piles. In his experimental setup he had a single pile with a diameter of 13.1 mm. This pile passed through a soft, under-consolidated clay layer that is 180 mm thick. This clay layer was overlain and underlain by sand layers. The pile was placed into the bottom sand layer and extended some 10 mm into this layer. Thus, the pile tip develops significant end bearing. The top sand layer overlying the clay layer acts as a surcharge and promotes consolidation of the soft clay layer generating the negative skin friction along the pile shaft. The distribution of drag load in an 80-*g* centrifuge test is shown in Figure 12.12.

In this figure it can be seen that the pile has recorded tensile loads in the upper regions of the pile. These change to compressive loads beyond a depth of about 145 mm. The neutral plane is identified in Figure 12.12 by a solid circle. The pile records tensile forces above the neutral plane while it records compressive forces below the neutral plane.

This research has shown that the centrifuge modelling technique can be used to investigate the problem of negative skin friction that acts on piles passing through consolidating clay layers.

12.8 LARGE-DIAMETER MONOPILES

With the urgent need for renewable energy sources and the societal reluctance to have onshore wind farms, offshore wind farms emerged as a sustainable, alternative form of energy production. The preferred foundations for these large offshore wind farms are large-diameter monopiles which can range from 4 m to 7 m in diameter and can extend 30 m to 50 m below the seabed. Often they pass through thick clay layers, especially offshore the United Kingdom. The continental coasts offshore of Germany,

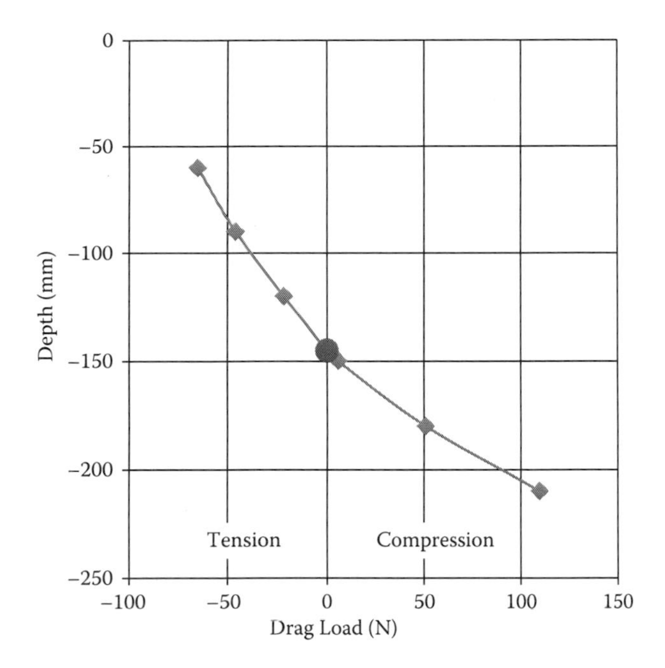

Figure 12.12 Negative skin friction in a pile (data from Lee, 2001).

Denmark, and Holland are predominantly sandy and monopile foundations are installed in these sand deposits. Given the long design life of the offshore wind turbines, the monopile foundations need to survive extreme wind and wave loads that may occur over their design life.

In Cambridge, research is underway to establish the monotonic and cyclic behaviour of monopile foundations installed in clay layers as well as in sand deposits. Given the large size of the problem, centrifuge modelling was carried out at 100 *g*. A 40-mm tubular model pile is used to simulate a 4-m-diameter monopile in the field. The two-dimensional actuator presented in Section 7.3 was used to initially install the model monopile in-flight and then subject it to lateral loads.

Figure 12.12 shows the jacking load required to push the model monopile into a normally consolidated clay layer of 35-m thickness. The clay was preconsolidated to a pressure of 500 kPa and is brought into equilibrium by allowing the excess pore pressures to fully dissipate during the centrifuge flight. The data in this figure are presented at prototype scale. Only data for the final 4 m of penetration (equivalent to one times the diameter or 1 D) is presented, just before the pile reaches the required depth. Figure 12.13 shows a peak force of nearly 7 MN is required to install the monopile to the required depth of 20 m.

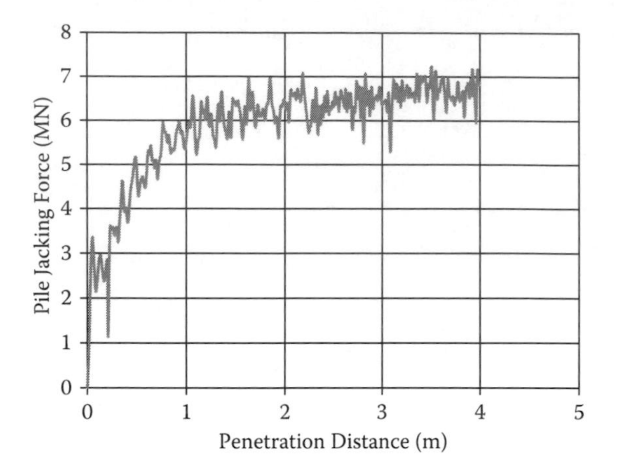

Figure 12.13 Installation forces for a 4-m-diameter monopole.

Once the monopile was installed, the clay around the monopile was allowed to reconsolidate. Following this, it was subjected to a monotonic lateral load until the pile had suffered large lateral displacements of 1.5 D. In Figure 12.14 the lateral response of the monopile is presented. While there is no clearly defined peak, the ultimate lateral capacity of this pile can be conservatively taken as 1.75 MN. This corresponds to a lateral displacement of about 1 m (or 0.25 D). Normally the design restriction of 0.2° of rotation at the seabed comes well before the lateral capacity of the monopile is reached. More such experiments in clays with different undrained shear strengths were conducted and these are discussed by Lau et al. (2014).

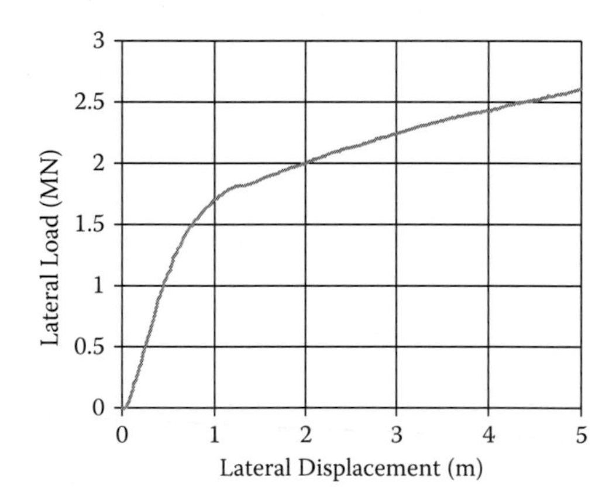

Figure 12.14 Lateral response of a 4-m-diameter monopole.

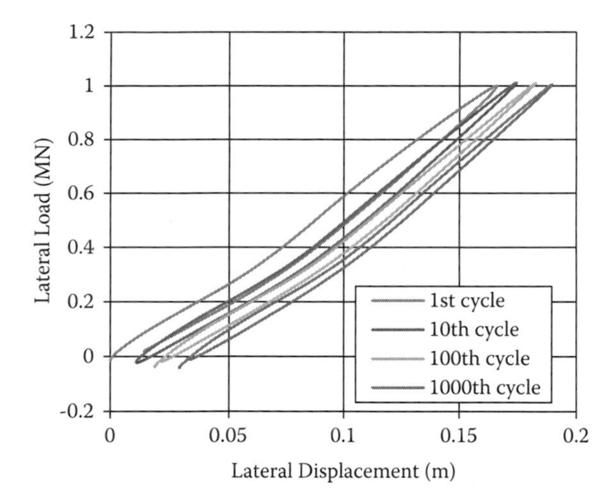

Figure 12.15 Cyclic response of a 4-m-diameter monopile in sand.

As an example, the centrifuge testing of monopiles driven into a dense sand layer is also considered here. In this case the 4-m-diameter monopile is subjected to horizontal cyclic loading. This is a realistic loading scenario for offshore monopoles, which are subjected to wind and wave loading. Figure 12.15 shows the response of the monopile to cyclic loading. In this figure the load-unload cycles are presented for the first, tenth, one hundredth, and one thousandth cycles. We can see that even after a large number of cycles the lateral displacement is still accumulating. Kirkwood and Haigh (2013) present more details of this study.

12.9 SUMMARY

Pile foundations are an important form of deep foundations. They are used worldwide and to support a wide variety of structures. In this chapter we have looked at use of centrifuge modelling in studying the behaviour of pile foundations under different loading conditions. Pile foundations can behave differently depending on if they were driven or cast in situ. In carrying out centrifuge modelling of pile foundations, efforts have been made to develop techniques to achieve pile driving in-flight. Some examples of this were considered in this chapter. A distinction was also made on some aspects of centrifuge techniques. Unlike in real field cases, the increased gravity field is applied suddenly by increasing the centrifuge speed. Piles that are installed at 1-*g* conditions and then subjected to this increased gravity field are called wished-in piles. In some respects they are similar to cast in situ or bored piles. By jacking in the piles in-flight, we can simulate driven piles.

In this chapter we have considered examples of centrifuge modelling of laterally loaded piles. While the techniques of applying the lateral load and the accuracy of the control has improved over the years, the emphasis in this chapter was on showing the quality of experimental data that can be obtained in centrifuge modelling. In the example considered it was shown that the bending moment distribution along the depth of the pile and lateral deflections suffered by the pile under various lateral loads were considered. Using this data the traditional p-y curves can be obtained at different pile depths. Similarly an example of the centrifuge modelling of a long, tension pile was also considered. In this case the distribution of the shaft friction along the pile depth can be obtained. An example of the negative skin friction generated along the pile shaft was considered for a pile passing through a soft, consolidating clay layer that was sandwiched between sand layers. Again the distribution of drag loads generated on the pile shaft could be determined. Finally centrifuge modelling of large-diameter monopiles used for offshore wind turbines was considered. While these are by no means an exhaustive list of the application of centrifuge modelling of pile foundations, these examples serve to highlight the types of problems that can be studied using centrifuge modelling.

Chapter 13

Modelling the construction sequences

13.1 ADVANCED CENTRIFUGE MODELLING

The behaviour of soil is complex. It is well known that its stress-strain behaviour is nonlinear at even moderate amounts of applied strain. Further, it can suffer volume changes and/or pore pressure changes when sheared. The volume changes may partly result in plastic, volumetric strains and shear strains. The behaviour of soil is also governed by the previous stress history it has seen. We normally calculate the over-consolidation ratio (OCR) based on the preconsolidation pressure the soil has seen before and its current stress state.

To illustrate the importance of stress history let us consider the simple case of a clay layer subjected to a surcharge loading. The behaviour of soil when it is subjected to larger surcharge pressures than it has seen in its history will be governed by the normal consolidation line (NCL). The settlement it will suffer for a given surcharge pressure can be calculated if the slope of the NCL (λ) is known. On the other hand consider a clay layer of the same thickness that has an OCR of 10, that is, it has seen a vertical stress 10 times larger than the current surcharge pressure. In this case the settlements will be much smaller. These settlements can be calculated by knowing the slope of the reconsolidation line (κ). For many clays, the values of λ and κ can differ by a factor of 3. Therefore, the settlement suffered by a clay layer with high OCR values will always be smaller. In other words, the stress history of the soil plays an important role in determining the behaviour of the soil.

The dependence of soil behaviour on previous stress history has consequences for centrifuge modelling. The stress history experienced by the prototype soil must be reproduced on the soil samples being tested in a centrifuge test, so that accurate soil behaviour is reproduced. Similarly, the stress paths taken by the soil can be very important when considering the construction sequence of geotechnical structures. Such cases are more complex than considering the OCR of a given soil at site and reproducing it in the centrifuge model. Many of the applications discussed in previous chapters such as the retaining walls in Chapter 11 did not consider the stress paths taken by the soil in the centrifuge models, although in some cases we have seen that the OCR of the soil was reproduced in the centrifuge tests. Some of the applications where it is important to consider the stress paths is considered in this chapter. Further, the advanced techniques required while attempting the centrifuge modelling of such problems is highlighted.

13.2 CONSTRUCTION SEQUENCE MODELLING

In many geotechnical engineering problems it is important to simulate the actual construction sequence that will be adopted in the field. The stress path taken by the soil in the prototype must be carefully considered and faithfully reproduced in the centrifuge test. Let us consider the construction of cantilever retaining walls discussed in Chapter 11. In that chapter we assumed that the model retaining walls can be placed in the soil at 1 *g* and then the gravity can be increased to simulate a prototype retaining wall. We did not try to model the actual construction sequence used in the field. Many retaining walls in the field are constructed by driving the sheet pile wall into the ground and then excavating the soil on one side. Similar processes are used while constructing diaphragm walls in clayey soils. This staged construction sequence will lead to a gradual buildup of horizontal stresses in the soil to active and passive earth pressures on either side of the wall and cause an increase in the shear force and bending moment in the wall. Neither those nor the wall deflections can be captured accurately if we make the centrifuge model at 1 *g* and then suddenly increase the *g* level. In fact, finite element analyses have been carried out to show the importance of modelling the excavation sequence in the construction of the retaining walls (e.g., Chandrasekaran and King, 1974). It is therefore important in some cases to simulate the staged construction sequences in the centrifuge. This will be explained in more detail in Section 13.3.

Similar to the construction of retaining walls and diaphragm walls, there are many other problems where the construction sequence can be important. Another example of this is the construction of new tunnels in the ground. If the tunnel alignment comes close to existing foundations such as piles then the load-carrying capacity of the foundations may be affected. At the very least, the foundations may suffer additional settlements due to the construction of the tunnel. Similar to this problem is new tunnels passing below existing pipelines or other tunnels. Again the construction of the new tunnel can induce additional stresses and/or deformations in the old structures. These types of problems are becoming common around the world with increased infrastructure needs in cities. For example, in the city of London the construction of the Cross Rail tunnels is currently underway and engineers must ensure that the existing foundations and the structures they support do not suffer excessive settlements.

Centrifuge modelling can be used to simulate many of these complex interaction problems. However, in each of these problems the construction sequence must be carefully simulated for the centrifuge tests to yield accurate results.

13.3 MODELLING OF STAGED EXCAVATIONS IN FRONT OF A RETAINING WALL

In order to simulate the behaviour of a retaining wall the gradual excavation of soil on one side of the wall should be modeled. Before the advances in robotics, this type of staged excavation could be modeled by use of a heavy fluid on one side of the wall that could be drained away in stages as shown in Figure 13.1. In this figure the triangles indicate the drop in the level of the heavy fluid with time to simulate the staged excavation. The heavy fluid that was used in the early days of centrifuge modelling was zinc chloride. Zinc chloride has a raw unit weight of about 30 kN/m^3. It could be mixed with water to make a solution of the required unit weight. Normally the unit weight of the heavy fluid is matched with the unit weight of the soil it replaces. The coefficient of earth pressure is estimated to be about 1 to 1.2 for diaphragm walls in over-consolidated clays constructed in a slurry trench. This matches the pressure exerted by the zinc chloride solution. As the *g* level increases, the wall will see the increasing horizontal earth pressure from the soil on the retained side and this is matched by increasing fluid pressure on the excavated side. Once the excess pore pressures reach equilibrium, the excavation process can be simulated by lowering the level of the zinc chloride solution gradually to match the rate of excavation.

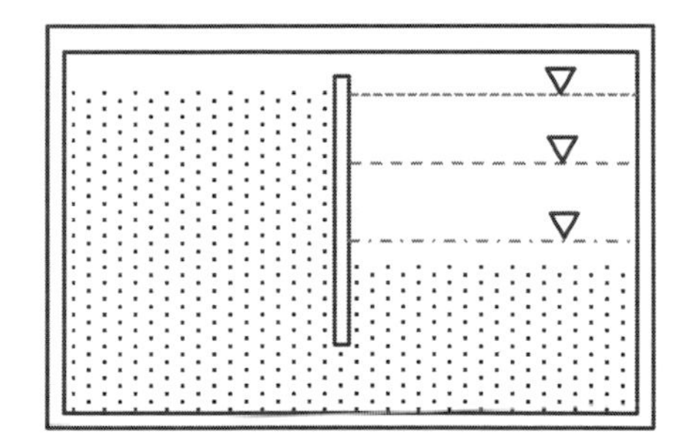

Figure 13.1 Simulation of staged excavation by drainage of heavy fluid.

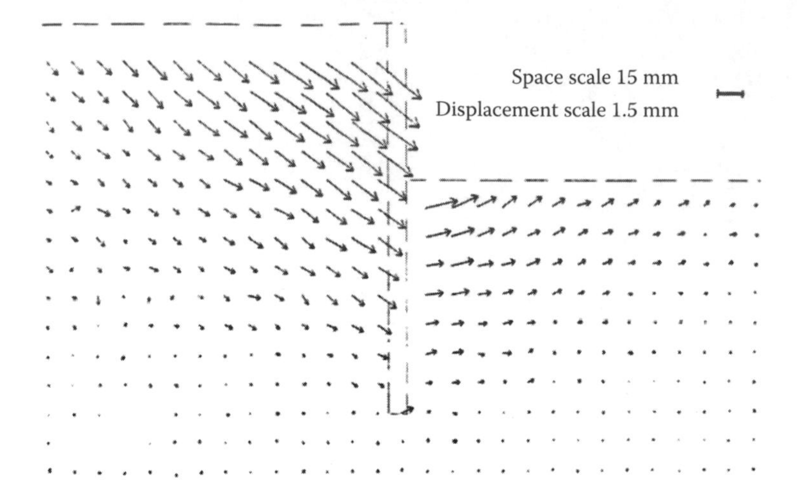

Figure 13.2 Soil deformations during excavation (after Powrie, 1986).

This procedure was used by Powrie (1986) to investigate the behaviour of diaphragm walls in clay. Initially he investigated the deformation of rigid, cantilever walls in over-consolidated clay by conducting the centrifuge tests at 125 *g*. The wall was made from Dural and had a thickness of 9.5 mm. He was able to obtain soil deformations following the excavation phase by taking photographs of the package in-flight using a Hasselblad camera at different times and comparing the position of the markers placed on the clay. An example of the soil deformations obtained in one of the diaphragm wall tests with a wall penetration of 15 m is shown in Figure 13.2.

Powrie (1986) also investigated propped retaining walls in clays that are used in the field when the penetration depths are limited. The model walls for this study were made from Dural and had a thickness of 4.76 mm. An example of a diaphragm wall propped at the top is considered here. The soil deformations obtained in one such test where the wall penetration depth was 15 m is presented in Figure 13.3. From this figure it is clear that the propped wall suffers rotation about the top prop and the soil deformations are more profound in this case. A prop force of about 450 kN/m was measured soon after excavation in this test.

The thinner walls used in this test were strain gauged to measure the bending moments in the wall. Figure 13.4 shows the bending moment distribution in the wall immediately after the excavation and after consolidating the soil for about 13.4 years (prototype scale). In this figure it can be seen that the peak bending moment in the wall increases substantially during the post-consolidation phase. The prop force also increased to about 650 kN/m in this phase.

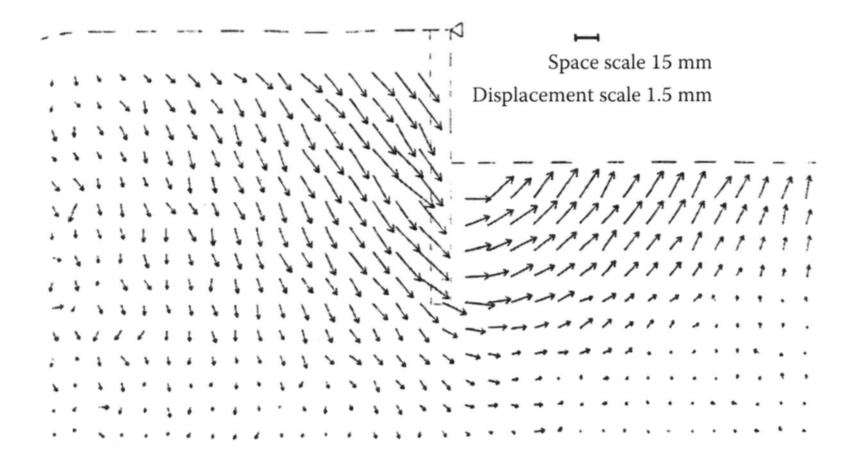

Figure 13.3 Soil deformations during excavation for a propped wall (after Powrie, 1986).

Although zinc chloride solution was used in many centrifuge tests, its use in modern-day centrifuge tests is reduced because of safety concerns. A new heavy fluid formed from sodium polytungstate solution can be used as a replacement for the zinc chloride solution. This fluid has unit weight of 28 kN/m^3 and can be adjusted to the required unit weight by mixing with water. It has been used recently by Choy (2004) and Elshafie (2008).

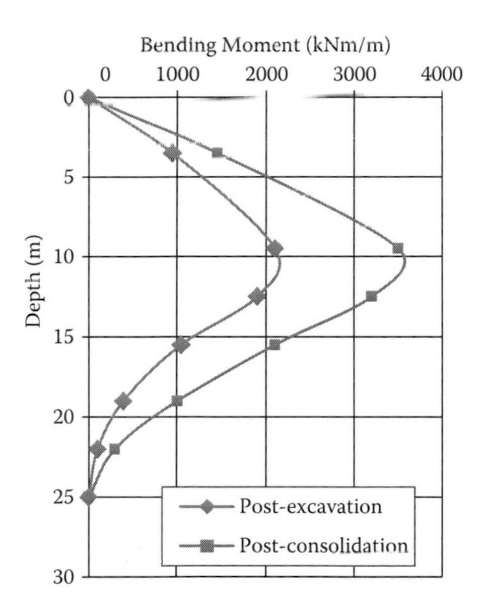

Figure 13.4 Bending moment distribution in a propped diaphragm wall (data from Powrie, 1986).

13.4 MODELLING THE INTERACTION BETWEEN A RETAINING WALL AND A BUILDING

The problem of buildings in close proximity to retaining walls involves modelling the complex interaction between these structures. Elshafie (2008) investigated the presence of a building behind a retaining wall. He considered buildings with different bearing pressures and bending stiffnesses. The main objective of the research was to investigate the influence of the presence of the buildings on the behaviour of the wall as the soil in front of it is being excavated. A typical cross-section of his centrifuge model is shown in Figure 13.5.

Initially a centrifuge test was conducted with no building but with just the wall at 75 *g*. This test is similar to the one seen in Figure 13.2 with the difference that the heavy fluid used in these tests was the sodium polytungstate solution. The deformations in the soil behind the wall were obtained used digital imaging and particle image velocimetry (PIV) analyses, as discussed in Section 8.3.2. In Figure 13.6 the displacement vectors recorded behind the retaining wall following a 12 m excavation in front of the wall are shown. These are plotted directly from the data obtained at the model scale.

For comparison, the displacement vectors behind the same retaining wall with a building that exerts a vertical pressure of 40 kPa behind the retaining wall is presented in Figure 13.7. In this figure it can be seen that the deformations are much larger when the building is present behind the wall. Further the settlements below the building are nonuniform, giving rise to both the settlement and rotation of the building. Elshafie (2008) investigated the effect of changing the bending stiffness of the wall and the building and how these influence the soil deformations behind the retaining wall. In addition, he also considered the case of individual strip footings supporting a portal frame structure behind the retaining wall. He observed differential settlement between the individual footings causing the portal frame to be subjected to additional moment loading at the slab level.

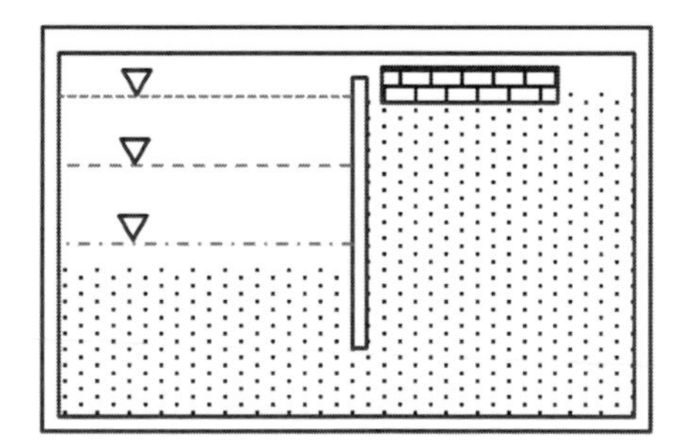

Figure 13.5 Building behind a retaining wall.

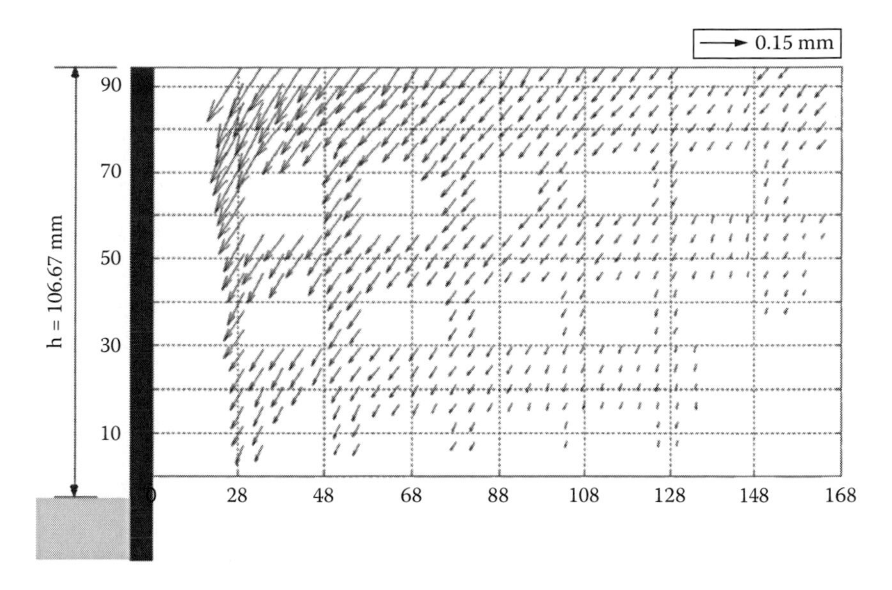

Figure 13.6 Displacement vectors behind a retaining wall (after Elshafie, 2008).

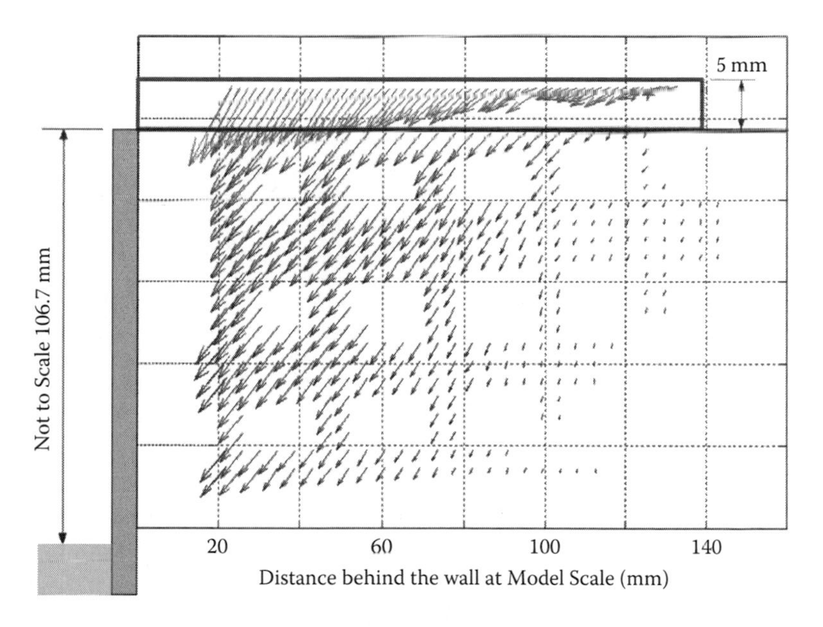

Figure 13.7 Displacement vectors behind a retaining wall with a building (after Elshafie, 2008).

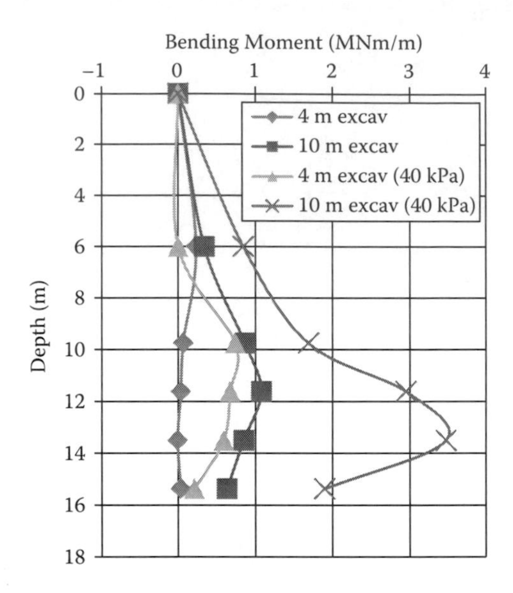

Figure 13.8 Bending moment distribution in the retaining wall (data from Elshafie, 2008).

Further bending moments in the wall were recorded during these centrifuge tests. The bending moments recorded in two centrifuge tests with and without the building are shown at prototype scale in Figure 13.8.

In this figure we can see that the bending moment increases with increasing excavation depth. When the building is present behind the retaining wall the bending moments are much larger for both depths of excavation. Also the position of maximum bending moment moves much deeper with the presence of the building behind the wall.

This research has shown the importance of considering the interaction between the building and the retaining wall.

13.5 INFLUENCE OF DIAPHRAGM WALLS ON A PILE FOUNDATION

With increased construction activities in congested cities worldwide, there is an increasing need to create underground spaces close to existing buildings and their foundations. Diaphragm walls are often used to create underground space for new foundations. Placing concrete diaphragm walls in situ can, however, affect the existing foundations such as the pile foundation shown in Figure 13.9. The horizontal stresses in the soil around an existing pile foundation can drop during the construction of the diaphragm

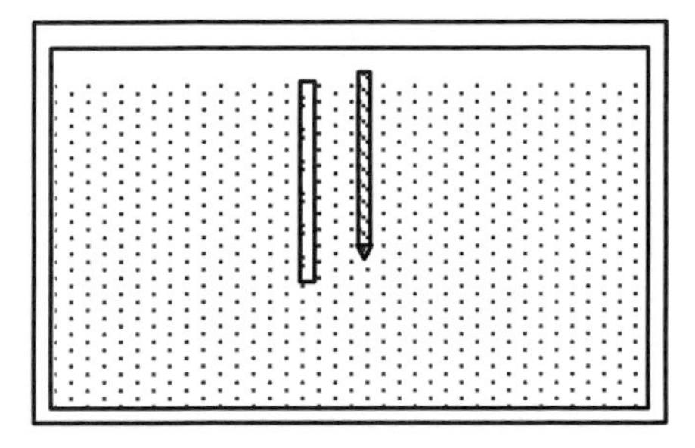

Figure 13.9 Schematic cross-section of the diaphragm wall next to a single pile.

wall. This can lead to a partial loss of pile capacity leading to additional settlements of the pile foundations.

Choy (2004) investigated the installation effects of diaphragm walls in fine sands on adjacent pile foundations. He carried out a series of centrifuge tests at 75 *g*. His model diaphragm walls were created by having the heavy fluid sodium polytungstate, described earlier in Section 13.4, filled into a latex bag. The trench construction is simulated by pumping out a controlled volume of the fluid by opening a valve. The model pile had a prototype diameter of 0.9 m and was placed adjacent to the model diaphragm wall. It was instrumented to measure the axial force and bending moments. It extended to a depth of 18.75 m at prototype scale. Further, it was driven in-flight at 75 *g* using the one-dimensional actuator so that representative horizontal stresses are generated along the pile shaft. The diaphragm wall had a panel width of 6 m and extended to a depth of 30 m below the ground surface. Concreting of the model diaphragm wall was achieved in-flight by replacing the heavy fluid in the latex bag with a mixture of sand, fine steel powder, and a rapid hardening adhesive.

Construction of the diaphragm wall in-flight by lowering the slurry level and replacing it with model concrete caused bending moments in the pile. Typical variation of the bending moment in the pile when the slurry level is lowered to different depths is presented in Figure 13.10 at prototype scale.

Clearly the bending moment in the pile increases with the construction of the diaphragm wall and increases with a drop in slurry level. It was possible to observe the soil settlements on both sides of the diaphragm wall as it was tested in the middle of an 850-mm tub. On one side it was the free field and on the other side the pile was present. This allowed for comparison of failure surfaces generated in the soil on either side of the diaphragm wall. An example of the failure surfaces is presented in Figure 13.11.

In this figure it can be seen that the failure surface extends to a larger distance on the free-field side compared to the side where the pile is present. This suggests that the larger lateral stresses induced by the pile influence

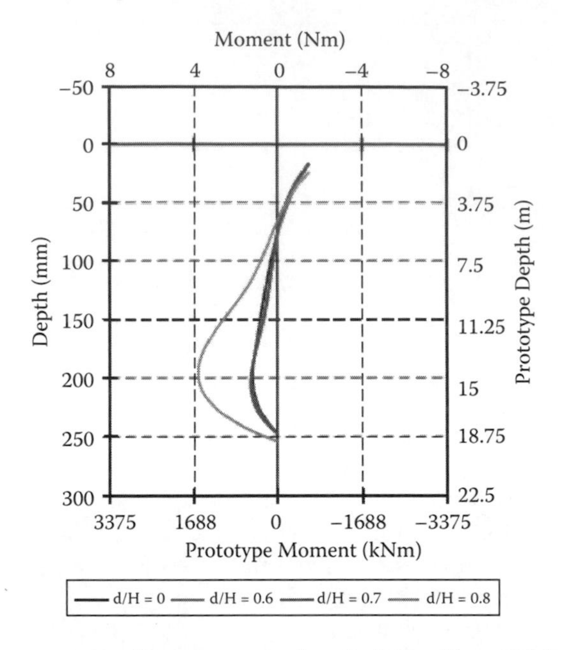

Figure 13.10 Changes in bending moment in the pile (after Choy, 2004).

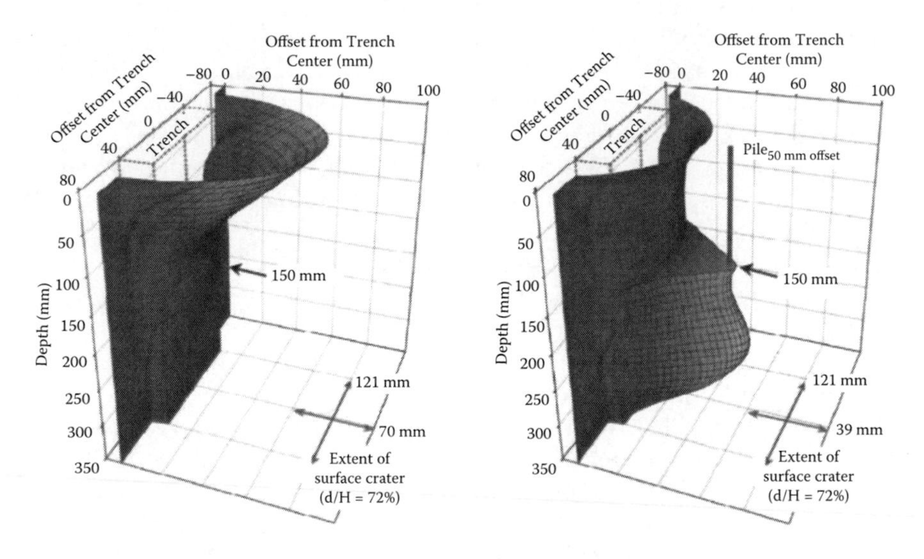

Figure 13.11 Failure surfaces around the trench with and without the pile foundation (after Choy, 2004).

the soil behaviour when the diaphragm wall is being constructed. Also in Figure 13.11 it can be seen that the influence of the diaphragm wall is greatest at the surface. More details of this can be found in Choy (2004).

13.6 MODELLING OF PROPPED RETAINING WALL

One of the recent advances in centrifuge modelling was the use of the two-dimenisonal actuator to excavate in front of a retaining wall. The two-dimensional actuator, described in Section 7.3.2, can be adapted to carry out excavations by attaching a scraping tool to its end. It can be programmed to excavate a given thickness of soil layer in each pass, thereby excavating the soil in a way to simulate the field excavations.

One of the first research projects to utilize the two-dimensional actuator to excavate soil was Lam (2010). He created excavations in front of a flexible retaining wall and deployed props at three different levels to support the wall. A typical schematic of his centrifuge model is presented in Figure 13.12. These were a complex series of experiments. The retaining wall was installed in the preconsolidated clay. Following this the centrifuge model was subjected to 60 *g* and the clay was allowed to reconsolidate.

The two-dimensional actuator was then activated to excavate the soil to a predetermined level. After this the first level of props close to the top of the wall was deployed. The excavation was then continued and the second and third level of props were deployed in stages. Figure 13.12 shows the top two props deployed and the position of the third prop (in dashed lines) that will be deployed after future excavation. Using this technique, Lam (2010) investigated the building up of bending moments in the wall and subsequent changes in the bending moment distribution following deployment of props. He also monitored the bending moments in the long term after the deployment of the final prop.

Lam (2010) discusses the changes in bending moments in the wall during excavation of the soil to a depth of 5.8 m in stages. He investigated

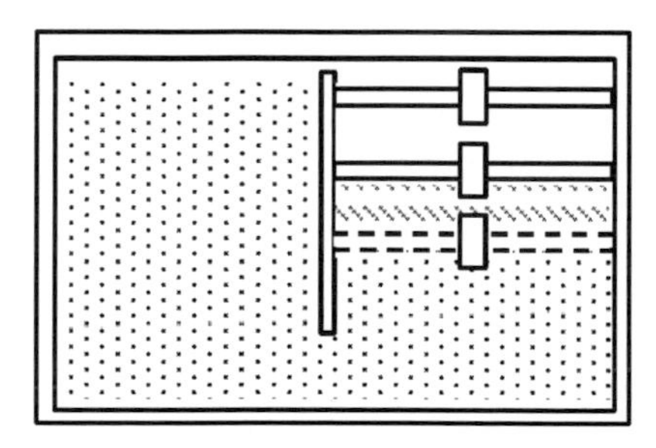

Figure 13.12 Schematic of a centrifuge model of a propped retaining wall.

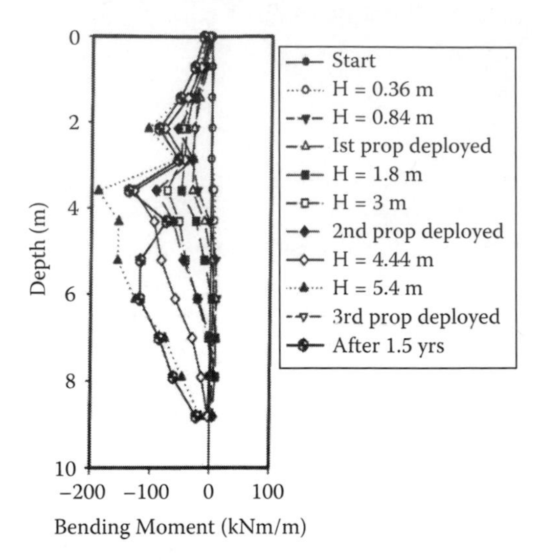

Figure 13.13 Variation of bending moment in the propped retaining wall (after Lam, 2010).

change in bending moment following the deployment of the props at three different levels. A typical example of the bending moment variation in the wall at different stages of excavation is presented in Figure 13.13 at prototype scale. In this figure it can be seen that the wall bending moments increase with increasing depth of excavation in front of the wall. Compared to the bending moments obtained for cases with no props (for example, see Figure 13.8), the deployment of props creates abrupt kinks in the bending moment distribution. Some reduction in the bending moments does occur following completion of construction once sufficient time has passed, as seen in Figure 13.13. Further, the deformations in the soil could be obtained using the PIV technique. An example of the accumulated soil deformations is presented in Figure 13.14, which clearly shows the active and passive wedge formations on either side of the wall.

Clearly the axial stiffness of the props is an important parameter that controls the magnitude of bending moments in the wall and the wall deformations. Lam (2010) investigated props with different stiffnesses. In Figure 13.15 the wall deflections for two prop stiffnesses are presented on the left. In this figure it can be seen that the wall deflections are much larger for the soft prop system, as one would expect. In the same figure the soil settlements behind the wall are also presented. Again it can be seen that the soil settlements are much larger for the soft prop system. It can also be seen that the settlement profile extends to some 14 m behind the wall. However,

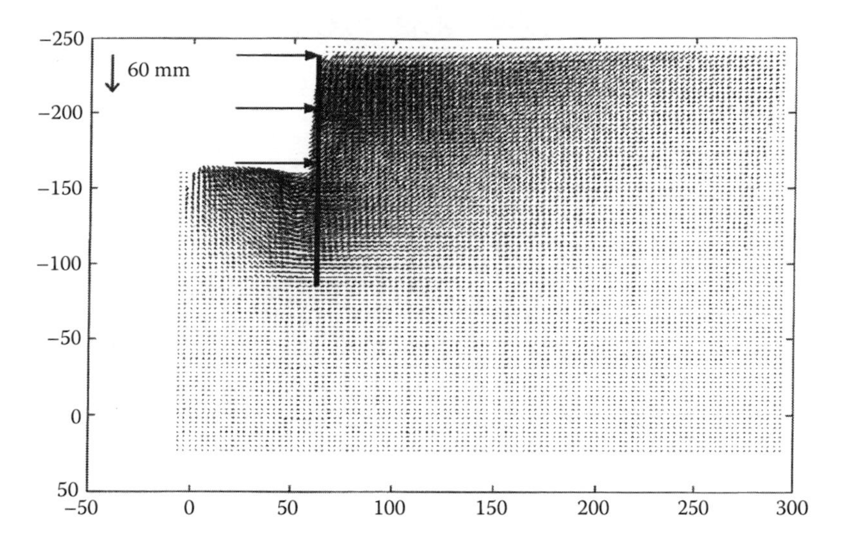

Figure 13.14 Soil deformations following excavation (after Lam, 2010).

the effect of the soft prop system on soil settlements diminishes at large distances away from the wall.

These series of centrifuge tests are complex as excavation is modeled in the centrifuge exactly as it would be carried out in the field. Such centrifuge tests are expected to give the closest representation of the problem of retaining wall construction in the field.

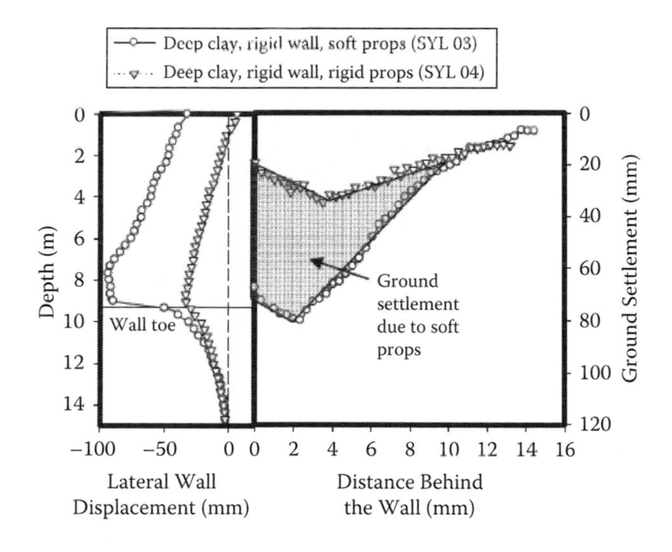

Figure 13.15 Effect of prop stiffness on wall and soil deformations (after Lam, 2010).

13.7 TUNNELING BELOW AN EXISTING PILE FOUNDATION

Construction of new tunnels is increasing, with several cities around the world looking to either extend existing underground tunnel networks or build new ones. Such tunnels often pass below the foundations of existing buildings. It is important that the construction of new tunnels does not cause excessive settlement or failure of existing buildings on the surface. It is therefore important for geotechnical engineers to understand the safe distance for the tunnels to pass below existing foundations without any adverse effects.

One of the early investigations that looked at the interaction between a new tunnel and an existing pile foundation in sand was reported by Jacobsz (2002). The schematic of the centrifuge model tests carried out by him is presented in Figure 13.16. These centrifuge tests were conducted at 75 *g*.

The piles were driven at 75 *g* using a one-dimensional actuator by about one pile diameter so that they mobilise the full shaft friction and end bearing. The model piles had a diameter of 12 mm and represented a prototype pile of 0.9 m diameter. They were subjected to axial loads using brass plates as dead weight on the top of the piles. The model pile was instrumented with several loads cells to measure axial forces in the pile. The outside diameter of the tunnel was 60 mm and represented a prototype tunnel of 4.5 m diameter. The model tunnel was made by having a latex tube around a brass mandrel and this tube was completely filled with water to begin with until the required *g* level was reached. The construction of the new tunnel was simulated in these centrifuge tests by pumping water out from the latex tube in a controlled manner through a valve system. Volume loss of up to 20 percent could be imposed on the tunnel using this system.

Jacobsz (2002) initially conducted a series of centrifuge tests without any pile foundations. He obtained the greenfield settlement troughs in sands by imposing known sets of volume losses. Figure 13.17 shows a typical example of the development of a settlement trough with increasing percentage of volume loss in the tunnel. These settlement troughs observed at the

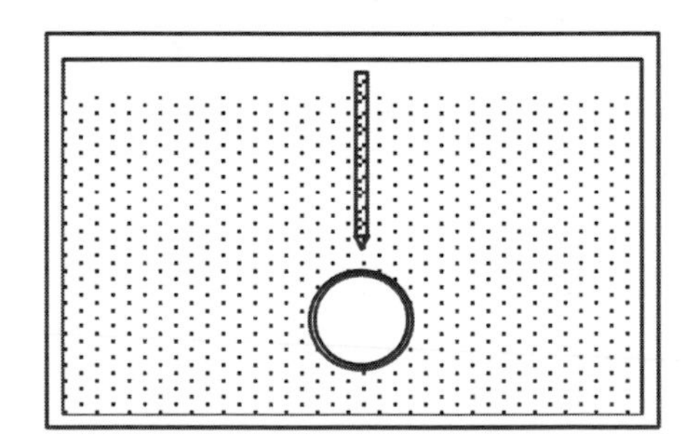

Figure 13.16 Schematic cross-section of a pile above a tunnel.

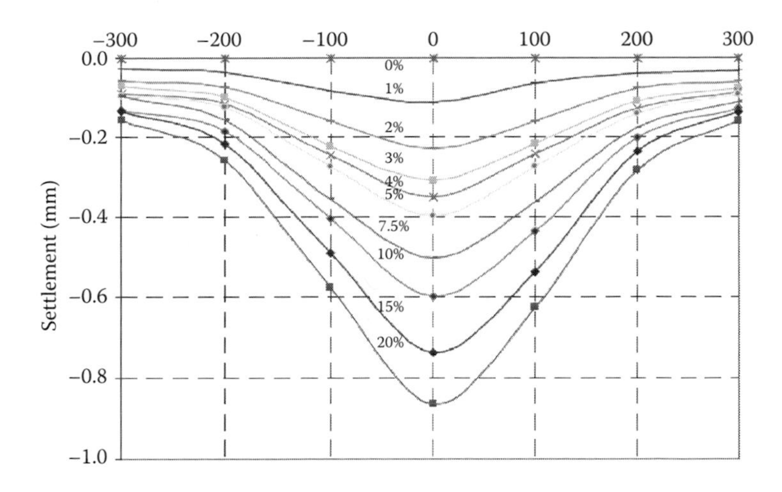

Figure 13.17 Settlement trough due to tunnel construction (after Jacobsz, 2002).

ground surface resemble the greenfield troughs expected when a tunnel is newly constructed and follow Gaussian curves about the center line of the tunnel (e.g., Fargnoli, Boldini, and Amorosi, 2013).

Once the greenfield settlements were established, Jacobsz (2002) could investigate the effect of tunneling passing below existing pile foundations. He investigated the effect of tunneling on piles located directly above the tunnel and also those that are offset by some distance. Here we will only consider the former. For the case of a pile with its tip about 4 pile diameters above the crown of the tunnel, the volume loss induces significant settlements. This is presented in Figure 13.18, which shows the pile settlement

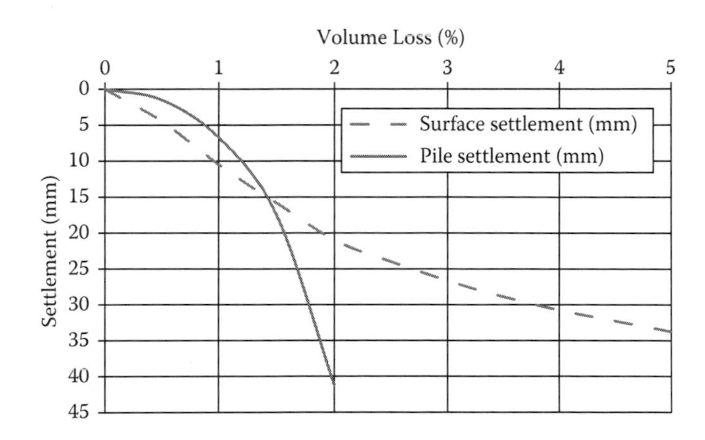

Figure 13.18 Settlement of piles following tunnel volume loss (data from Jacobsz, 2002).

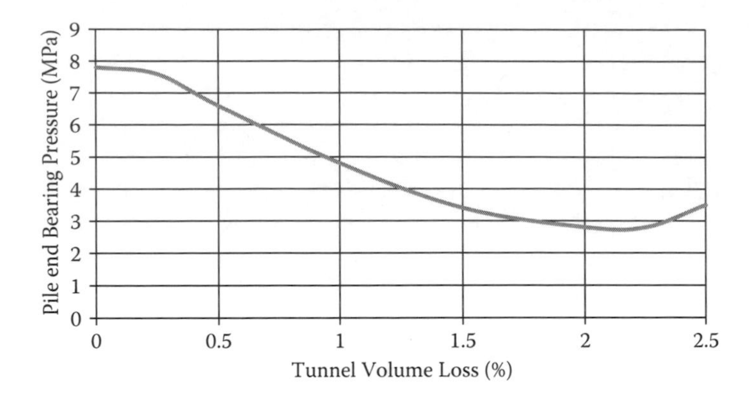

Figure 13.19 Loss of end bearing of piles following tunnel volume loss (data from Jacobsz, 2002).

induced by the volume loss due to tunneling and the settlement of the ground surface. In this figure we can see that the pile settlement is more dramatic compared to the ground settlement and reaches some 40 mm at prototype scale for a tunnel volume loss of 2 percent. At this volume loss the ground settlement is only 20 mm. Further, it is interesting to know the loss of load-carrying capacity of the pile due to the construction of the tunnel. The loss of end-bearing pressure of the pile with volume loss is presented in Figure 13.19. In this figure it can be seen that the initial end-bearing pressure of 7.9 MPa reduces to 2.9 MPa for a 2 percent volume loss in the tunnel. This is a significant drop in the end-bearing pressure of the pile. The centrifuge data does suggest a small recovery in the end-bearing pressure beyond 2 percent volume loss but this could be due to the pile coming too close to the crown of the tunnel at these levels of volume loss.

Although the construction of the tunnel in this research was simulated by volume loss, it does demonstrate the ability of centrifuge modelling to capture the pile settlements and loss of the pile's load-carrying capacity.

13.8 TUNNELING BELOW EXISTING PIPELINES

Construction of new tunnels affects not only existing foundations as discussed in Section 13.7 but also existing water and sewer pipelines that are normally buried at shallow depths. Again the interaction between the new tunnel and existing pipeline can be a complex one. It clearly depends on the flexibility and type of construction of the pipeline. Vorster (2005) has investigated the influence of tunneling on existing pipelines using centrifuge

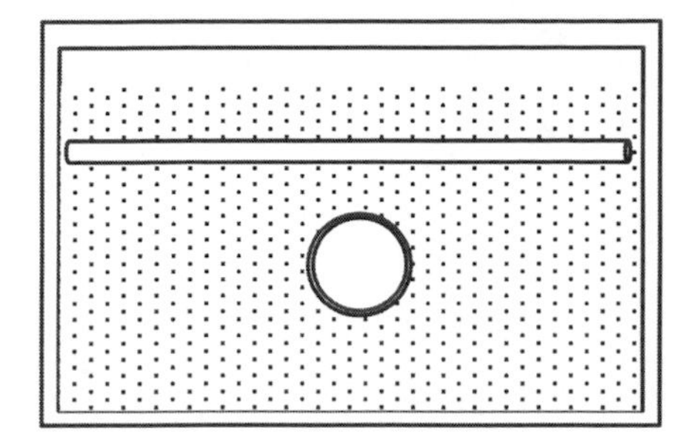

Figure 13.20 Schematic of the centrifuge model cross-section for tunneling below pipelines.

testing. The schematic representation of his centrifuge models is presented in Figure 13.20. The tunnel passes at right angles to the existing pipeline as shown in this figure. These centrifuge tests were also carried out at 75 *g*. A wide variety of pipelines with different flexibility was tested. The volume loss in the tunnel was achieved using similar method to that described in Section 13.7 and used earlier by Jacobsz (2002). Once volume loss is initiated we would expect the pipeline to suffer flexural bending. Depending on its flexural stiffness it may impede the greenfield settlement of the soil. Very flexible pipes of course will simply follow the soil settlements.

The diameter of the pipeline and the tunnel considered here are 1.2 m and 4.5 m, respectively. Figure 13.21 presents the relative settlement of the pipeline with respect to the tunnel for two cases. The case of a pipeline buried at a shallow depth that passes over a tunnel at great depth

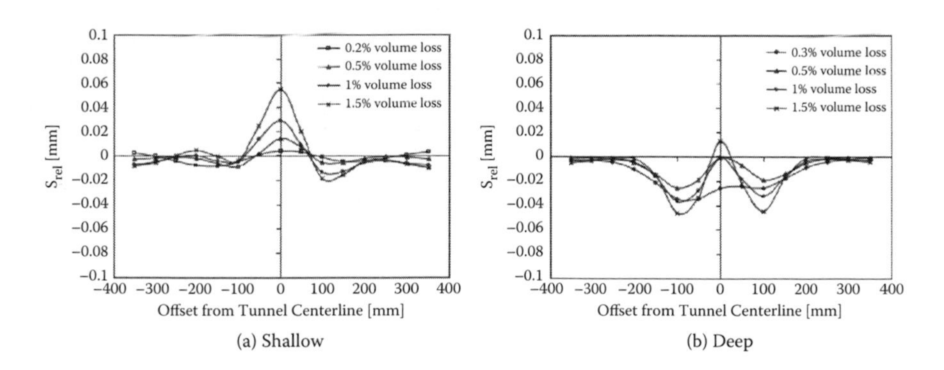

Figure 13.21 Relative settlement of the pipeline for shallow and deep cases (after Vorster, 2005).

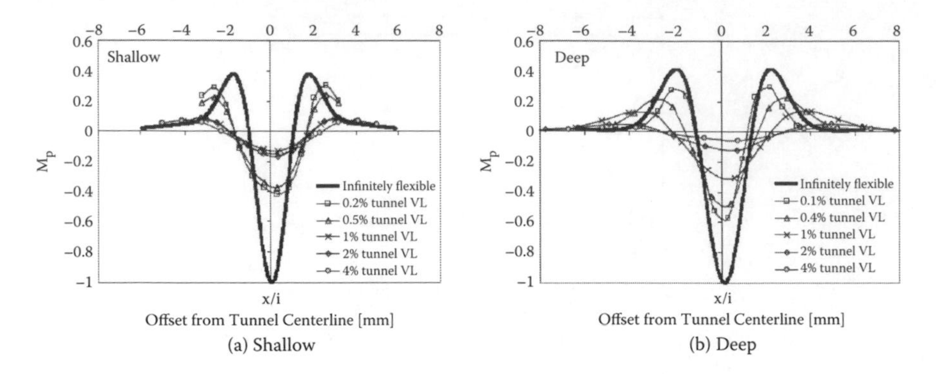

Figure 13.22 Normalised bending moment in pipelines for shallow and deep cases (after Vorster, 2005).

(9 m above the crown of the tunnel) is shown on the left. On the right is the case of a pipeline that is buried at great depth but passes closer to the tunnel (about 4.2 m above the crown).

In this figure it can be seen that the pipeline buried at shallow depth is able to resist the soil settlements created by the volume loss in the tunnel. This can be viewed as a positive down drag, that is, the soil above the pipeline is being supported by it. This is true for all volume losses considered, that is, up to 1.5 percent. On the other hand, when the pipeline is deeper and the volume loss is small it is unable to resist the down drag. This trend changes when the volume loss increases, as seen in Figure 13.21.

The model pipeline was also instrumented to measure bending moments generated due to the construction of the tunnel. In Figure 13.22 the normalised bending moments are presented for the shallow and deep pipeline cases. The measured bending moment is normalised by the maximum bending moment that would be generated in an infinitely flexible pipeline. In this figure it can be seen that for shallow pipelines the bending moment increases with increasing volume loss, as one would expect. For the deep pipeline case, the bending moments are larger compared to the shallow pipeline for any given volume loss in the tunnel.

This research has shown that significant interaction can occur between an existing pipeline and a new tunnel that is being constructed. Centrifuge modelling can provide insights into such an interaction and generate data for specific cases which could be used to validate numerical or analytical solutions for such problems.

Marshall (2009) investigated similar problems of pipelines and piles that lie above new tunnels that are being constructed. For the problem of

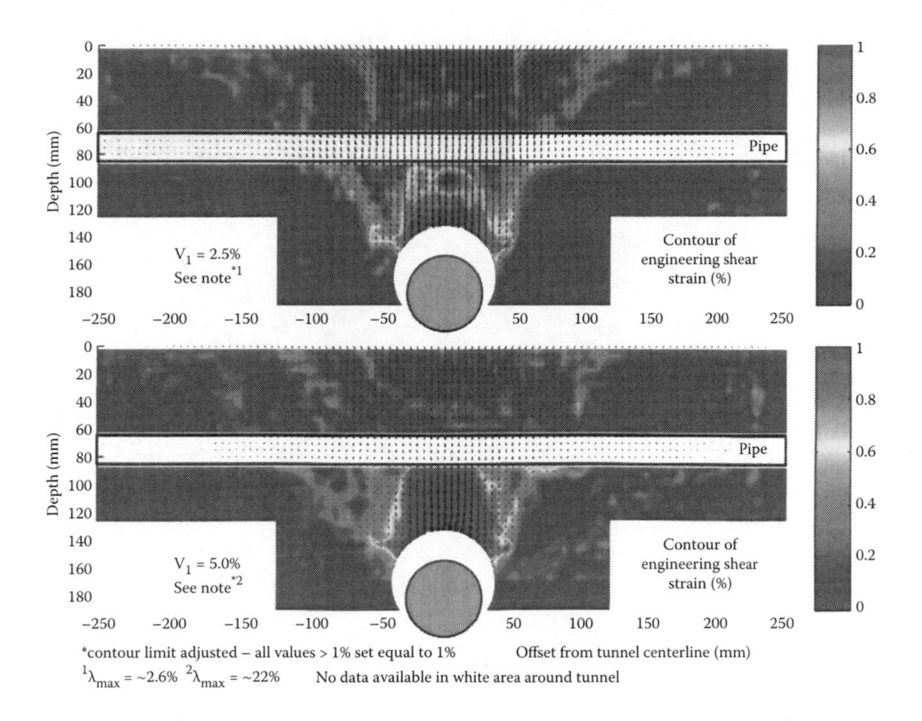

Figure 13.23 Shear strains in soil around a tunnel suffering volume loss (after Marshall, 2009).

pipeline crossing above a tunnel, he could obtain visual contours of shear strains in the soil as the tunnel suffers volume loss using the PIV analyses. Figure 13.23 shows examples of the contours of shear strains in the soil for two different volume losses. In this figure it can be seen that with increasing volume loss in the tunnel, the shear strains accumulate. Also the pipeline seems to resist the generation of shear strains in the soil above its elevation. This further strengthens the thought that the pipelines can support the soil to some extent but this is a function of the pipeline stiffness and the depth at which the pipeline is present.

13.9 TUNNELING BELOW A MASONRY STRUCTURE

Another problem with construction of new tunnels is the effect they have on masonry structures that lie above them at ground level. This was a major issue in London during the Jubilee line extension where the tunneling came close to Big Ben and the Palace of Westminster. In that case compensation grouting was undertaken to prevent any damage to the historic structures. The greenfield surface settlements due to tunneling are known to extend on either side of the tunnel line. If masonry buildings of historic or public importance are present in the zone of influence then they need to be protected against damage. However, the presence of the buildings may alter the extent to which the surface settlements may progress when a new tunnel is constructed. There is a need to understand the interaction between the masonry structures and the tunneling process. This can help in designing appropriate methods to protect the buildings from damage.

Farrell (2010) investigated the response of buildings to tunneling in sands. Again he used the same method as Jacobsz (2002) to create volume loss in a tunnel. The centrifuge tests were all carried out at 75 *g*. He investigated the response of buildings with different flexural stiffness to tunneling. An interesting aspect of his research was the creation of model buildings to represent masonry structures at prototype scale. He used micro-concrete and model bricks to create structures with varying flexural stiffness. The micro-concrete was made by using Portland cement and varying the water/cement ratio and sand aggregate to achieve the required stiffness. Similarly the masonry structures were created by using clay-fired model bricks that are scale models that are 1/12th or 1/50th of the dimensions of prototype bricks. The mortar was created by using silica gel.

An example of the settlement data obtained by Farrell (2010) is presented in Figure 13.24. In this figure the greenfield settlements are overlain onto the settlement profiles below different types of structures. The micro-concrete structure suffers slightly smaller settlements compared to greenfield settlements as seen in this figure. Also the settlements increase with increasing volume loss in the tunnel. Another point is that the settlement profile is quite smooth for this structure, which means that the structure is able to sustain the bending strains induced. For the 1/12th scale model brick masonry structure the settlements are comparable to the micro-concrete structure. Again the settlement profile is quite smooth. In the case of the 1/50th scale model brick masonry structure the settlement profile becomes discontinuous when the tunnel volume loss goes above 2 percent. This suggests that the bending strains in this building caused cracking. The generation of cracks causes the settlement profile to become discontinuous. However, the maximum settlements suffered by this building are again comparable to greenfield settlements. This may be due to the fact that on either side of the cracks, the building will follow the ground settlements.

This example demonstrates the use of centrifuge modelling to simulate the behaviour of masonry structures on the ground. Further research in this area is currently underway at Cambridge.

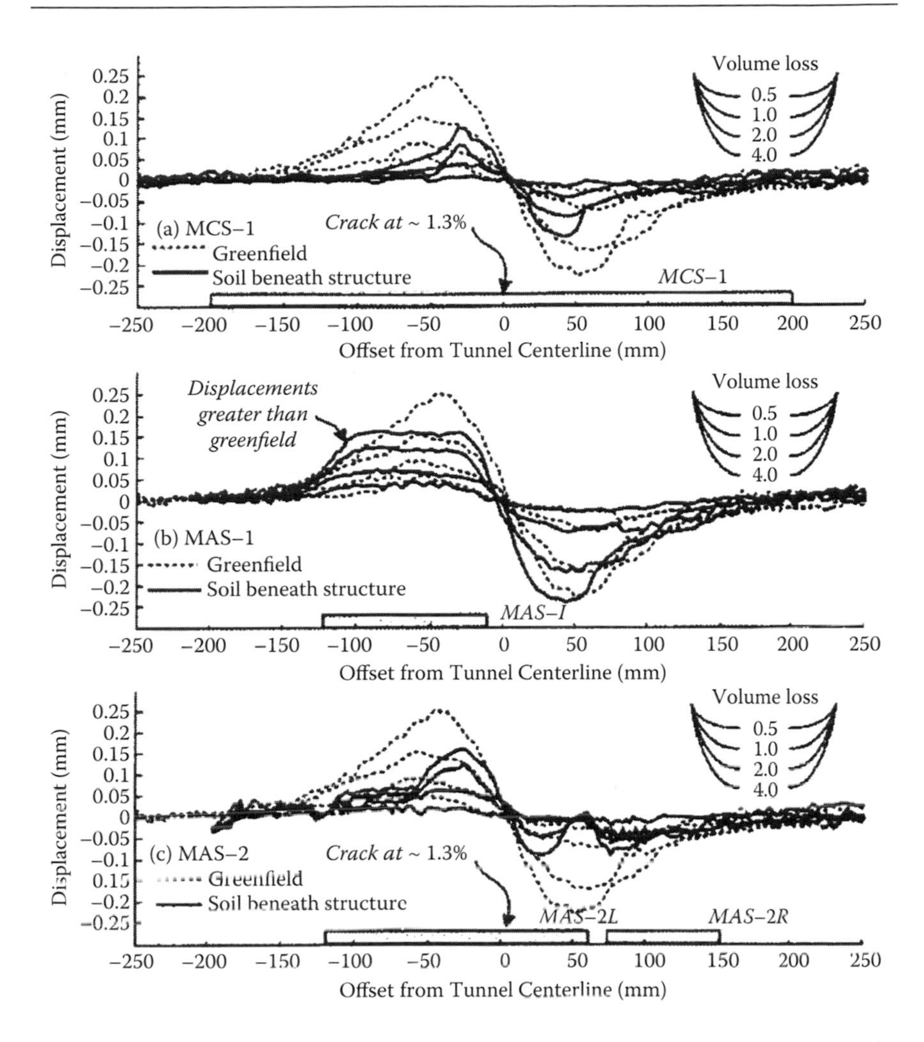

Figure 13.24 Settlement of micro-concrete and masonry structures (after Farrell, 2010).

13.10 SUMMARY

In this chapter we considered advanced problems that can be modeled using centrifuge modelling. The interaction between soil and structure, such as a retaining wall and the soil, or two geotechnical structures in soil such as a pile foundation above a tunnel, is a complex problem. Centrifuge modelling offers an attractive way to understand the mechanisms at play in such problems. Of course the understanding and insights provided by centrifuge modelling can help us develop simpler numerical tools that will capture the essential mechanisms at play in these problems.

The problems we considered fall broadly into two categories. In the first category we considered the retaining walls. We endeavored to model these geotechnical structures to capture the essential features used during their construction in the field. Initially, we looked at modelling the excavation sequence using heavy fluids. These fluids are drained at a controlled rate at high gravity to simulate the excavation rate in the field. A variation on this theme was the presence of other structures. such as buildings behind the wall or piles close to diaphragm walls. We investigated how the presence of these structures close to retaining walls affects their behaviour, a clear case of modelling interaction between them. In the second category we looked at modelling of new tunnel construction in close proximity to other geotechnical structures, such as piles, pipelines, or masonry structures. In each of these cases we have seen that centrifuge modelling is able to offer a way to investigate the interaction between the new tunnel and other structures.

With further advances in control gear and miniature actuators we should be able to replicate the construction processes in an increasingly realistic fashion in the future.

Chapter 14

Dynamic centrifuge modelling

14.1 MODELLING OF DYNAMIC EVENTS

Dynamics in civil engineering plays an important role in a wide variety of problems. The aerodynamics of a long span suspension bridge, or wind-, blast-, or earthquake-induced vibrations of a skyscraper are all examples of problems where the dynamic behaviour of structures is important. Small-scale physical models of suspension bridges or tall towers subjected to wind loading can be tested in a wind tunnel. However, if we are interested in the interaction between structures and the foundation soil during a dynamic event, we cannot test at a small scale. This point was explained in detail in Chapters 2 and 4.

The use of centrifuge modelling is particularly important for earthquake loading on civil engineering structures as it offers a unique opportunity to study the performance of structures before the earthquake actually happens. This offers various possibilities such as testing the earthquake resistance of existing structures or establishing the most effective ways of remediating earthquake risk. The only other way to obtain information on dynamic performance of structures subjected to earthquake loading is to instrument full-scale field structures and wait for an earthquake event to occur. Though this is being done to obtain information on structural performance in California and in Taiwan, this is often very expensive and time consuming. Also, using field instrumentation, it may not be possible to evaluate competing technologies to mitigate damage during earthquake loading. Dynamic centrifuge modelling, on the other hand, offers an effective methodology for such comparisons. Some examples of this are considered here. In this chapter although the emphasis is on earthquake engineering, it must be pointed out that other dynamic problems such as blast loading on underground structures, soil-structure interaction under wind loading, or modelling of ground-borne vibrations can all be studied using centrifuge modelling. The same principles we will develop for modelling of earthquake-related problems can be applied to these dynamic problems.

Before describing examples of dynamic problems, we shall look at some basic scaling laws that are applicable while attempting dynamic centrifuge modelling. We will then look at some of the specialist equipment that is required to carry out modelling of such problems. These will include earthquake actuators and model containers. We will also learn additional model-making techniques and saturation procedures normally used to model earthquake-induced liquefaction problems.

14.2 DYNAMIC SCALING LAWS

In Chapter 4, Table 4.1, we listed the scaling laws that relate the model parameters to the prototype for both "slow" and "dynamic" events. We, however, did not go into details of how we would classify a certain problem as a slow or dynamic event. We can classify an event as slow if the inertial effects in the direction normal to the gravity field can be neglected. For example, this may include a slow migration of groundwater in the soil model from one location to another. Similarly a lateral load applied to a pile or a retaining wall slowly so that the inertial forces are negligible can be classified as a slow event. In contrast, a dynamic event is one where the inertial effects are large and cannot be ignored. During an earthquake loading, the soil is subjected to vertically propagating primary (P) waves and horizontally polarized shear (S_h) waves. These waves are known to cause damage to structures on the ground surface when they reach them. The structures are subjected to large lateral forces under the action of these stress waves. The action of these forces and the inertia of the structures cannot be ignored, if we want to investigate the effect of such dynamic events. Similarly the effects of a blast or wind loading on structures can be rapid and again the inertial forces cannot be ignored in the modelling of such problems.

Just like we have derived the general scaling laws for slow events in Chapter 4, we need to derive scaling laws for dynamic events.

Let us consider a simple case of a layer of soil overlying the bedrock. Let us imagine that earthquake motion that can be applied as a sinusoidal displacement of amplitude x_o at the bedrock-soil layer interface due to some distant fault rupture. We can express this input motion as:

$$x = x_o \sin \omega t \tag{14.1}$$

where ω is the angular frequency. By differentiating this, we can get the velocity and acceleration applied to the soil layer at the bedrock level as shown in Equations 14.2 and 14.3.

$$\dot{x} = x_o \omega \cos \omega t \tag{14.2}$$

$$\ddot{x} = -x_o \omega^2 \sin \omega t \tag{14.3}$$

Let us now consider an equivalent centrifuge model with the model container's base as the bedrock and a layer of soil above it. Let us suppose that the geometric scale factor for this centrifuge model is N and it is subjected to a centrifugal acceleration of Ng. We wish to know what displacement we should apply to the base of the model container to simulate the earthquake motion considered above. We know from the general scaling laws in Chapter 4 that the displacement in the centrifuge model (which has the dimension of length), must be N times smaller than the prototype. We also know that the acceleration as a parameter must scale as N times larger in a centrifuge model compared to the prototype. This follows directly from increasing the g level on the centrifuge model to Ng, that is, we are scaling up the centrifugal acceleration by a factor of N. We can make both these requirements work if we increase the angular frequency of the sinusoidal input motion to the centrifuge model by N. Thus, the input motion to the centrifuge model will be:

$$x_m = \frac{x_o}{N} \sin N\omega t \tag{14.4}$$

Differentiating this equation, we get:

$$\dot{x}_m = \frac{x_o}{N} N\omega \cos N\omega t = x_o \, \omega \cos N\omega t \tag{14.5}$$

Differentiating again, we get:

$$\ddot{x}_m = -N \; x_o \omega^2 \sin N\omega t \tag{14.6}$$

In Equations 14.4 to 14.6, we see that the displacement amplitude of the model was scaled down by N and the acceleration amplitude was increased correctly by a factor of N. We have demonstrated the scaling of displacement, velocity, and acceleration of the centrifuge model using a very simple sinusoidal input motion. However, we can use an identical argument for more complex input motions that simulate realistic earthquake motions with variable amplitudes and richer frequency content. We have already seen in Chapter 9 that any time-varying function that is a stationary process can be expressed as a summation of sine and cosine series, using the Fourier components. If we replace the displacement function in Equation 14.1 as such a summation, we can apply the same argument as above to show that the displacements and accelerations in a centrifuge model must be scaled by a factor N. There is, however, an argument as to whether real earthquake motions are actually stationary. In dynamic centrifuge modelling we make the assumption that a sufficiently longtime window will be considered on either side of the input acceleration-time history (i.e., before and after the earthquake motion) that it can be assumed to repeat itself on either side of that time window.

14.2.1 Scaling law for frequency

Equations 14.4 to 14.6 have some important implications for dynamic centrifuge modelling. The first one of these is that frequency of loading applied to a centrifuge model must be N times larger than the prototype loading. In other words the scaling law for frequency is

$$\frac{f_{model}}{f_{prototype}} = \frac{N\omega}{\omega} = N \tag{14.7}$$

We know from historic records that the earthquake motions will have a typical frequency range of 1 to 5 Hz. If we were to model a given prototype at 50 g then according to Equation 14.7 the model input motions must have a frequency range of 50 to 250 Hz. This means that the earthquake actuators must be powerful and be able to apply the loading at these high frequencies. We will consider this aspect in more detail in Section 14.4 when dealing with earthquake actuators used in dynamic centrifuge modelling. Similar to modelling of earthquakes, frequency of loading must be increased for other types of dynamic loading such as wind or wave loading, if inertial effects are an important aspect of the problem being modeled.

14.2.2 Scaling law for velocity

Comparing Equations 14.2 and 14.5 we can see that velocity in the model is the same as that of the prototype, that is, the velocity scaling factor is 1. This has some interesting consequences in dynamic centrifuge modelling. Any wave propagation that occurs in the model will occur at the same wave speed as in the prototype, for example, the velocity of the shear waves will be the same. However, the centrifuge models are N times smaller than the prototype, which means that the time it takes for the waves to travel from the bedrock to the ground surface will be N times smaller in the centrifuge model.

14.2.3 Scaling law for time

Just as with the scaling for frequency, the time for dynamic events in centrifuge models also has to scale. We know that the time period of a dynamic event is determined as:

$$t = \frac{1}{f} \tag{14.8}$$

This implies that the scaling law for time in a dynamic event must be

$$\frac{t_{model}}{t_{prototype}} = \frac{1}{N} \tag{14.9}$$

Again this scaling law for time has important consequences for dynamic centrifuge modelling. Typical earthquake durations can last anywhere from 10 to 90 seconds. In a centrifuge test carried out at 50 g, the scaling law for time given in Equation 14.9 indicates that the earthquake duration will be 0.2 to 1.8 seconds. This means that to model dynamic events in a centrifuge the loading must be applied in a very short period of time. In addition there are serious implications for the data sampling rates by the instruments within the centrifuge model. If we imagine that we are logging 50 instruments from a centrifuge model, we need to collect data from all these instruments using a high sampling rate. This is necessary as we need sufficient data points for every cycle of loading to adequately reproduce the cyclic variations in the analog signals. Some of these aspects are covered in Chapter 9. Typically a sampling rate of 5 to 10 kHz per channel is required. This gives a throughput requirement of about 250 to 500 kHz to log 50 channels. Currently data acquisition boards with a throughput logging frequency of 1 MHz are available at a relatively low cost. Faster data acquisition boards with 100 MHz are also coming onto the market.

14.3 DISCREPANCIES BETWEEN GENERAL AND DYNAMIC SCALING LAWS

The above scaling laws for velocity and time in dynamic events are different to the general scaling laws considered in Chapter 4. Comparing the scaling laws for these two parameters for static and dynamic events, we can see that the velocity scales as a factor N between the model and the prototype in static events, while it scales as unity for dynamic events. Similarly we can see that time in static events scales as $\frac{1}{N^2}$ between the model and the prototype, while for dynamic events it scales as $\frac{1}{N}$. These discrepancies between the scaling laws for static and dynamic events need to be carefully addressed.

The implications of this mismatch depend on the problem being considered. For example, the seepage velocity of groundwater may scale as N between the model and prototype but the velocity of a wave will scale as unity. Similarly if we are trying to model earthquake-induced liquefaction in loose, saturated sands, the generation of the excess pore water pressure due to cyclic shear stresses from the earthquake will scale as $\frac{1}{N}$ but the dissipation of these excess pore pressures in the post-earthquake period will scale as $\frac{1}{N^2}$. Clearly we cannot use different scale factors for time while modelling the same event in a centrifuge.

This discrepancy in scaling laws for velocity and time can be rectified in one or two ways. These are considered next.

14.3.1 Use of viscous pore fluid

One way to avoid the discrepancy in scaling laws for velocity and time while modelling earthquake engineering problems in a centrifuge is to use a substitute pore fluid. The normal pore fluid in saturated soils will be water, which has a kinematic viscosity of 1 cSt (at 20°C). If we saturate our centrifuge models with a model pore fluid that has N times more viscosity then water, we can avoid the discrepancies in the scaling laws for static and dynamic events. A fluid that is N times more viscous than water will obviously flow more slowly through the soil pores. Darcy's law for flow of water through soils can be written as:

$$v_w = k \frac{\gamma_w}{\mu_w} i \tag{14.10}$$

where v_w is the flow velocity, k is the intrinsic permeability of the soil due to the pore structure, i is the hydraulic gradient driving the flow, and γ_w and μ_w are the unit weight and kinematic viscosity of water. Let us consider a centrifuge model that has the same pore structure. If we saturate this model with a pore fluid that is N times more viscous, then we can write the above equation as:

$$v_v = k \frac{\gamma_v}{\mu_v} i \tag{14.11}$$

where the subscript v has been used to indicate viscous pore fluid. Dividing Equation 14.10 by Equation 14.11 we can write:

$$\frac{v_w}{v_v} = \frac{\mu_v}{\mu_w} \frac{\gamma_w}{\gamma_v} = N$$

$$v_v = \frac{1}{N} v_w \tag{14.12}$$

From Equation 14.12 we can see that the flow velocity is reduced by a factor N when we use the viscous pore fluid, provided that the substitute pore fluid has the same unit weight as water. Thus, using a viscous pore fluid we can reduce the fluid flow velocity in a centrifuge model so that the scaling law for fluid flow becomes unity. This matches with the scaling for velocity in dynamic events. Similarly by using the viscous pore fluid we will increase the post-earthquake pore pressure dissipation time in the centrifuge model by a factor of N thereby changing the scaling law for time in dynamic events from $\frac{1}{N}$ to $\frac{1}{N^2}$. This ensures that in earthquake-induced liquefaction problems the rate of generation of excess pore pressure due to dynamic loading matches the rate of dissipation due to the consolidation processes.

Substituting the pore fluid in the centrifuge models with a viscous fluid is therefore clearly desirable in a certain class of problems. The required range of the viscosities for such a fluid is between 50 and 100 cSt to cover centrifuge tests from 50 *g* to 100 *g*. One of the early fluids that was used for this purpose was silicone oil. This oil is available in a range of viscosities. Further intermediate viscosities could be made up by mixing silicone oils of different viscosities, as required. At Cambridge silicone oil was used as the saturation fluid of choice until 1996. However, more recently hydroxypropyl methyl cellulose (HPMC) is used as it is relatively cheap and easily available. This is available as a powder and can be mixed with de-aired and de-ionized water to make a saturation fluid of required viscosity. The relationship between the amount of methyl cellulose powder that needs be mixed with water and viscosity of the resulting fluid is given by Stewart, Chen, and Kutter (1998).

$$\mu_{20} = 6.92 \quad C^{2.54} \tag{14.13}$$

where μ_{20} is the viscosity at 20° centigrade and C is the quantity of methyl cellulose powder expressed as a percent by weight. Small quantities of benzoic acid need to be added to the solution to protect it from bacterial growth.

Saturating soils with high-viscosity fluid needs some special considerations and these are discussed in Section 14.5.

14.3.2 Change of soil permeability

Another alternative to use of a substitute high-viscosity pore fluid is to change the soil permeability (k). Hazen's equation relates the soil permeability to the particle sizes as:

$$k = \alpha D_{10}^2 \tag{14.14}$$

where α is a constant and D_{10} is the particle size that corresponds to the size at which 10 percent of the particles are smaller. Therefore, it is possible to choose a soil with finer particles to reduce the permeability of the soil by a factor of N. Let us suppose that we wish to conduct a centrifuge test at 100 *g*. Let us suppose that the D_{10} size in our prototype soil is 0.2 mm. In order to reduce the permeability of soil by a factor of 100, we need to use a soil with a D_{10} size of 0.02 mm in our centrifuge model. The advantage of reducing the permeability of soil by modifying the size of the soil particles is that we can use water as the pore fluid. However, as we are changing the particle sizes we must ensure that the soil behaviour is not altered by this change; that is, the soil used in the centrifuge model has the same stress-strain behaviour as the soil in the prototype we are trying to model. We may have to ensure that the mineralogy of the soil is not altered, for example if the prototype has silica sand, we may have to model this using fine rock flour. Clearly we should not expand this argument and reduce the soil particle sizes to fine silts or clay as the water absorbency may become important and start to dominate the soil behaviour at those small particle sizes. Although this method of changing the particles sizes is a valid one, in practice very few centrifuge modellers attempt this, mostly due to the difficulty of establishing identical stress-strain behaviour between the soil with reduced particle size in the centrifuge model and the prototype soil.

There is also an argument to be made with regard to the use of a high-viscosity fluid to substitute for the pore water as described in Section 14.3.1. As we are looking at dynamic problems there is a concern about the amount of viscous damping that will arise by using these fluids in the centrifuge tests. Bolton and Wilson (1989) investigated the increased amount of viscous damping in the soil element using resonant column tests and quantified this increase. Damping in soil during dynamic events comes predominantly from three sources: (1) material damping in the soil, (2) geometric or radiation damping, and (3) viscous damping in the pore fluid.

The relative magnitudes of these components that make up the soil damping are quite important. Madabhushi (1994) carried out conjugate centrifuge tests on his tower structures on a liquefiable soil bed, only changing the pore fluid from 80 cSt silicone oil to water. In these tests no discernible difference in damping was observed between the two tests. It was postulated that this was due to the relative magnitude of material damping in the soil, which "drowns" the other two components of damping. Later Ellis et al. (2000) carried out resonant column tests and confirmed that the viscous damping does increase with use of silicone oil, but its relative magnitude was much smaller than the material damping in the soil at large strains. Thus, viscous damping only plays a dominant role for small strain vibrations, that is, small magnitude earthquake loading.

Many researchers have shown that the use of viscous pore fluid is both desirable and important while modelling liquefaction problems by performing conjugate centrifuge tests, that is, identical in every respect other than the pore fluid used for saturation. Peiris, Madabhushi, and Schofield (1998) investigated settlement of rock-fill dams into liquefied soil deposits and showed that the amount of settlement of the dam is underestimated significantly if water is used as the pore fluid. More recently Chian and Madabhushi (2010) carried out conjugate centrifuge tests to investigate the floatation of tunnels in liquefiable soils and showed that the pore fluid viscosity plays an important role in determining the amount of uplift the buried tunnels suffer.

14.4 EARTHQUAKE SIMULATION IN CENTRIFUGES

Earthquake events in nature involve release of a significant amount of energy in a short period of time. To model such events in a centrifuge we require actuators that are able to deliver large quantities of energy quickly. The time scales for centrifuge model earthquakes is even shorter due to the time scaling law discussed in Section 14.2.3. This makes building an earthquake actuator for centrifuges quite challenging. Further, the mechanical design of the entire centrifuge must be taken into account while attempting modelling of earthquakes. The magnitude of time-varying force that needs to be applied to the centrifuge models is large and can typically be in the range of 10 to 100 kN. The centrifuge swing basket and the hinges supporting the swing basket must be able to resist such large forces. Further, these forces will ultimately be transferred to the main bearing of the centrifuge as it will be applied as an unbalanced force. The bearing assembly must be able to safely carry these forces.

Broadly the earthquake actuators can be classed as:

- Mechanical actuators
- Servo-hydraulic actuators
- Electro-magnetic (E-M) actuators
- Piezo-electric actuators

Examples of the first two are emphasized in this chapter. However, there are other examples of E-M and piezo-electric earthquake actuators that have been developed. For E-M-based earthquake actuators the main limitation is the electric current required to drive the actuators. A working example of such an actuator is present at Shimizu Corporation in Japan. A smaller version of an E-M actuator was also developed at Cambridge and was used mostly for undergraduate research (Madabhushi, Collison, and Wilmshurst, 1997). For piezo-electric actuators the main limitations are the dynamic force they can generate and the electric current required to drive these actuators.

14.4.1 Mechanical earthquake actuators

Early attempts at simulating earthquakes in a centrifuge aimed at developing mechanical actuators. Some of the earliest work in this area was carried out by Morris (1979) at Cambridge. He used a leaf spring assembly that he could preload and that was released during the centrifuge flight. The spring system caused the centrifuge model to vibrate. This system excited free vibrations in the centrifuge model at a given frequency. Kutter (1982) developed a very successful earthquake actuator called the "bumpy road," which utilised a sinusoidal track built along the perimeter of the centrifuge chamber to cause sinusoidal vibrations in the centrifuge model. This was a simple, mechanical earthquake actuator that worked by deploying a radial wheel that came into contact with the sinusoidal track on the wall. The in-out movement of the radial wheel was translated into the lateral shaking of the centrifuge package through a bell-crank mechanism. The magnitude of earthquakes could be adjusted in-flight by changing the levering distance in the bell-crank mechanism. However, the frequency and duration of the model earthquakes were related to the angular velocity (and hence the *g* level) of the centrifuge.

Madabhushi, Schofield, and Lesley (1998) describe the development of an earthquake actuator called the Stored Angular Momentum (SAM) actuator. This actuator has been very successful and has been in use for over a decade. A schematic diagram of the SAM actuator and a view of the same are presented in Figure 14.1. In this actuator the energy required for the earthquake is stored in a set of flywheels which are rotated by a simple three-phase motor. The flywheels turn a reciprocating rod through a cross-head. The main component of this actuator is a fast-acting hydraulic clutch that engages rapidly to commence earthquake motion of the centrifuge model. The SAM actuator is able to operate in 100-*g* centrifuge tests and fire earthquakes of desired sinusoidal frequency and duration. It can also fire swept-sine motion that rolls through a range of frequencies and can help pick up resonant frequencies in soil-structure systems. The user is able to choose the magnitude and duration of the earthquake and can change them in-flight between successive earthquakes.

An example of a typical earthquake input generated by the SAM actuator is shown in Figure 14.2 at prototype scale. In this figure it can be seen that the

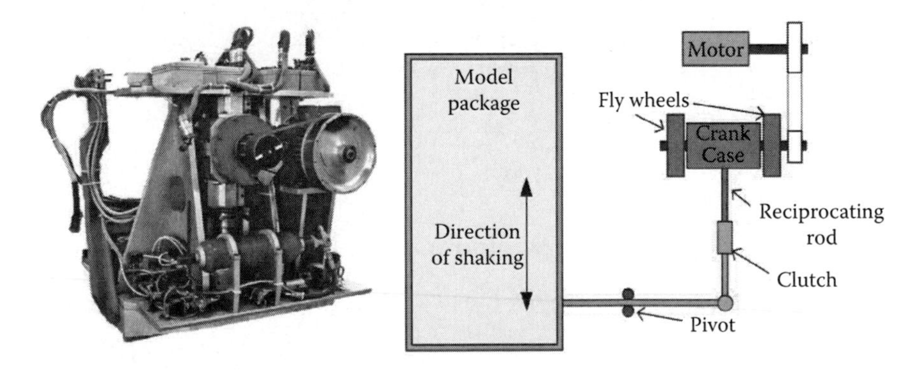

Figure 14.1 Schematic diagram and a view of the SAM earthquake actuator.

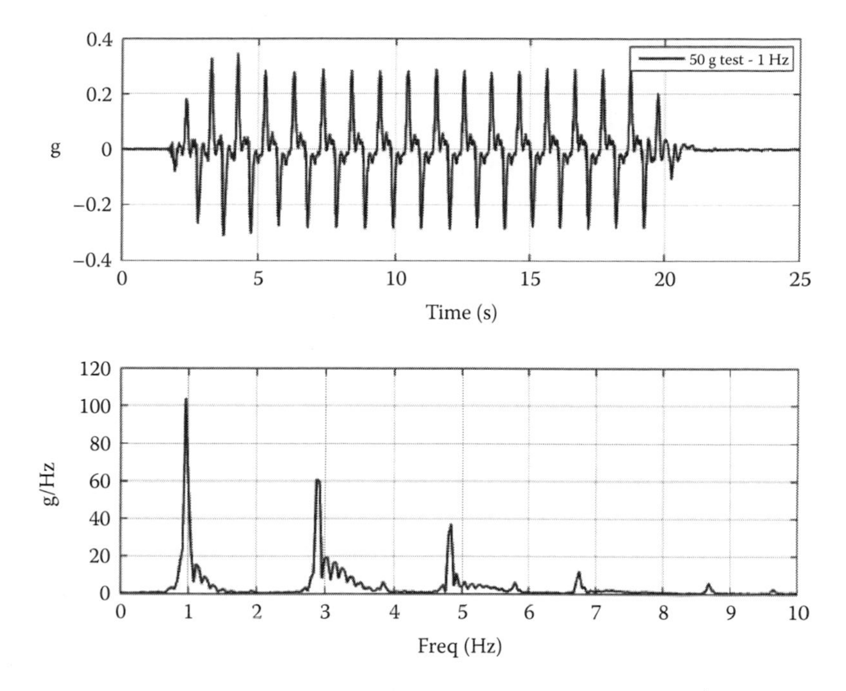

Figure 14.2 Example of a sinusoidal earthquake generated by the SAM actuator.

model earthquake generates ±0.35 *g* of acceleration that has about 16 cycles and lasts about 20 seconds. This particular earthquake was recorded during a 50-*g* centrifuge test. We can back-calculate the dynamic force generated by the SAM actuator in this earthquake as 42.9 kN. In Figure 14.2 the fast Fourier transform (FFT) of the input acceleration trace is presented. As explained in Chapter 9, the FFT expresses the input motion in a frequency domain. Thus, the horizontal axis is now given in terms of frequency and the vertical axis has units of g/Hz. In this figure we can see that the predominant frequency of the earthquake is 1 Hz. However, we can also see that the actuator generates higher harmonic motions close to 3 and 5 Hz.

As the SAM actuator works based on the stored energy in the flywheels, it is possible to roll down the frequencies by switching off the motor during an earthquake motion. This generates an input motion that drops in both amplitude and frequency. As it rolls down through a range of frequencies it is possible to pick up the amplification in the soil-structure systems and identify resonant frequencies. An example of the swept-sine wave input motion is presented in Figure 14.3 at prototype scale that was recorded during a 50-*g* centrifuge test. In this figure the SAM actuator was running at 60 Hz, that is, a prototype frequency of 1.2 Hz, when the power to the motor was switched off. It rolls down from 1.2 Hz to 0 Hz. The FFT for this

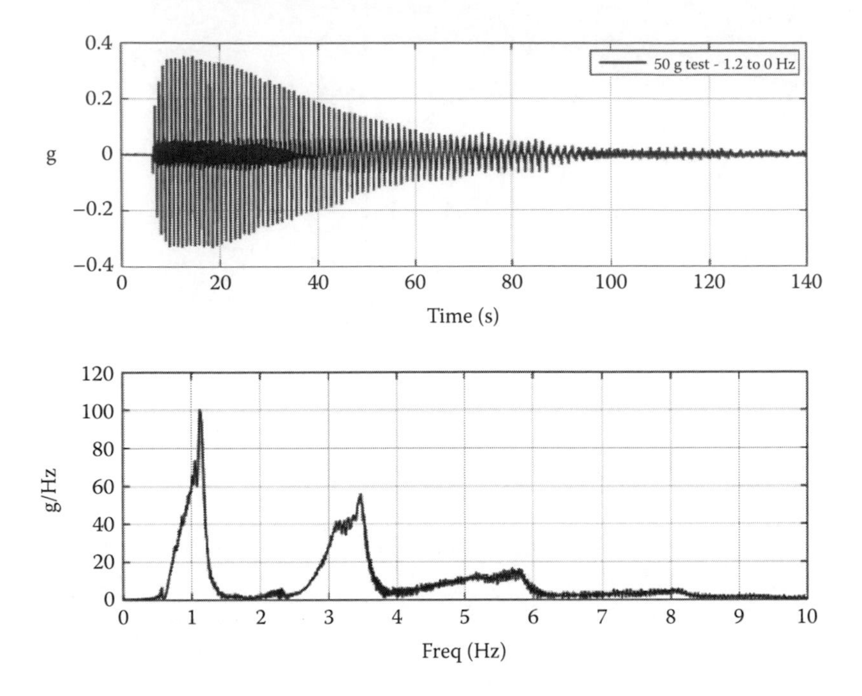

Figure 14.3 Example of a swept-sine wave earthquake generated by the SAM actuator.

input trace is also presented in Figure 14.3. In this figure we can see that the rolling down of the SAM actuator not only shows components between 1.2 Hz and 0 Hz but excites ranges of frequencies between 2.5 Hz and 3.75 Hz and between 4.25 Hz and 6 Hz. Thus, any soil-structure system resonances in the centrifuge models in those ranges of frequencies can also be detected.

Similar mechanical actuators have been developed elsewhere and were used extensively, for example at Tokyo Institute of Technology in Japan (Kimura, Takemura, and Saitoh, 1988). Mechanical actuators are relatively cheap to fabricate and easy to operate and maintain.

14.4.2 Servo-hydraulic earthquake actuators

Another way of generating earthquakes on a centrifuge is by using servo-hydraulics. Early attempts at developing a servo-hydraulic shaker at Caltech were described by Scott (1981, 1994). The concept of servo-hydraulic actuators comes from large shaking tables used in many structural dynamics labs. The earthquake actuators at the University of California, Davis and RPI in the United States, the earthquake modelling facility at C-Core in Newfoundland, Canada, the one at Hong Kong University of Science and Technology, and the IFSTTAR centrifuge center in France are all based on servo-hydraulics. These actuators rely on stored energy in hydraulic oil that is pressurized to 20–30 MPa (200–300 bar). They use this energy to shake the centrifuge model. Unlike the mechanical actuators the servo-hydraulic actuators can follow a given earthquake motion such as the El Centro or Kobe earthquake. This is achieved by controlling the flow of the high-pressure hydraulic fluid through a servo-valve which can open proportionally to the earthquake signal, thereby changing the force-time history being generated by the actuator. The servo-valve is the most important component of these actuators and it is often made specifically to have a good response at high frequencies even when subjected to additional loading due to high gravity in a centrifuge.

The main advantage of the servo-hydraulic shaker is that it offers researchers the opportunity to simulate real earthquake motions. They are able to vary the amplitude and frequency content of earthquake motion applied to the centrifuge models.

In Cambridge a new servo-hydraulic shaker has been developed that was commissioned in late 2011. Madabhushi et al. (2012) describe the construction and performance of this actuator. A view of this earthquake actuator is shown in Figure 14.4. It uses many of the features of the Turner beam

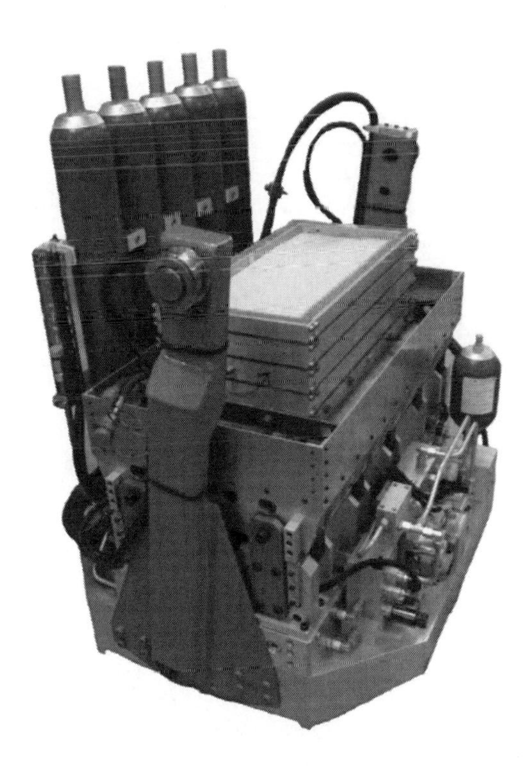

Figure 14.4 A view of the new servo-hydraulic earthquake actuator.

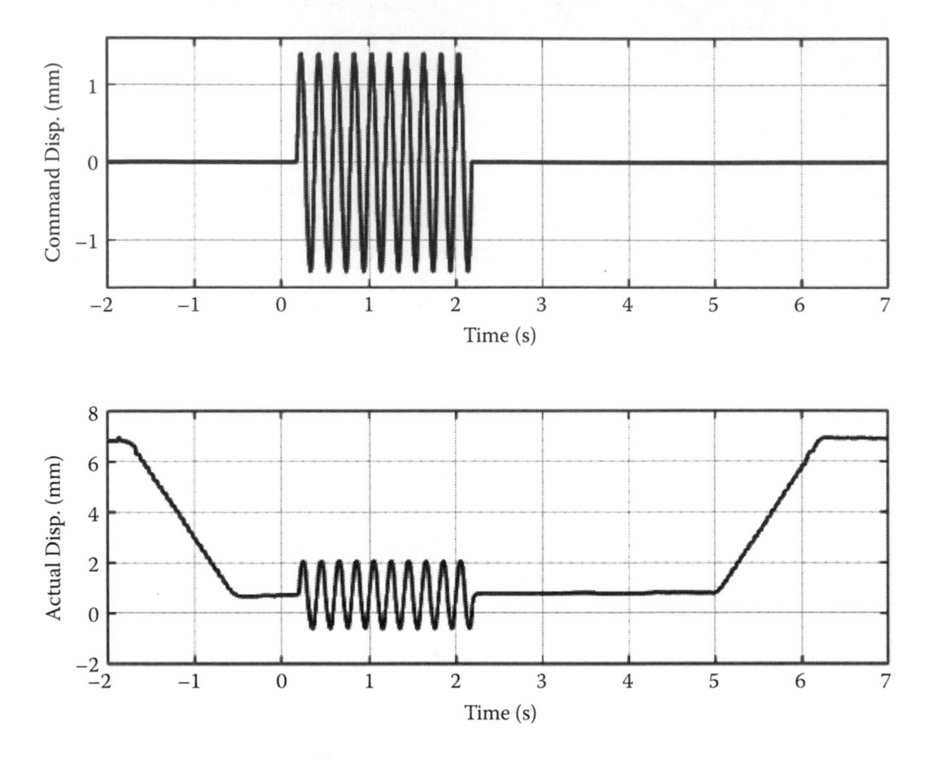

Figure 14.5 Command displacement-time history.

centrifuge described in Section 5.2.1. For example, the main reaction to the earthquake shaking force imparted to the centrifuge model will be provided by the main body of the beam centrifuge. The entire shaker assembly is mounted on a self-contained swing which can be loaded and unloaded like any other centrifuge package tested on the centrifuge. The soil model containers can be mounted on the shaking table as seen in Figure 14.4. The hydraulic power pack that supplies the high-pressure fluid is outside the centrifuge and is supplied to the earthquake actuator through high-pressure fluid slip rings.

The servo-hydraulic actuators work by taking in a displacement-time history and moving the shaking table proportional to this record. However, the default parking position of the actuator is to one end of the stroke. This means that they need to be centralized before shaking can commence. This process is normally automated as illustrated in Figure 14.5. Let us suppose we wish to apply a sinusoidal motion of 10 cycles with a displacement amplitude of up to ±1.5 mm. The actuator starts to move from its parked position of about +7 mm to a near-central position and then the sinusoidal shaking is applied as seen in Figure 14.5. After the end of shaking it

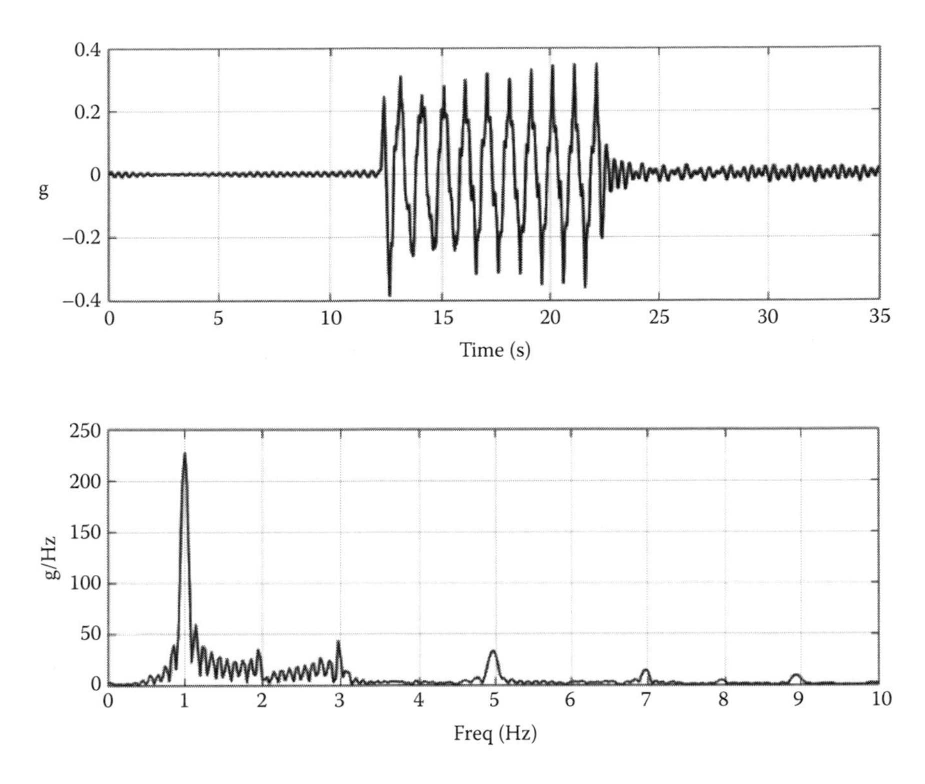

Figure 14.6 Sinusoidal motion generated by the servo-hydraulic shaker.

moves to the parked position again. This centralizing procedure is important so that the actuator does not hit the end stops when a strong shaking is applied. Further, it is important there is enough lead and lag time on either side of earthquake motion to allow time for these pre- and post-earthquake translations, without impacting onto the end stops.

A typical example of a sinusoidal earthquake motion generated by the servo-hydraulic actuator during a 50-*g* centrifuge test for the level sand bed model shown in Figure 14.4 is shown in Figure 14.6. This figure is presented at prototype scale. It must be noted that this is the actual acceleration measured at the base of the model in response to a displacement-time history applied as shown in Figure 14.5. The acceleration-time history is not as smooth and it picks up some of the system resonant frequencies. The FFT of the acceleration trace is also presented in Figure 14.6. Clearly there are some high frequency components present although their relative magnitudes are smaller than for the SAM actuator motion shown earlier in Figure 14.2.

One of the main advantages of using a servo-hydraulic earthquake actuator is that we can simulate more realistic motions as mentioned earlier. Figure 14.7 shows a prototype scale example of earthquake motion generated

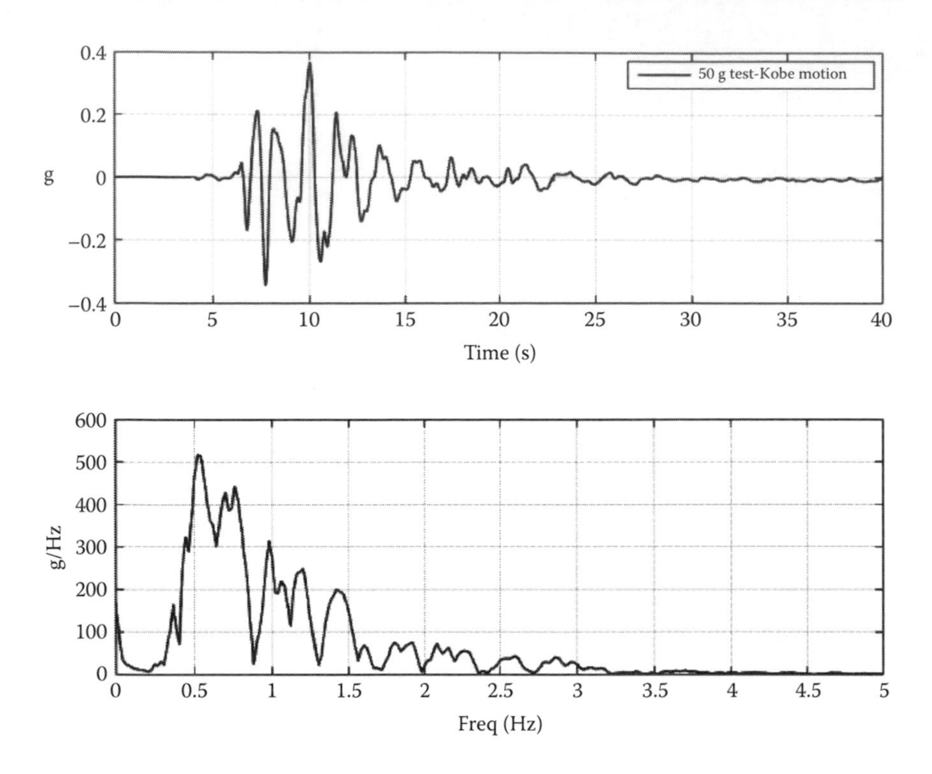

Figure 14.7 Kobe earthquake motion generated by the servo-hydraulic actuator.

by the servo-hydraulic actuator to simulate a motion recorded during the Kobe earthquake of 1995. Figure 14.7 also shows the FFT of the motion. This test was one of the early attempts at using this actuator and was conducted at 50 *g*. Clearly, the actuator is able to replicate this motion well with good high frequency response. Other historic earthquake motions such as the El Centro or Northridge motion can be simulated in a similar way.

One of the important considerations for the safe operation of a servo-hydraulic actuator on a centrifuge is the provision to carry the reaction forces. As explained earlier, the servo-hydraulic actuator in Cambridge transfers the reaction forces into the main beam centrifuge due to the torsion bar arrangement and the centrifuge package engaging with the end plates of the centrifuge as explained in Chapter 5. However, most modern centrifuges do not have such a provision and therefore the reaction forces can load the swing basket and hinges excessively. This can be avoided by having an arrangement of reaction masses on either side of the centrifuge model which shake in the opposite direction to the centrifuge model. Such a servo-hydraulic shaker with reaction masses on the sides of the shaking table is shown in Figure 14.8. These types of actuators are being used

Figure 14.8 The IFSTTAR earthquake actuator. (From J.L. Chazelas, S.P.G. Madabhushi, and R. Phillips (2007). In: Proc. IV International Conference on Soil Dynamics and Earthquake Engineering. Thessalonica, Greece. With permission.)

in several centrifuge centers, such as IFSTTAR in France, the University of Dundee, and the C-Core facility in Newfoundland, Canada. The main advantage of these actuators is that they do not transfer any additional loads onto the main beam centrifuge or its bearings. Therefore, they can be used in conjunction with centrifuges that have not been originally designed to conduct earthquake centrifuge tests.

14.4.3 Model containers

In the field the soil strata are considered semi-infinite, that is, they extend laterally to large distances. By their very nature centrifuge models have limited lateral extent. This is not a problem for static tests but when we are attempting dynamic centrifuge tests this can result in unwarranted P waves generated by the end walls as the whole soil model is shaken laterally. Further, every vertical plane in a soil layer will be subjected to complementary shear stresses when the bedrock imposes a horizontal shear stress during earthquake loading. This is true in a centrifuge model as well, except at the boundary of the model container. In order to simulate field conditions accurately, we must have model containers that allow horizontal shear deformations in the soil layer while being able to carry the complementary shear stresses in the vertical direction.

In the early days of centrifuge modelling, in about the 1980s, model containers with rigid ends were used. To minimize the impact of P waves from the end walls and to prevent any stress wave reflections by the boundaries back into the soil, absorbing materials like Duxseal were used. Duxseal is a clay-like putty material that was supposed to absorb incident stress waves. Steedman and Madabhushi (1991) have shown that the Duxseal is able to absorb about 65 percent of incident wave energy for both P and S_h waves.

This concept of having absorbing boundaries works reasonably well for some problems. However, where dynamic soil-structure interaction problems are being studied the boundaries may cloud the results. A new concept of designing the end walls as compound structures that undergo the same shear deformation as the soil layer was developed by Zeng and Schofield (1996). Such model containers are called equivalent shear beam (ESB) model containers. They aim to match the shear deformation of the soil layer in a given earthquake in a step-wise fashion by having alternate layers of rubber and aluminum rings. One of the first model containers built in this fashion is shown in Figure 14.9. This model container was relatively small, having a depth of 220 mm, and could model a soil layer of about 10 m depth in a 50-*g* centrifuge test.

The small ESB model container was used to model a wide variety of problems. Shear sheets that were made from thin aluminum sheets onto which sand was glued were used next to the end walls to carry the complementary shear stresses. Teymur and Madabhushi (2003) investigated the boundary effects in this model container using in-flight cone penetrometers and concluded that the middle third of this container is ideal for any type of centrifuge testing with earthquake loading.

In order to model deeper soil deposits a much larger ESB model container was built (Brennan and Madabhushi, 2002). This model container could simulate soil deposits of 22 m deep in a 50-*g* centrifuge test. A view of this

Figure 14.9 A view of the small equivalent shear beam model container.

Figure 14.10 A view of the deep equivalent shear beam model container.

large ESB model container is shown in Figure 14.10. In this figure the lid used during the saturation of soil with viscous fluid is also seen. This model container has been extensively used in a wide variety of centrifuge tests. A newer ESB model container that is of much lighter construction is currently being fabricated at the Schofield Centre.

It must be pointed out that these model containers are only able to match the shear deformation in soil under a given earthquake. During soil liquefaction, the stiffness of the soil changes with the generation of excess pore pressures and once that happens the boundaries no longer match the soil stiffness. There will be some wave reflections in the soil as it transforms to this liquefied state. Another disadvantage of the ESB model containers is that the lateral deformations in the soil can be limited, for example, if we are modelling a sloping ground it can spread laterally by a large distance once it is liquefied. The ESB model containers prevent such lateral spreading from occurring.

To avoid this problem and model the liquefaction problems more accurately, a laminar model container is used. The concept of the laminar model container is that it has zero lateral stiffness of its own and therefore its deformation is driven by the soil deformation. This concept has been around for a long time (e.g., Scott, 1994). The laminar model container is built by having individual laminas that are separated by cylindrical bearings and therefore can move freely relative to one another. They are very useful in simulating liquefaction-induced lateral spreading problems. A view of the laminar model container is presented in Figure 14.11. Brennan, Madabhushi, and Houghton (2006) have compared the relative performance of the ESB and laminar model containers. The lamina are normally restrained to move in the direction of the earthquake loading by having PTFE side supports as seen in Figure 14.11. However, laminar model containers that are able to

Figure 14.11 A view of the laminar model container.

freely move in two directions have also been developed to accommodate biaxial shaking of soil models. However, these model containers need an inner rubber bag, normally made from latex, to prevent leakage of water.

Although the ESB and laminar model containers are very useful in simulating the field boundary conditions accurately, their design prevents any digital imaging through the side walls. With the development of particle image velocimetry (PIV) techniques and the availability of high-speed, high-resolution digital cameras it is desirable to have a box with transparent sides. To fill this gap, a model container with Perspex sides and a 45° mirror has been developed at Cambridge. A view of this model container is presented in Figure 14.12. Duxseal is used along the end walls to minimize the stress wave reflections from the boundary. The 45° mirror allows

Figure 14.12 A view of the model container with a 45° mirror.

mounting of the digital camera looking vertically down, providing the ability to view the cross-section of the centrifuge model.

With advances in digital imagining equipment, we can now have compact cameras which can be mounted directly in front of the transparent side, normal to the direction of high gravity, allowing the mirror to be dispensed with and digital images to be obtained directly.

14.5 SATURATION OF CENTRIFUGE MODELS

The need to use viscous fluids to saturate centrifuge models used in dynamic tests was explained earlier. In this section we will see how this is achieved in practice. The viscous fluids are normally used to saturated sandy soils, particularly to model earthquake-induced liquefaction problems. The hydraulic conductivity of sands will be in the range of 10^{-2} to 10^{-4} m/s. If we wish to saturate such soils for example with a fluid that has a viscosity of 50 cSt, then we need special measures to achieve good saturation. In this context it must be pointed out that the degree of saturation must be as close to 100 percent as possible, as the bulk modulus of the fluid drops rapidly with a drop in the degree of saturation.

The viscous fluid used in centrifuge models is either silicone oil or methyl cellulose (HPMC) as described earlier. It is important to have the pore fluid at the right viscosity depending on the *g* level at which the centrifuge test will be conducted. Using the scaling laws considered earlier, we would require a pore fluid of 50 cSt in a 50-*g* centrifuge test, 100 cSt in a 100-*g* centrifuge test, and so on. In order to obtain pore fluid of desired viscosity we need to mix HPMC powder at different concentrations as dictated by Equation 14.13. Recently a study was conducted by Adamidis and Madabhushi (2013) to investigate the validity of this expression for the HPMC obtained in the United Kingdom and establish the stability of the methyl cellulose solution over a long period of time. Figure 14.13 presents the data collected over a period of 48 days. It can be seen that the methyl cellulose solution holds its viscosity without much change over this period. We can also see that the data after 1 day shows slightly smaller viscosities but these settle down well after that initial period. This may be due to dissolution of any small pockets of HPMC powder that remain undissolved in that initial period. In Figure 14.13 Equation 14.13 is also plotted. It appears that this equation predicts smaller viscosities particularly at higher concentrations of HPMC. Adamidis and Madabhushi (2013) proposed a slight modification to Equation 14.13 as shown in Figure 14.15, which fits the experimental data much better.

$$\mu_{20} = 6.92\ C^{2.9} \tag{14.15}$$

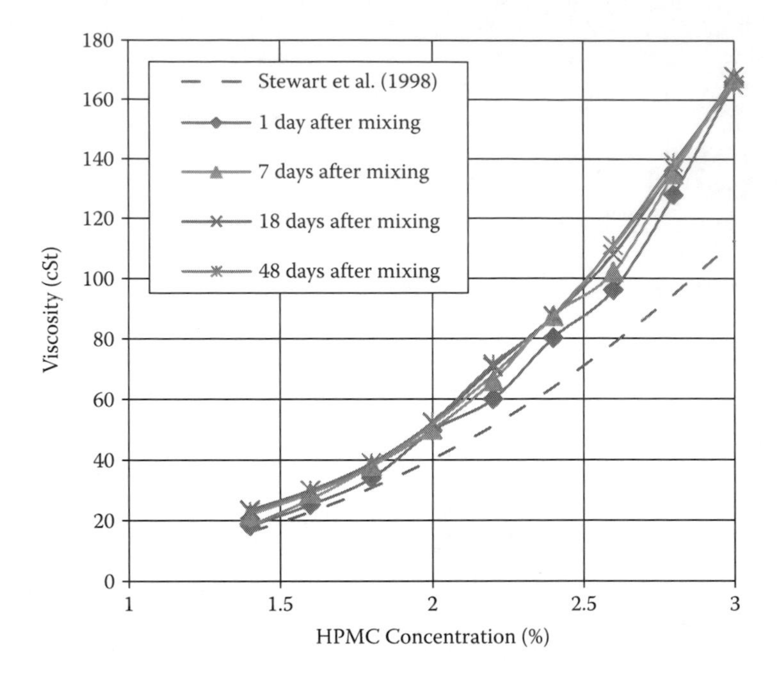

Figure 14.13 Variation of viscosity with HPMC concentration.

This equation also fits the manufacturers data that a 2 percent HPMC solution should have a viscosity of 50 cSt.

Until recently the centrifuge models were saturated by admitting the fluid from the base of the model and applying a suction at the top to draw the fluid through the soil body. This was achieved by having a lid on top of the model container that is connected to a vacuum pump. The soil models were evacuated for a period of an hour or so and then flushed with carbon dioxide gas. This process of evacuation and flushing with carbon dioxide gas was repeated several times. The idea behind this is to replace all the air in the soil body with carbon dioxide gas, which is soluble in water. The de-aired methyl cellulose (HPMC) was then admitted into the centrifuge model through the base of the model through multiple ports. Filter plates made from Vyon sheets cover the entry wells at the base of the model, to prevent any clogging of the ports. As we apply a vacuum of about –95 kPa at the top of the centrifuge model, the viscous pore fluid migrates through the soil body.

The rate at which we admit the pore fluid is important. If we admit the fluid too quickly the upward hydraulic gradients may be large enough to

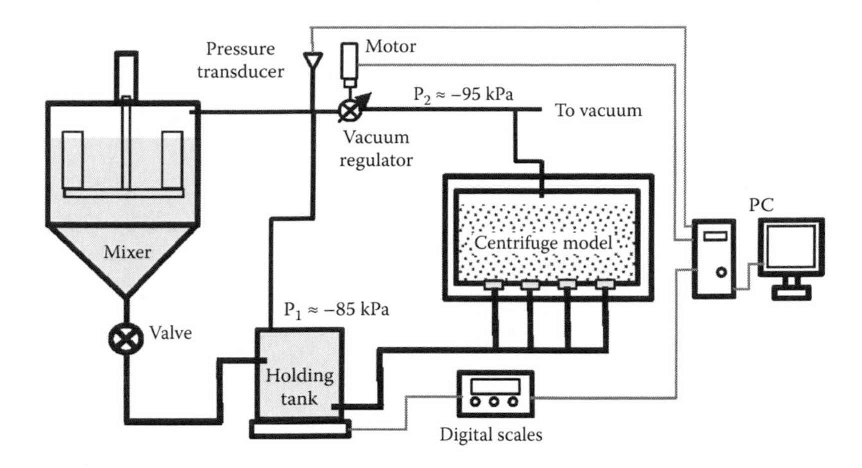

Figure 14.14 Schematic diagram showing the modified Cam-Sat system.

fluidize the soil bed. On the other hand, too slow a flow rate means that the saturation process can take days to complete. The flow rate of the fluid into the centrifuge model is controlled by needle valves and the fluid mass flux is kept below 0.5 kg/hour by manually controlling the needle valves.

The above procedure though very successful does take 10 to 15 hours to saturate a centrifuge model of reasonable size. It also requires constant monitoring to ensure a successful saturation. In order to make the process smoother a novel automatic saturation system was developed at Cambridge, called the Cam-Sat system (Stringer and Madabhushi, 2009). This system was designed for any viscous pore fluid and for soils with a range of hydraulic conductivities.

In the Cam-Sat system the saturation process is automated. The fluid mass flux is used as one of the main control parameters. The schematic of this system is presented in Figure 14.14. This is slightly modified from the system presented in Section 7.2.2. The system includes a methyl cellulose mixer that produces the viscous fluid under vacuum pressure. The model container is also kept under vacuum pressure after a few cycles of soil flushing with the carbon dioxide gas. A differential vacuum pressure regulator is used to maintain the pressure difference between the holding tank and the centrifuge model container, as shown in Figure 14.14, and provides the driving force to move the viscous fluid into the centrifuge model.

The fluid mass flux is constantly monitored using a digital balance underneath the holding tank. The output from the digital balance is fed into the computer to work out the changes in fluid flux with time. The computer also monitors the vacuum pressure difference between the centrifuge model

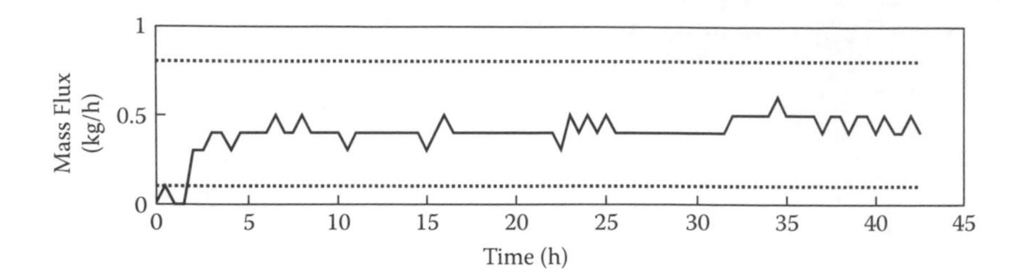

Figure 14.15 Control of fluid mass flux using the Cam-Sat system.

and the holding tank through a pressure transducer. It adjusts the vacuum regulator through a motor to maintain the fluid flux within threshold values as shown in Figure 14.15. In this figure we can see that the system is able to maintain the fluid flux between the thresholds over a long period of time. The system can also warn users by email if the fluid flux is approaching one of the thresholds.

A view of the centrifuge model during saturation is shown in Figure 14.16. The soil in this model was a fine sand called Hostun sand. In this figure it can be seen that the saturation front is more or less horizontal as it propagates upwards. This system has been integrated with regular centrifuge model making and saturation at the Schofield Centre and is being used now for a wide variety of earthquake engineering problems.

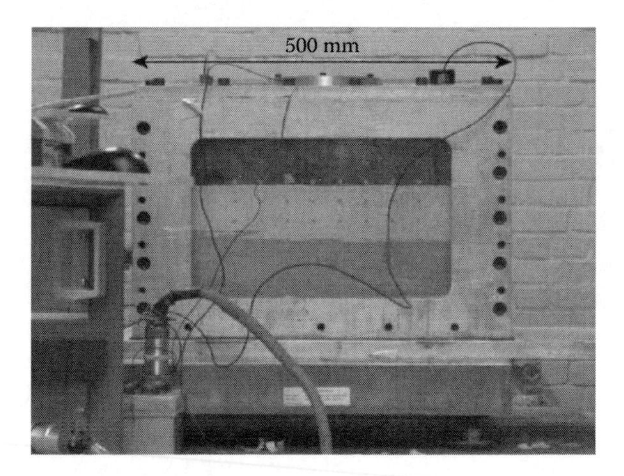

Figure 14.16 A view of the cross-section of a centrifuge model during saturation.

14.6 CENTRIFUGE MODELLING OF THE DYNAMIC SOIL–STRUCTURE INTERACTION PROBLEMS

One of the advantages of using centrifuge modelling for geotechnical earthquake engineering problems is that soil liquefaction can be modeled accurately. As a result many researchers have used dynamic centrifuge modelling to study liquefaction problems. However, the same modelling can be used to investigate dynamic soil-structure interaction (SSI) problems. In practice the SSI effects are often ignored at design stages of a structure. This is because these are difficult to quantify and there is a general perception that ignoring SSI effects in design leads to an additional factor of safety. However, this may not always be true and ignoring SSI effects under some circumstances can lead to a reduction in the factors of safety. Dynamic centrifuge modelling can offer an effective method of quantifying the SSI effects for various structures subjected to earthquake loading.

14.6.1 Tower–soil interaction

Madabhushi and Schofield (1993) report on a series of centrifuge tests carried out on tower structures that were subjected to earthquake loading. They show that degradation of soil stiffness following partial liquefaction has the propensity to cause a drop in the natural frequency of the tower-soil system to one of the earthquake frequencies driving the system. This can then lead to resonant vibrations in the tower structure. Although the predominant aim of this research was to look at liquefiable foundations below tower structures, Madabhushi (1991) also reports on a large number of tests carried out with towers founded on dry sand beds. This was one of the early studies on SSI effects associated with tower structures. The shear strain contours mobilised in the dry sand layer below the structures were obtained. Examples of these for low and high natural frequency structures which apply the same bearing pressure on the sand are presented in Figure 14.17 for a moderate earthquake magnitude that caused bedrock accelerations of 0.12 *g* in an 80-*g* centrifuge test.

In this figure we can see that the low frequency structure produces shear strain of 0.18 percent at the edge of the zone of influence. The high frequency structure influences the sand layer to a greater depth and produces a shear strain of 0.23 percent at the edge of the zone of influence. This figure shows the importance of considering dynamic soil-structure interaction as the natural frequency of the structure is shown to influence the foundation soil behaviour.

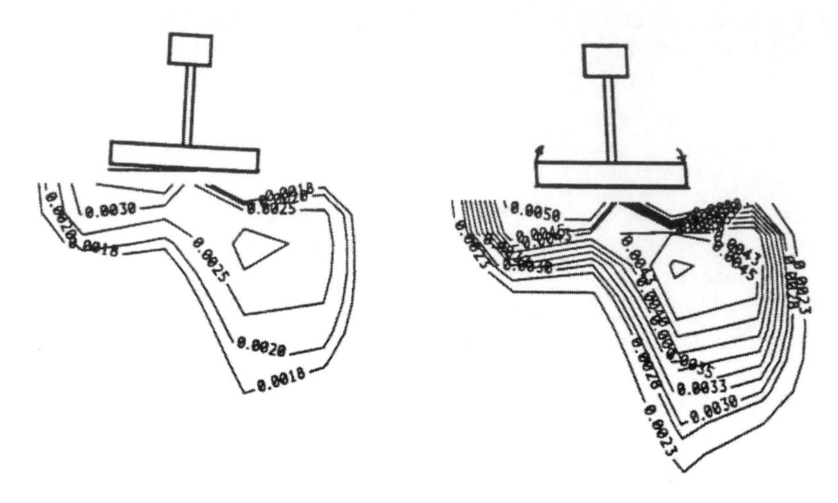

Figure 14.17 Shear strains generated below tower structures.

14.6.2 Tunnel–soil interaction

Another example where SSI effects can play an important role is the dynamic behaviour of underground structures like tunnels. In the design of tunnel structures in earthquake-prone regions of the world, the interaction effects are largely not considered. Cilingir and Madabhushi (2011a, 2011b, 2011c) investigated the seismic behaviour of circular and square tunnels using dynamic centrifuge modelling. They utilised the model container with the 45° mirror shown in Figure 14.12. An example of the square tunnel model is shown in Figure 14.18.

Figure 14.18 Centrifuge model of a square tunnel in sand.

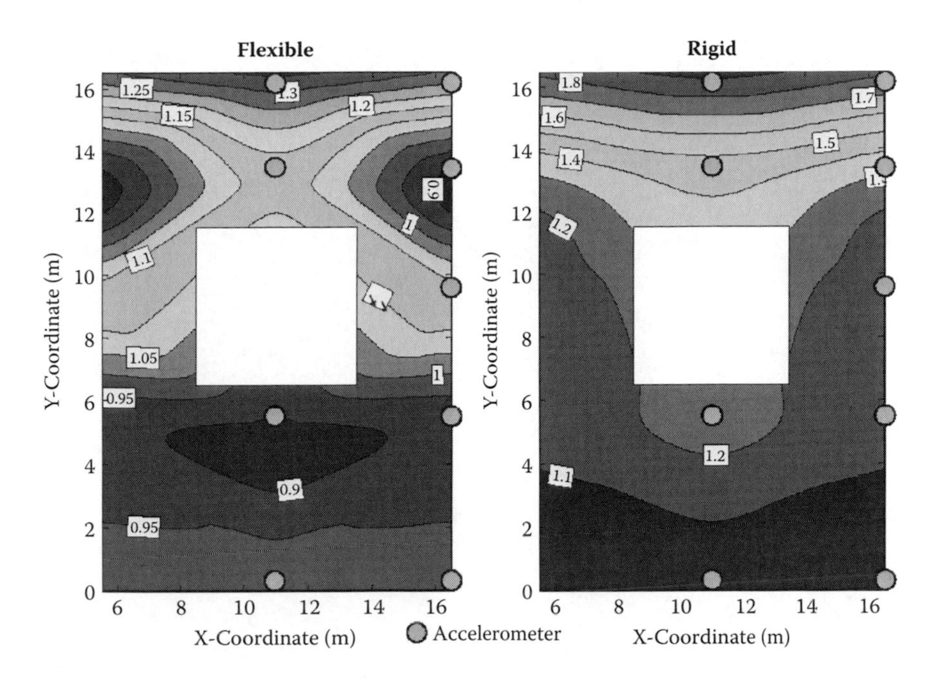

Figure 14.19 Acceleration contours around square tunnels.

The test program included testing of circular and square tunnels with different wall thicknesses to change the flexibility of the tunnel. Tunnels buried at different depths were also tested. In Figure 14.19 the acceleration contours are plotted around square tunnels with flexible and rigid walls. The contours clearly change with flexible-walled tunnels interacting more with the soil around it and changing the stress wave propagation in the soil. Rigid-walled tunnels interact to a lesser extent.

In this test series, PIV analysis was attempted for the first time in a dynamic centrifuge test (Cilingir and Madabhushi, 2011b). High-speed imaging was carried out using the Phantom camera with images taken at 1000 frames per second. An example of PIV analysis is shown in Figure 14.20 for the case of a square tunnel buried at a shallow depth compared with a test in which the same tunnel was buried much deeper.

Another innovation that was tried in this test series is the use of miniature pressure transducers around the tunnels to measure the dynamic earth pressures. An example of the variation in earth pressure during an earthquake loading is shown in Figure 14.21 along with the base acceleration applied to the model. The peak acceleration applied was about 0.2 *g* in this earthquake and the earth pressure is seen to vary with the cyclic loading. In the first few cycles the earth pressure increases continually. This phase is termed the transient phase. Following this the earth pressure

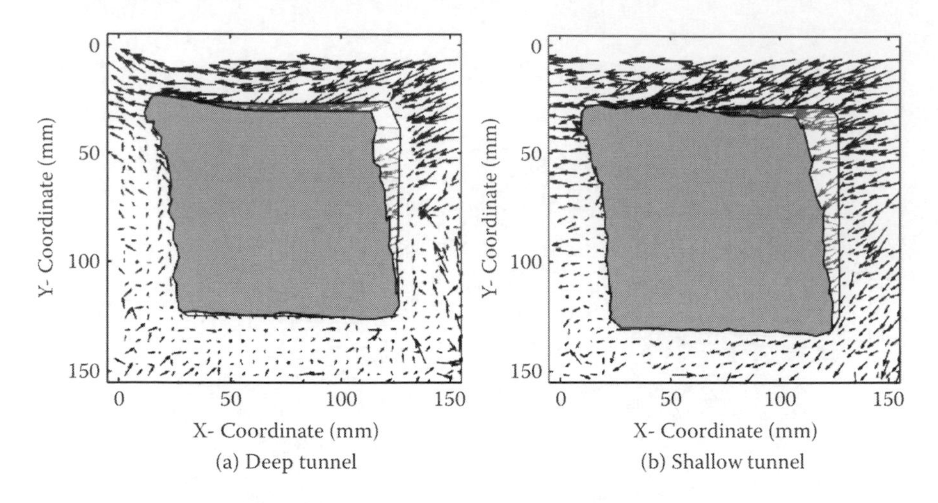

Figure 14.20 PIV analysis of square tunnels.

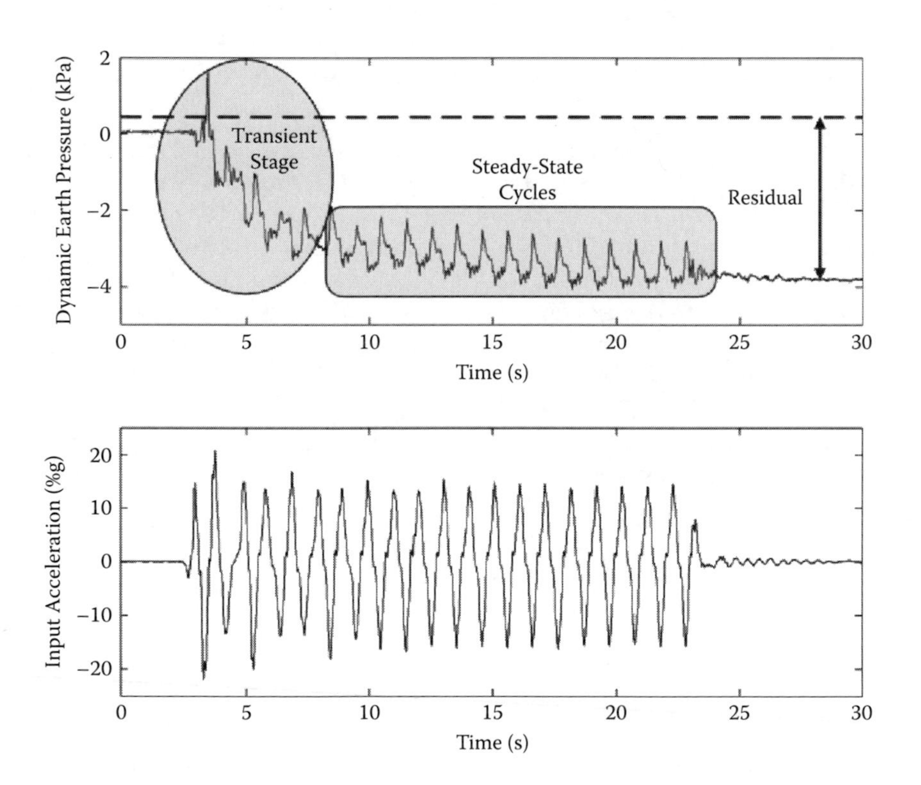

Figure 14.21 Dynamic earth pressure measurement.

varies about a mean value and this phase is termed a steady-state phase. One can also see that the earth pressures maintain a residual value after the end of the earthquake.

The above examples of dynamic behaviour of tunnels show clear interaction between the tunnel structures and the soil. It was also seen that using dynamic centrifuge modelling we can gain insights into the changes in stress wave propagation around tunnel structures and variations in earth pressures acting on tunnels. Further carrying out PIV analyses offers ways of visualizing the interaction between the tunnel and the surrounding soil.

14.6.3 Rocking of shallow foundations

Centrifuge modelling of shallow foundations was discussed extensively in Chapter 10. In this section we shall briefly revisit that topic and consider shallow foundations subjected to earthquake loading. Recently Heron (2013) considered the rocking of shallow foundations when subjected to earthquake loading. He considered shallow foundations of different aspect ratios exerting different bearing pressures. This series of experiments was carried out as part of the European Union-funded FP7 project called SERIES. Similar experiments were also carried out at the IFSTTAR centrifuge in France to provide a direct comparison between centrifuge tests carried out at different centers (Cilingir et al., 2012).

Shallow foundations are known to suffer rocking vibrations. Rocking was also detected in the raft foundations below tower structures that were considered in Section 14.6.1. However, without high-speed imaging it is impossible to pick up the behaviour of rocking foundations under cyclic loading particularly the lift-off they suffer when the lateral accelerations increase beyond a threshold value. The amount of the foundation that suffers lift-off depends on the relative density of the soil, with loose sands deforming more easily and not allowing much lift-off. On the other hand, denser sand deposits offer more rigid support to the structure, encouraging more lift-off. In this series of experiments high-speed digital imaging was carried out with a new camera that was quite compact in size. This meant that there was no need to use the 45° mirror and the camera could be mounted directly in front of the Perspex window shown in Figure 14.12. An example of the PIV analyses carried out on a shallow foundation is shown in Figure 14.22. In this figure the accumulated displacement vectors in the sand are shown for the whole earthquake. This figure reveals that very little sliding of the foundation occurred during the earthquake loading. Most of the displacement vectors indicate the rocking of the foundation and the soil tending to move away from beneath the structure. The figure also shows a very symmetric pattern of deformations in the soil about the central axis of the structure. Further the PIV analysis of these dynamic centrifuge tests has improved in quality with fewer wild

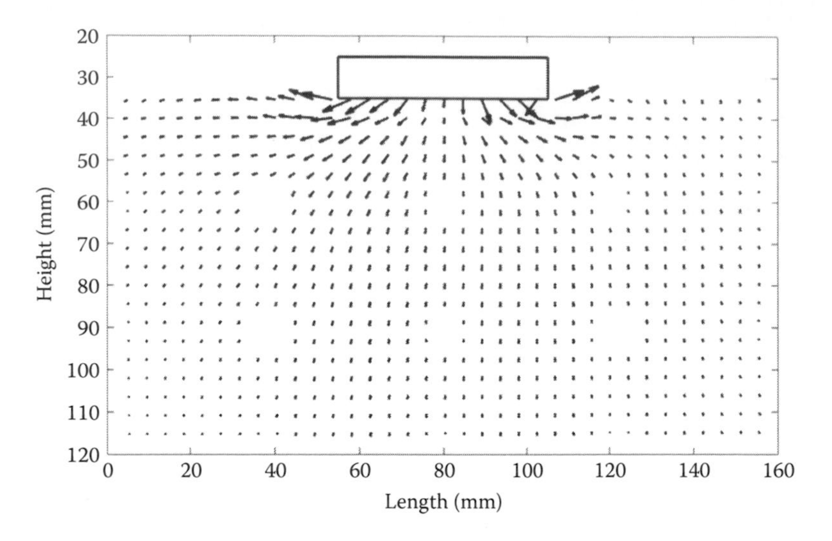

Figure 14.22 PIV analysis of a shallow foundation.

vectors, due to the orientation of the camera, improvement in its resolution, and lighting arrangements.

The examples presented in this section all confirm that dynamic centrifuge modelling can be used successfully to study soil-structure interaction problems. With the development of PIV analysis we can now visualize the failure mechanisms that may develop around various geotechnical structures that are subjected to earthquake loading.

14.7 CENTRIFUGE MODELLING OF LIQUEFACTION PROBLEMS

As mentioned earlier centrifuge modelling offers a unique opportunity to model soil liquefaction problems. Although there are finite element codes that are able to predict liquefaction, such as Swandyne (Chan, 1989), these codes require experimental data to validate their predictions. A large comparison exercise, "Verification of Numerical Procedures for the Analysis of Soil Liquefaction Problems—VELACS," that compared centrifuge test data and numerical predictions was funded by the National Science Foundation in 1992 (Arulanandan and Scott, 1993). Currently a more up-to-date version of this exercise, called the "Liquefaction Experiments and Programs—LEAP" project, that uses modern centrifuge modelling techniques and instrumentation is being planned with participants from Japan, the United Kingdom, and the United States. A similar exercise on tunnel-soil interaction that used dynamic centrifuge test data and compared them with numerical predictions, called "Round Robin Tunnel Testing—RRTT," was recently concluded (Bilotta, Silvestri, and Madabhushi, 2014). This exercise was conducted using the centrifuge test program carried out as part of the RELUIS project that was a collaboration of a consortium of Italian universities and industry and the University of Cambridge.

Over the past decade a number of researchers have carried out centrifuge experiments to model soil liquefaction. In this section some interesting experimental studies that were conducted at Cambridge are explained briefly. These examples are given with the view to explaining what can be achieved using this modelling technique rather than to provide a comprehensive discourse on the range of centrifuge tests that were performed.

14.7.1 Piles passing through liquefiable soils

Modelling of pile foundations was considered in Chapter 12. In this section we shall see the modelling of pile foundations in liquefiable soils. Pile foundations transfer structural loads onto competent soil strata. If the surface layers are formed by loose, saturated sandy soils the pile foundations are driven through these until they go into more competent strata like dense sands or stiff clays. If the surface layers form a sloping ground as shown in Figure 14.23 then they are likely to suffer lateral spreading following earthquake-induced liquefaction. Such laterally spreading soils can impose large lateral loads onto piles and were responsible for damage to pile foundations in many of the recent earthquakes. Considerable research has taken place on pile foundations in liquefiable soils. Centrifuge modelling has played an important role in developing an understanding of the magnitude of additional lateral loads on piles due to the lateral spreading

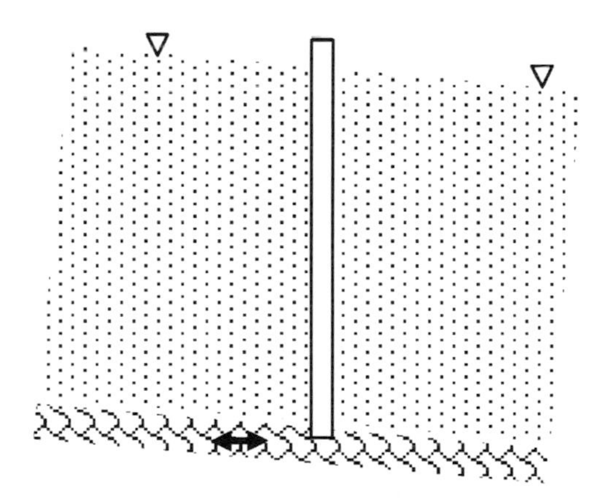

Figure 14.23 Pile foundations in sloping ground.

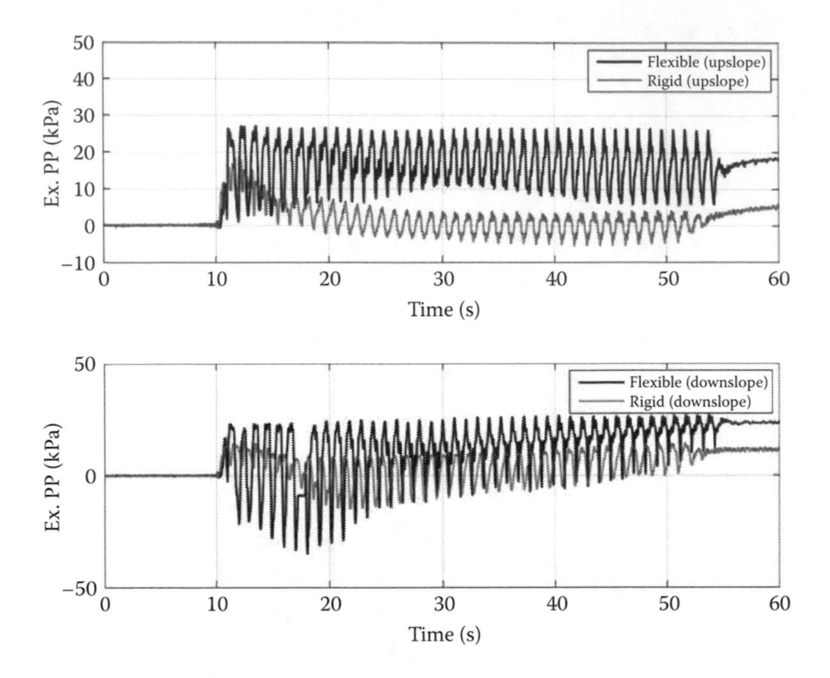

Figure 14.24 Excess pore pressures near pile foundations.

effects. Also new mechanisms of failure for pile foundations in level, liquefiable ground were postulated based on centrifuge test data (Madabhushi, Knappett, and Haigh, 2009).

At Cambridge the piles in laterally spreading soils were first investigated by Haigh (2002). He researched the effects of lateral spreading on flexible and rigid piles. The interaction between the pile and the soil plays an important role in this problem.

In Figure 14.24 the excess pore pressures generated near the upslope and downslope of flexible and rigid piles are presented. In this figure we can see that on the upslope side the excess pore pressure generated near a flexible pile is much larger than for a rigid pile.

It appears that a rigid pile is able to resist the downward movement of the soil and therefore reduces the cyclic shear strains in the soil during earthquake loading. This results in a lower excess pore pressure. In contrast the flexible pile allows more soil movement, which results in larger excess pore pressures. On the downslope side of the piles the soil is able to move away from the pile and therefore the excess pore pressure is influenced to a lesser degree by the pile stiffness. In Figure 14.24 we can see that the excess pore pressures generated for both rigid and flexible piles are somewhat similar. The slightly larger excess pore pressure for a flexible pile can be attributed to the additional cyclic strains generated in the soil due to the extra

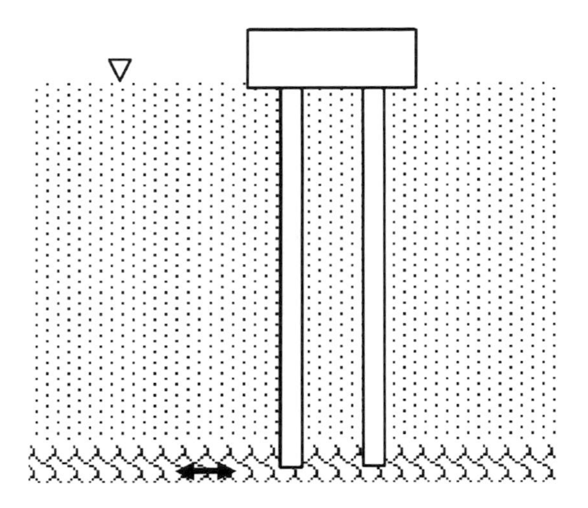

Figure 14.25 Pile group passing through a liquefiable soil layer.

movement of the flexible pile compared to a rigid pile. More details of this research are given in Haigh and Madabhushi (2005).

In the case of pile foundations passing through level, liquefiable ground earthquake loading and soil liquefaction can present a different kind of problem. Piles are generally very slender and in normal ground they do not suffer buckling because of the lateral support they elicit from the soil surrounding their shaft. When soil suffers liquefaction owing to earthquake loading there could be a temporary loss in this lateral support and the piles could suffer buckling during that period. This premise was first tested by Bhattacharya, Madabhushi, and Bolton (2004) for a single pile and later by Knappett and Madabhushi (2009) for pile groups like the one shown in Figure 14.25. It must be pointed out that buckling of piles is only an issue when the end tips of the piles are rock-socketed as seen in Figure 14.25. If the pile tips end in dense sands then the piles are more liable to settle rather than buckle. This aspect is considered in a later example.

For rock-socketed piles that pass through liquefiable soils, buckling is a viable failure mechanism. Dynamic centrifuge modelling offers a means of verifying such a hypothesis without waiting for real failures of piles in earthquake events. Also parametric studies can be conducted to investigate the importance of slenderness ratio, load eccentricity, relative pile-soil stiffness, etc.

An example of the lateral displacement suffered by the pile group shown in Figure 14.25 in a 50-*g* centrifuge test is shown in Figure 14.26, presented at prototype scale. In this figure we can see that the lateral displacements increase initially during the earthquake (marked as **a**) and reach about 0.4 m of total deflection. Following the end of the earthquake this stays

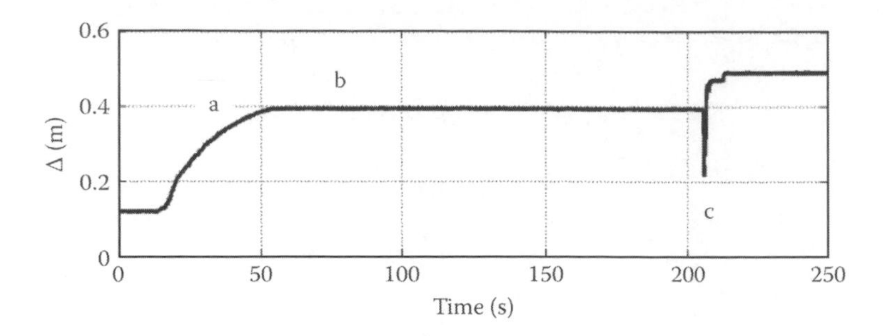

Figure 14.26 Lateral displacement of the pile group.

constant (marked as **b**) for nearly 150 seconds. It is only after this period, some 200 seconds after the earthquake struck, that the pile group suffers buckling (marked as **c**). This is a good example of where failures occur in the post-earthquake period rather than during the earthquake shaking.

After the centrifuge test the pile group was excavated and photographed as shown in Figure 14.27. In this picture we can see the buckling of each of the piles clearly and the rotation suffered by the pile cap. Knappett and Madabhushi (2009) give more details of these types of failures for pile groups with different pile spacing.

This type of failure is quite easy to design against. Once the geotechnical engineer knows it is a plausible failure mechanism he or she can easily check for the Euler buckling loads and design against it.

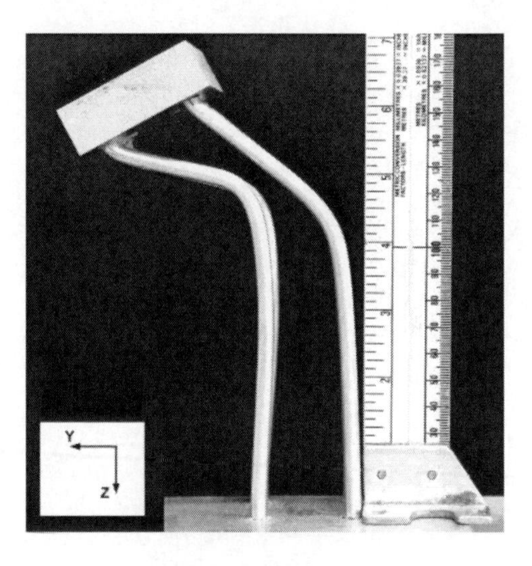

Figure 14.27 A view of the buckled pile group.

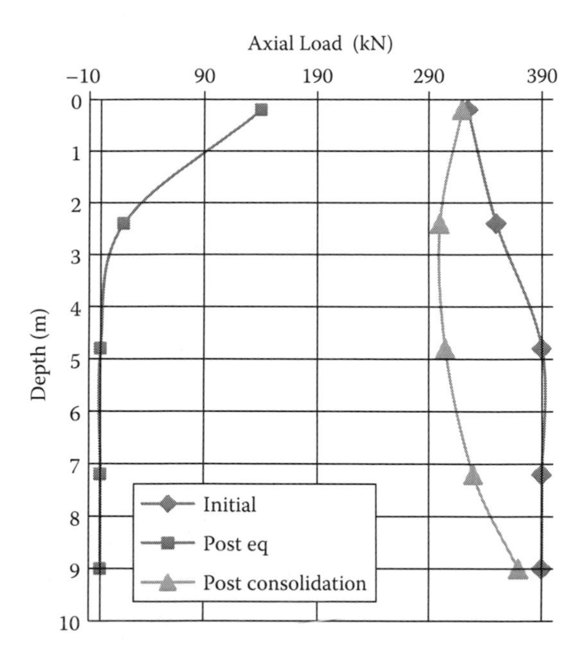

Figure 14.28 Axial load distribution in a pile.

As explained earlier, if the pile foundations end in dense sands or stiff clays then buckling may not be the failure mechanism. Pile settlements can become more important. Stringer and Madabhushi (2012, 2013) investigated the axial behaviour of pile foundations particularly when the pile cap was resting on the ground surface. The model piles were instrumented to measure the axial loads at different points along the pile. In Figure 14.28 example data of the variation of the axial loads along the length of the pile are presented. In this figure we can see that before any earthquake loading the axial loads vary between 310 and 390 kN. Following soil liquefaction this drops alarmingly to about 120 kN at the pile head and reduces down to 0 at deeper levels. After the earthquake loading is completed the liquefied soil slowly reconsolidates allowing the piles to recover their shaft capacity. In Figure 14.28 we can see that post-consolidation the pile load distribution has almost recovered to its initial values.

During the period between earthquake loading and the reconsolidation the pile group needs to be supported. Vertical bearing pressures were monitored below the pile cap as shown in Figure 14.29. In this figure we can see that the pile cap bearing pressures increase significantly during the earthquake loading confirming that the pile cap provides temporary support to the pile group in this phase. We can carry out equilibrium checks on

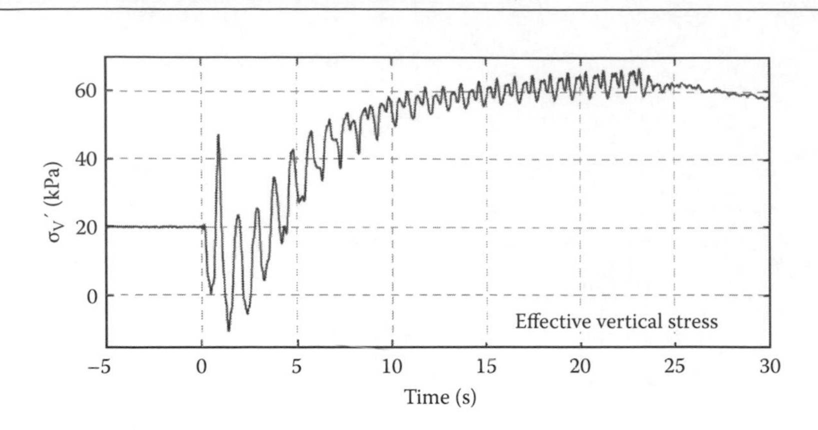

Figure 14.29 Changes in effective stress below the pile cap.

this data. If we take the increased bearing pressure of about 62.5 kPa seen in Figure 14.29, under a 5 m × 5m pile cap (100 mm × 100 mm model scale), this gives an additional vertical support of 1562 kN. Given that there are four piles in this two × two pile group, this is equivalent to 390 kN per pile. This matches perfectly with the values in Figure 14.28. The piles lost their axial load-carrying capacity of 390 kN and thus the pile cap had to generate this additional bearing pressure to compensate for this loss.

From a design point of view, this type of centrifuge test reveals the importance of the pile cap bearing capacity which is routinely ignored in the design. It also shows the importance of good structural detailing that connects the pile cap and the piles.

14.7.2 Nuclear reactor building interaction with soil

The performance of nuclear infrastructure during and after an earthquake event is one of the safety critical requirements in nuclear power generation. The core reactor buildings and appurtenant buildings that house support equipment need to be adequately checked for their seismic performance. Often the bearing pressures exerted by such buildings is quite high (~150 kPa or more). For this reason where possible the raft foundations for these structures is located directly on the bedrock. However, there may be sites where bedrock is quite deep and engineers are required to locate them on improved ground.

Ghosh and Madabhushi (2007) investigated the SSI of heavy structures on stratified soils as shown in Figure 14.30. This centrifuge test program was carried out at 50 *g* and the model structures exerted a bearing pressure of 150 kPa on the soil. Horizontal stratifications were considered as naturally occurring site conditions. Vertical stratification was considered to account for in situ densification carried out as a ground improvement prior to founding a heavy structure.

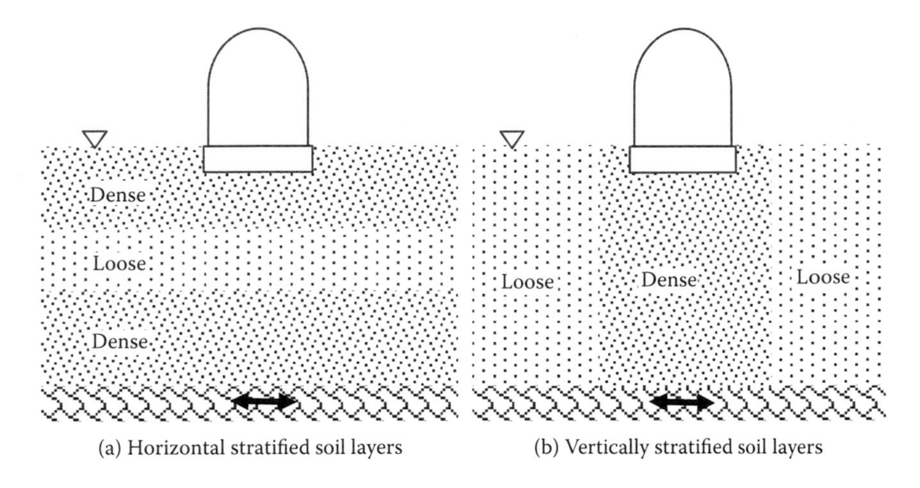

Figure 14.30 A nuclear reactor building founded on stratified soil layers.

The excess pore pressures generated in the soil below the raft foundation of these heavy structures is presented in Figure 14.31. In this figure we can see that the excess pore pressures generated in a densified layer are very different to those in an equivalent homogeneous layer. The dense sand is able to dilate strongly creating large suctions in the excess pore pressures during earthquake loading. This shows a strong interaction between the structure

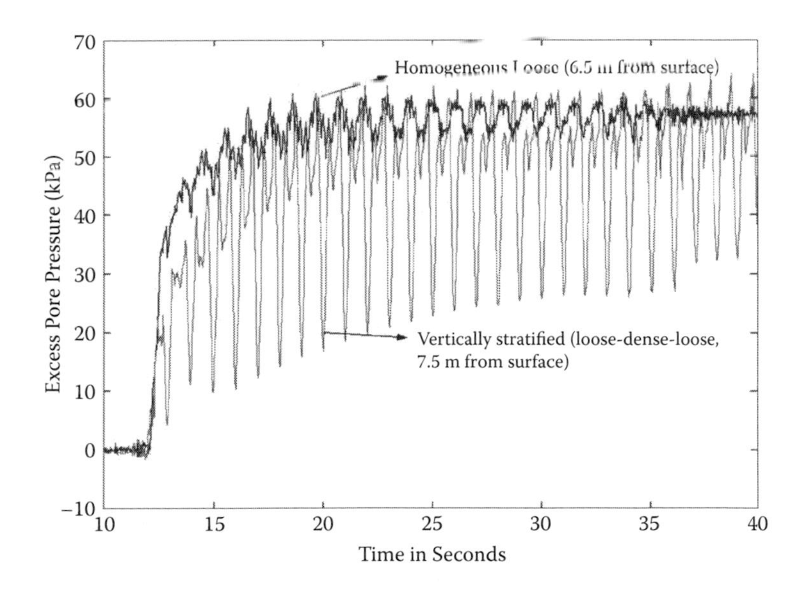

Figure 14.31 Excess pore pressures generated below the structure.

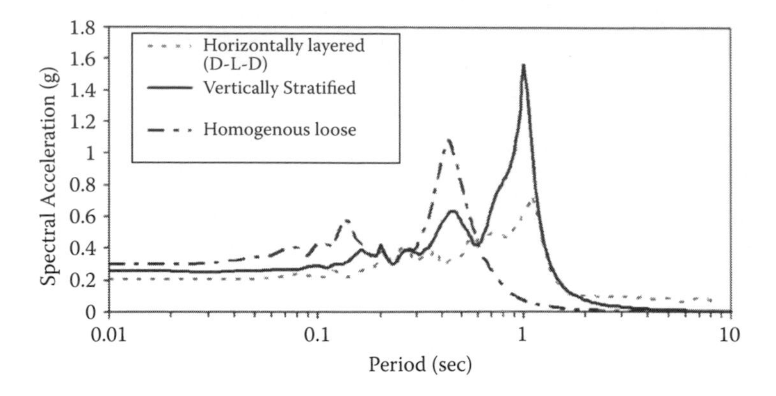

Figure 14.32 Response spectra of the reactor building.

and the soil and how the density of the soil deposit can influence the overall structural behaviour.

Figure 14.32 shows the response spectra obtained from an acceleration measurement on the structure during the centrifuge test. In this figure we can see that the response spectra for the horizontally and vertically stratified soils are very different from the one for a homogeneous soil deposit. The amplification of the spectral acceleration is greatest for the vertically stratified case. The horizontally stratified soil layers seem to isolate the structure due to excess pore pressure generation and show a significant drop in the spectral acceleration.

The settlements suffered by the structure are shown for various soil configurations in Figure 14.33. In this figure we can see that the structure on the homogeneous soil settles the most. The structure on vertically stratified soil settles the least, with the one on horizontally stratified soil lying in between. Again we would expect this result as the densified zone reduces the settlement of the structure. It can also be seen that most of the settlement occurs during the shaking with the structures only settling marginally in the post-earthquake period.

In these centrifuge tests we can see that quantitative data can be obtained using this modelling technique for the performance of the structure both in terms of its dynamic response and the settlement. The cost of centrifuge modelling in such cases is quite small compared to its importance and the safety critical nature of such infrastructure.

14.7.3 Floatation of tunnels

Underground structures such as tunnels, basements, or storage tanks are buoyant structures. If they are located in liquefiable soils they can suffer floatation following an earthquake event (Figure 14.34). Numerous examples

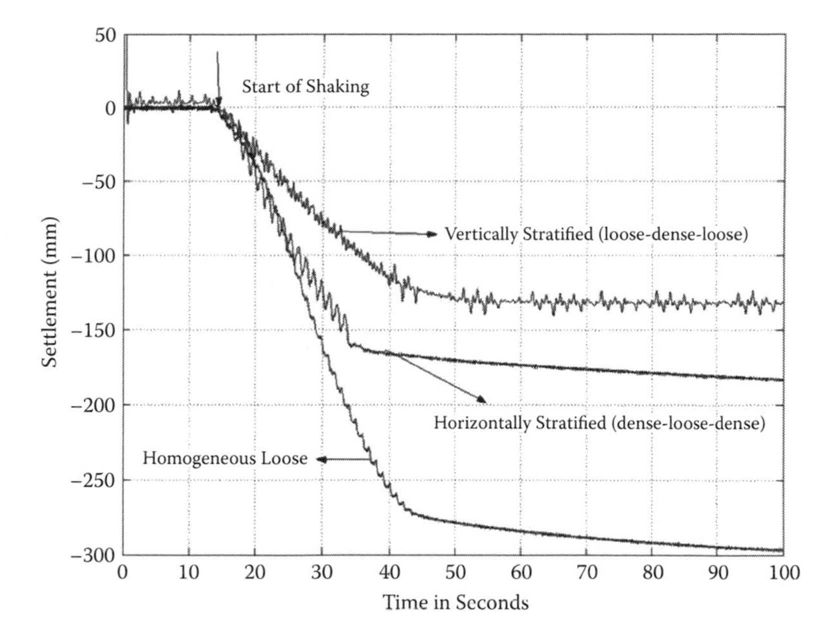

Figure 14.33 Settlement of the reactor building.

exist from recent earthquakes where sewer pipes, empty water tanks, etc., have suffered floatation following liquefaction. This problem was recently investigated by Chian and Madabhushi (2012). They considered tunnels of various diameters that were buried at different depths. These centrifuge tests were conducted at 50 *g* using the box with transparent sides presented

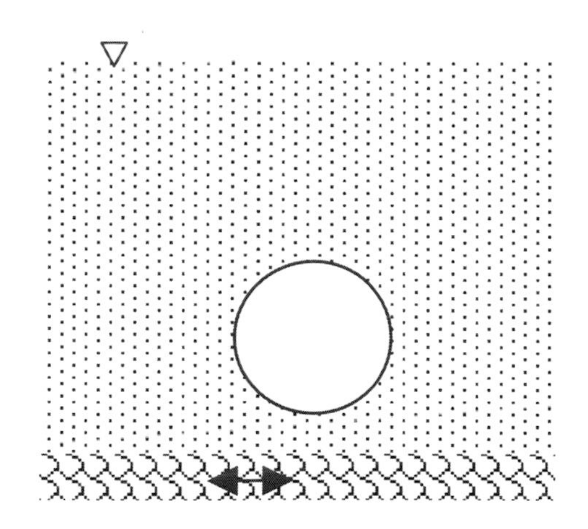

Figure 14.34 Schematic diagram of a buried tunnel in liquefiable soil.

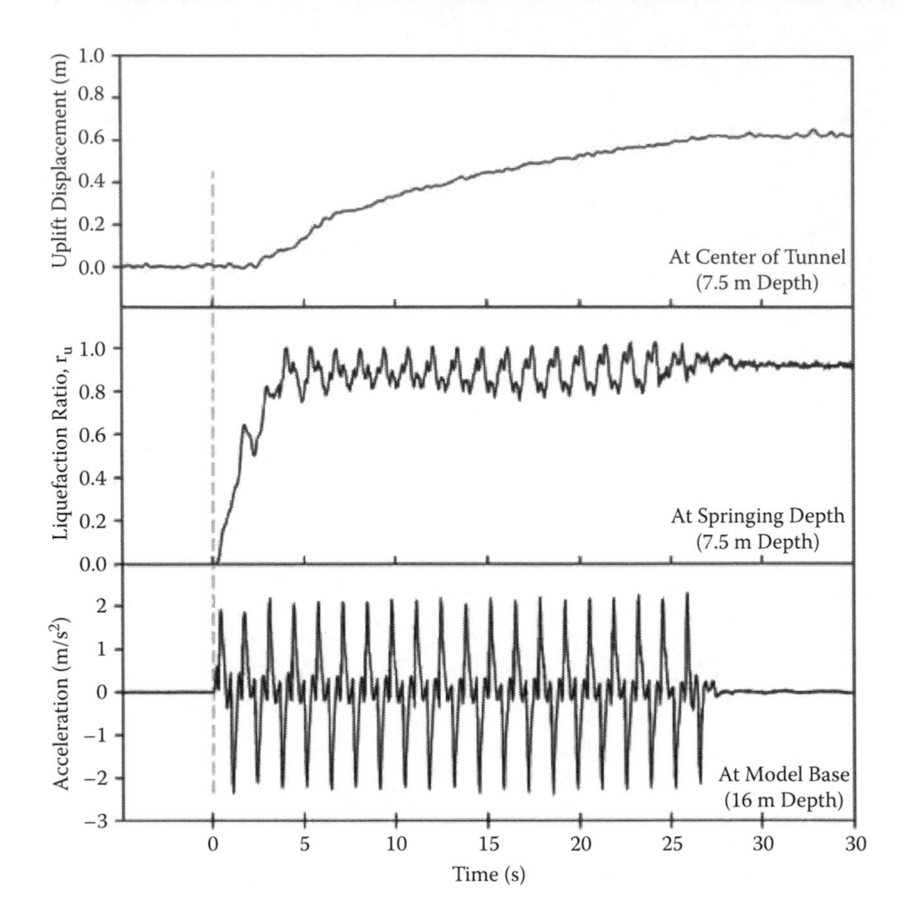

Figure 14.35 Example data from centrifuge tests on tunnel floatation.

in Figure 14.12. The tunnels themselves were rigid. High-speed imaging was utilised and PIV analysis of the images was carried out subsequently.

Figure 14.35 shows an example of the experimental data obtained from this test series. In this figure we can see that excess pore pressures are generated within the first few cycles of earthquake loading. The uplift displacement of the tunnel starts soon after and continues throughout the earthquake loading. This was measured at the crown of the tunnel and reaches a maximum value of 0.6 m by the end of the earthquake.

An example of the PIV analysis on tunnel floatation is presented in Figure 14.36. In this figure we can identify the soil deformations around the tunnel to accommodate its uplift displacements. A near circular mechanism develops that allows for the soil in region A to move into region C. Similarly, soil away from the tunnel moves into region B to fill the space

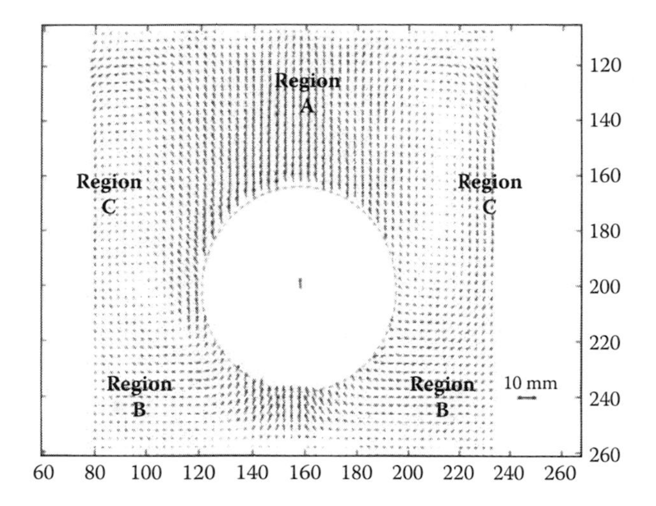

Figure 14.36 PIV analysis of tunnel floatation.

below the tunnel invert. Identifying such mechanisms can help us better design the tunnels against floatation.

PIV analysis also provides interesting data on the actual movement of the tunnel during the earthquake loading. Figure 14.37 shows an example of the tunnel movements obtained from PIV analysis. In this figure we can see that the tunnel suffers quite a lot of horizontal displacement in every cycle while it

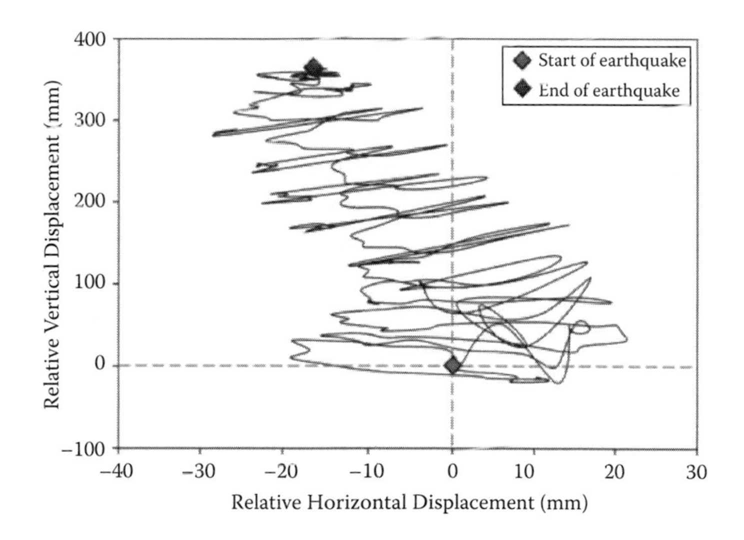

Figure 14.37 Horizontal and vertical displacement of the tunnel.

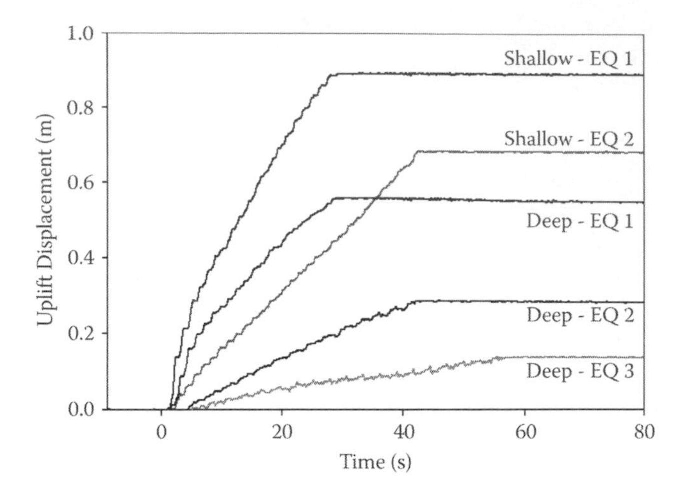

Figure 14.38 Uplift of deep and shallow tunnels.

accumulates the vertical uplift cycle by cycle. Further, we see that the amplitude of the horizontal displacements is reduced but the vertical displacement suffered in each cycle remains much the same but does accumulate.

This research involved testing of tunnels buried at different depths. In Figure 14.38 the uplift displacements for a shallow and deep tunnel are presented. In this figure we can see that as expected the shallow tunnels suffer larger uplift than deep tunnels. Also when subjected to repeated earthquakes the uplift displacements of the tunnel are reduced. In each case uplift displacement for earthquakes 2 and 3 are smaller than for earthquake 1. This is due to the densification suffered by the soil above the tunnel in each earthquake, which increases the uplift resistance of the tunnel.

14.7.4 Vertical drains to protect against liquefaction

So far we have seen centrifuge modelling used to investigate the behaviour of geotechnical structures in liquefaction scenarios. We can also use centrifuge modelling to investigate the efficacy of various protection measures normally used to offer liquefaction resistance to structures such as in situ densification or use of drainage to relieve excess pore pressures.

The performance of vertical drains to relieve excess pore pressures following soil liquefaction was investigated by Brennan and Madabhushi (2005). It is relatively straightforward to create centrifuge models with vertical drains in a loose, saturated liquefiable soil body. These centrifuge tests were carried out at 50 *g*. Model drains were made from coarse-grained sands. The installation effects of drains were, however, not modeled in this research.

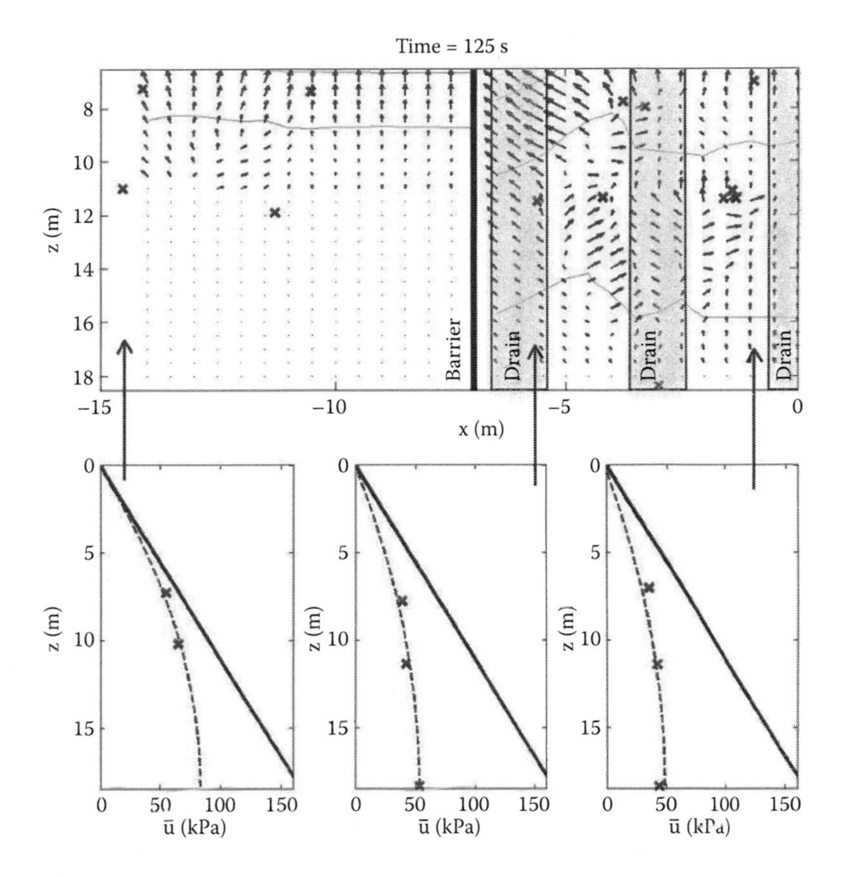

Figure 14.39 Fluid velocity vectors 100 seconds after the end of the earthquake.

Drains offer an easier path for the pore fluid to move to the surface when excess pore pressures are generated due to soil liquefaction. Figure 14.39 shows an example of the centrifuge data obtained during this test series.

The centrifuge model had an impermeable barrier separating the regions where drains were present and the free field. In this figure we can see fluid velocity vectors after 100 seconds (model scale) following an earthquake that liquefied the soil bed. In the top half of this figure we can see the velocity vectors indicating the fluid movement into the drains and then upward in the right-hand region. In the free field the movement of the pore fluid is more or less vertical toward the soil surface. The same information can be seen using excess pore pressure isochrones shown in the bottom of the figure. Here we see that the excess pore pressures in the region where drains were present dropped to about 50 kPa in and around the drains. In the free field the excess pore pressures are still close to 80 kPa. This example shows not only that drains are effective in reducing the excess pore pressures but also how effective they are. Various drain configurations can be tested to optimize their design for a given site.

14.7.5 Propped retaining walls

As a final example of modelling of liquefaction problems, let us return to the case of a propped retaining wall. We considered these structures under static loading in Chapter 11. Here we shall subject the same model to earthquake loading. Also it will be assumed that the water table coincides with the excavation level and the soil is liquefiable. The schematic cross-section of the model is shown in Figure 14.40. As the retaining walls are propped we expect them to rotate about the prop level and displace laterally into the excavated space. Soil liquefaction presents an interesting aspect to this problem. As the soil below the water table liquefies, the retaining walls rotate and cause heaving of the soil in the excavation. This could be quite critical if the earthquake happens during the construction phase where the bottom prop is not yet placed.

Tricarico et al. (2013) have investigated this problem by conducting centrifuge tests at 40 *g*. An example of the data obtained from these centrifuge tests is presented in Figure 14.41 at prototype scale. An earthquake that applied a lateral acceleration of 0.2 *g* at the bedrock level was applied to the model using the SAM actuator. Figure 14.41 presents the accelerations recorded at the top of the left and right walls. Due to the prop these accelerations are forced to be identical as confirmed by these recordings. At the same time, the prop loads increase as shown in Figure 14.41. A peak load of nearly 1 MN/m is recorded in the prop. This is a substantial load and the props have to be designed to carry such loads safely. It is also interesting to note that the frequency of variation of the axial loads in the prop is about

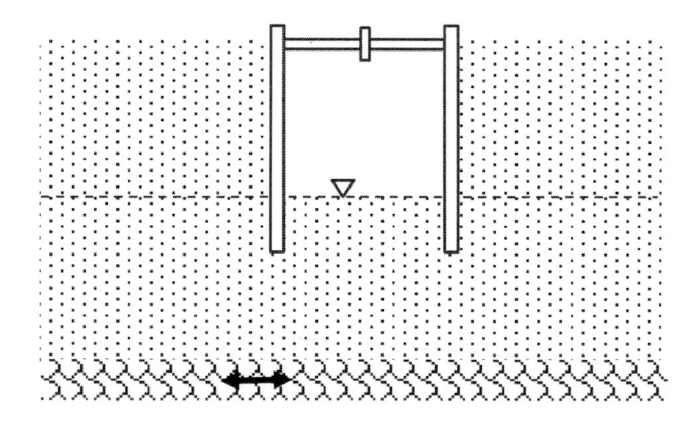

Figure 14.40 Schematic of propped retaining wall problem.

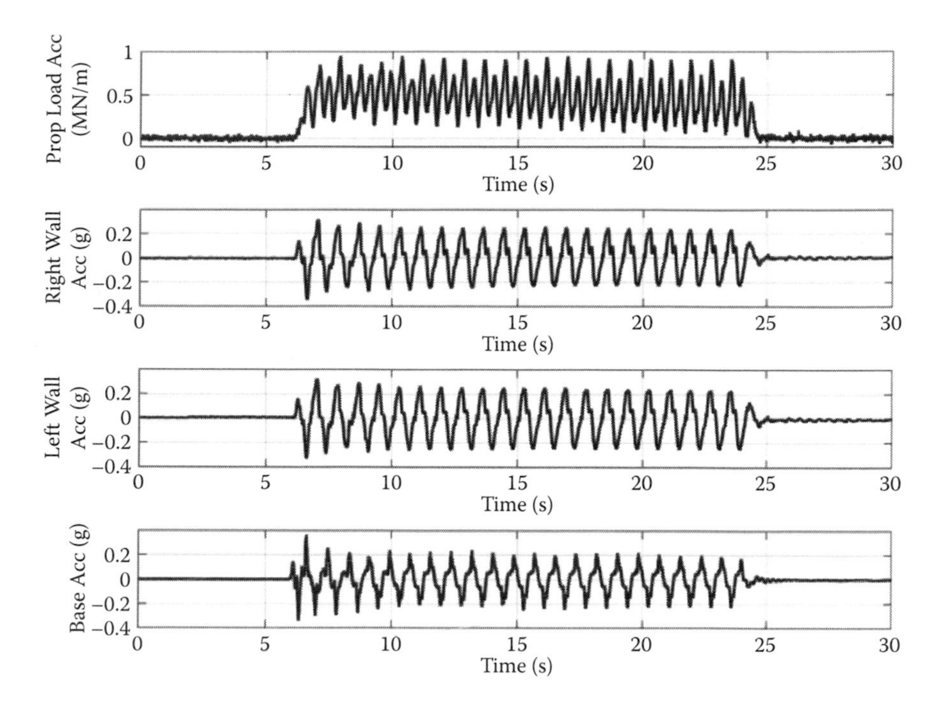

Figure 14.41 Example of data from a propped retaining wall problem.

twice that of the earthquake. This is because the prop has to prevent the left wall moving into the right wall in one half cycle of the earthquake and it has to prevent the right wall moving into the left one in the second half of the cycle. This results in a double frequency variation in the axial load in the prop.

14.8 SUMMARY

In this chapter we considered the modelling of dynamic events in a geotechnical centrifuge. The dynamic loads on structures in nature can come from a variety of sources, such as wind, waves, blasts, or earthquakes. We have developed additional scaling laws to model dynamic events and considered some of the special modelling techniques we need to employ to achieve congruency between general and dynamic scaling laws. More emphasis was placed throughout this chapter on modelling of earthquake loading. We considered various additional equipment that are required to carry out earthquake modelling in a centrifuge. Examples of the earthquake actuators currently used along with the specialist model containers for minimizing boundary effects were presented.

In the final sections of this chapter we considered two classes of examples in earthquake centrifuge modelling. The first one was soil structure interaction problems and we considered simple structures and tunnels as examples. In the second class of examples we considered liquefaction problems. A wide variety of examples of these problems was presented. Worldwide there has been more research on liquefaction problems owing to the complex nature of liquefaction phenomena and continued observation of the damage this caused in recent earthquakes.

In each of the examples we considered in this chapter it was possible to clearly identify the insights provided by centrifuge testing. Centrifuge test data can also be used to validate predictions from numerical codes or design procedures used by the industry.

Although not considered here, centrifuge modelling can be used for other dynamic loading, such as wind, waves, or even blast loading. These are challenging but interesting problems and no doubt centrifuge modelling will be used by civil engineers to investigate these problems in the future.

References

Abdoun, T., and V. Bennett, 2008. A New Wireless MEMS-Based System for Real-Time Deformation Monitoring. *Geotechnical News*, 26 (1), 36–40.

Adamidis, O., and Madabhushi, S.P.G., 2013. Stability of HPMC Solution. Unpublished Internal Report, University of Cambridge, United Kingdom.

Allersma, H.G.B., 1990. Online Measurement of Soil Deformation in Centrifuge Tests by Means of Image Processing. In *Proc. 9th Int. Conf. on Experimental Mechanics*. Copenhagen, Denmark, 1739–1748.

Arulanandan, K., and R.F. Scott, 1993. Verification of Numerical Procedures for the Analysis of Soil Liquefaction Problems. In: *Proc. of VELACS Conference*. Davis, California: Balkema, 1–15.

Barton, Y.O., 1982. Laterally Loaded Model Piles in Sand: Centrifuge Tests and Finite Element Analyses. PhD thesis, University of Cambridge, United Kingdom.

Bhattacharya, S., S.P.G. Madabhushi, and M.D. Bolton, 2004. An Alternative Mechanism of Pile Failure during Seismic Liquefaction. *Géotechnique*, 54 (3), 203–213.

Bilotta, E., F. Silvestri, and S.P.G. Madabhushi, 2014. Comparison of Centrifuge Test Data and Numerical Predictions on Seismic Performance of Tunnels. *Acta Geotechnica* (Special issue on the Round Robin Tunnel Tests, Springer).

Bolton, M.D., M.W. Gui, J. Garnier, J.F. Corte, G. Bagge, J. Laue, and R. Renzi, 1999. Centrifuge Cone Penetration Tests in Sand. *Géotechnique*, 49 (4), 543–552.

Bolton, M.D., and J.M.R. Wilson, 1989. An Experimental and Theoretical Comparison between Static and Dynamic Torsional Soil Tests. *Géotechnique*, 39 (4), 585–599.

Borowitz, S., and L.A. Bornstein, 1968. *A Contemporary View of Elementary Physics*. New York: McGraw Hill.

Bourne-Webb, P.J., D.M. Potts, D. Koenig, and D. Rowbottom, 2011. Analysis of Model Sheet Pile Walls with Plastic Hinges. *Géotechnique*, 61 (6), 487–499.

Brandenberg, S.J., S. Choi, B.L. Kutter, D.W. Wilson, and J.C. Santamarina, 2006. A Bender Element System for Measuring Shear Wave Velocities in Centrifuge Models. In: *Physical Modelling in Geotechnics: ICPMG '06*. Hong Kong: Balkema, 165–170.

Bransby, P.L., and G.W.E. Milligan, 1975. Soil Deformations near Cantilever Sheet Pile Walls. *Géotechnique*, 25 (2), 175–195.

Brennan, A.J., and S.P.G. Madabhushi, 2002. Design and Performance of a New Deep Model Container for Dynamic Centrifuge Testing. In: R. Phillips, P.J. Guo, and R. Popescu, eds. *Proc. International Conference on Physical Modelling in Geotechnics*. St. Johns, Newfoundland: Balkema, 183–188.

Brennan, A.J., and S.P.G. Madabhushi, 2005. Liquefaction and Drainage in Stratified Soil. *ASCE Journal of Geotechnical and Geo-environmental Engineering*, 131 (7), 876–885.

Brennan, A.J., S.P.G. Madabhushi, and N.E. Houghton, 2006. Comparing Laminar and ESB Containers for Dynamic Centrifuge Modelling. In: *Physical Modelling in Geotechnics: ICPMG '06*. Hong Kong: Balkema.

Broms, B., 1966. Methods of Calculating the Ultimate Bearing Capacity of Piles: A Summary. *Sols-Soils*, 5 (18–19), 21–31.

Chan, A.H.C., 1989. A Generalised Fully Coupled Effective Stress Based Computer Procedure for Problems in Geomechanics. SWANDYNE User Manual, Swansea, United Kingdom.

Chan, K.C., 1975. Stresses and Strains Induced in Soft Clay by a Strip Footing. PhD thesis, University of Cambridge, United Kingdom.

Chandrasekaran, V.S., 2001. Numerical and Centrifuge Modelling of Soil-Structure Interaction. *Indian Geotechnical Journal*, 31 (1), 1–32.

Chandrasekaran, V.S., and G.J. King, 1974. Simulation of Excavation Using Finite Elements. *ASCE Journal of Geotechnical and Geo-environmental Engineering*, 100 (GT9), 1086–1089.

Chazelas, J.L., S.P.G. Madabhushi, and R. Phillips, 2007. Reflections on the Importance of the Quality of the Input Motion in Seismic Centrifuge Tests. In: *Proc. IV International Conference on Soil Dynamics and Earthquake Engineering*. Thessalonica, Greece.

Chian, S.C., and S.P.G. Madabhushi, 2010. Influence of Fluid Viscosity on the Response of Buried Structures in Earthquakes. In: S.M. Springman, J. Laue, and L. Seward, eds. *Proc. 7th International Conference on Physical Modelling in Geotechnics*. Zurich, Switzerland: Balkema.

Chian, S.C., and S.P.G. Madabhushi, 2012. Effect of Buried Depth and Diameter on Uplift of Underground Structures in Liquefied Soils. *Journal of Soil Dynamics and Earthquake Engineering*, 41 (1), 181–190.

Chian, S.C., M.E. Stringer, and S.P.G. Madabhushi, 2010. Use of Automatic Sand Pourer for Loose Sand Models. In: *Proc. 7th International Conference on Physical Modelling in Geotechnics*. Zurich, Switzerland: Balkema.

Choy, C.K., 2004. Installation Effects of Diaphragm Walls on Adjacent Piled Foundations. PhD thesis, University of Cambridge, United Kingdom.

Cilingir, U., S.K. Haigh, C.M. Heron, S.P.G. Madabhushi, J.-L. Chazelas, and S. Escoffier, 2012. Cross-Facility Validation of Dynamic Centrifuge Testing. In: *Role of Seismic Testing Facilities in Performance-Based Earthquake Engineering: Geotechnical, Geological, and Earthquake Engineering*, Vol. 22, 83–98.

Cilingir, U., S.K. Haigh, S.P.G. Madabhushi, and X. Zeng, 2011. Seismic Behaviour of Anchored Quay Walls with Dry Backfill. *Geomechanics and Geoengineering: An International Journal*, 6 (3), 227–235.

Cilingir, U., and S.P.G. Madabhushi, 2011a. Effect of Depth on the Seismic Response of Circular Tunnels. *Canadian Geotechnical Journal*, 48 (1), 117–127.

Cilingir, U., and S.P.G. Madabhushi, 2011b. A Model Study on the Effects of Input Motion on the Seismic Behaviour of Tunnels. *Journal of Soil Dynamics and Earthquake Engineering*, 31 (1), 452–462.

Cilingir, U., and S.P.G. Madabhushi, 2011c. Effect of Depth on the Seismic Response of Square Tunnels. *Soils & Foundations*, 51 (3), 449–457.

Clegg, D.P., 1981. Model Piles in Stiff Clays. PhD thesis, University of Cambridge, United Kingdom.

Ellis, E., K. Soga, F. Bransby, and M. Sato, 2000. Resonant Column Testing of Sands with Different Viscosity Pore Fluids. *ASCE Journal of Geotechnical and Geoenvironmental Engineering*, 126 (1), 10–17.

Elshafie, M.Z.E.B., 2008. Effect of Building Stiffness on Excavation-Induced Displacement. PhD thesis, University of Cambridge, United Kingdom.

Fargnoli, V., D. Boldini, and A. Amorosi, 2013. TBM Tunnelling-Induced Settlements in Coarse-Grained Soils: The Case of the New Milan Underground Line 5. *Journal of Tunnelling and Underground Space Technology*, 38 (1), 336–347.

Farrell, R.P., 2010. Tunnelling in Sands and the Response of Buildings. PhD thesis, University of Cambridge, United Kingdom.

Fourier, J.B., 1822. *Théorie Analytique de la Chaleur*. Paris: Chez Firmin Didot, père et fils.

Ghosh, B., and S.P.G. Madabhushi, 2002. An Efficient Tool for Measuring Shear Wave Velocity in the Centrifuge. In: R. Phillips, P.J. Guo, and R. Popescu, eds. *Proc. International Conference on Physical Modelling in Geotechnics*. St. Johns, Newfoundland: Balkema.

Ghosh, B. and S.P.G. Madabhushi, 2007. Centrifuge Modelling of Seismic Soil-Structure Interaction Effects. *Nuclear Engineering and Design*, 237 (1), 887–896.

Haigh, S.K., 2002. Effects of Earthquake-Induced Liquefaction on Pile Foundations in Sloping Ground. PhD thesis, University of Cambridge, United Kingdom.

Haigh, S.K., N.E. Houghton, S.Y. Lam, Z. Li, and P.J. Wallbridge, 2010. Development of a 2D Servo-Actuator for Novel Centrifuge Modelling. In: S. Springman, J. Laue, and L. Seward, eds. *Physical Modelling in Geotechnics: Proc. 7th International Conference on Physical Modelling in Geotechnics, June 28–July 1. Zurich, Switzerland*. Boca Raton, FL: CRC Press, 239–244.

Haigh, S.K., and S.P.G. Madabhushi, 2005. The Effects of Pile Flexibility on Pile Loading in Laterally Spreading Slopes. In: R.W. Boulanger and K. Tokimatsu, eds. ASCE-GI Special Publication STP 145. 24–37.

Halliday, D., R. Resnick, and J. Walker, 2004. *Fundamentals of Physics*. Vol. 6. New York: John Wiley & Sons.

Heron, C.M., 2013. The Dynamic Soil-Structure Interaction of Shallow Foundations on Dry Sand Beds. PhD thesis, University of Cambridge, United Kingdom.

Hibbeler, R.C., 2007. *Engineering Mechanics: Dynamics*. 11th ed. Singapore: Prentice-Hall, Inc.

Jacobsz, S.W., 2002. The Effect of Tunnelling on Piled Foundations. PhD thesis, University of Cambridge, United Kingdom.

Khokher, Y.R., and S.P.G. Madabhushi, 2010. Dynamic Earth Pressures and Earth Pressure Cell Measurements. In: *Proc. 5th International Conference on Recent Advances in Geotechnical Earthquake Engineering and Soil Dynamics*. San Diego, CA.

Kim, N.R., and D.S. Kim, 2011. A Shear Wave Velocity Tomography System for Geotechnical Centrifuge Testing. *ASTM Geotechnical Testing Journal*, 33 (6), 434–444.

Kimura, T., O. Kusakabe, and K. Saitoh, 1985. Geotechnical Model Tests of Bearing Capacity Problems in a Centrifuge. *Géotechnique*, 35 (1), 33–45.

Kimura, T., J. Takemura, and K. Saitoh, 1998. Development of a Simple Mechanical Shaker Using a Cam Shaft. In: *Proc. Centrifuge '98*. Tokyo, Japan: Balkema, 107–110.

Kirkwood, P. and S.K. Haigh, 2013. Centrifuge Testing of Monopiles for Offshore Wind Turbines. In: *Proc. ISOPE 2013*. Anchorage, Alaska.

Knappett, J.A., and R.F. Craig, 2012. *Craig's Soil Mechanics*. 8th edition. Abingdon, Oxon, UK: Spon Press.

Knappett, J.A., S.K. Haigh, and S.P.G. Madabhushi, 2006. Mechanisms of Failure for Shallow Foundations under Earthquake Loading. *Journal of Soil Dynamics and Earthquake Engineering*, 26 (2), 91–102.

Knappett, J.A., and S.P.G. Madabhushi, 2009. Influence of Axial Load on Lateral Pile Response in Liquefiable Soils, Part I: Physical Modelling. *Géotechnique*, 59 (7), 571–581.

Knappett, J.A., C. Reid, S. Kinmond, and K. O'Reilly, 2011. Small-Scale Modelling of Reinforced Concrete Structural Elements for Use in a Geotechnical Centrifuge. *Journal of Structural Engineering*, 137 (11), 1263–1271.

Kreyszig, E., 1967. *Advanced Engineering Mathematics*. 2nd edition. New York: John Wiley & Sons.

Kutter, B.L., 1982. Centrifugal Modelling of the Response of Clay Embankments to Earthquakes. PhD thesis, University of Cambridge, United Kingdom.

Kutter, B.L., I.M. Idriss, T. Khonke, J. Lakeland, X.S. Li, W. Sluis, X. Zeng, R.C. Tauscher, Y. Goto, and I. Kubodera, 1994. Design of a Large Earthquake Simulator at UC Davis. In: *Proc. International Conference Centrifuge 1994*. Singapore: Balkema, 169–175.

Kutter, B.L., X.S. Li, W. Sluis, and J.A. Cheney, 1991. Performance and Instrumentation of the Large Centrifuge at Davis. In: *Proc. International Conference Centrifuge 1991*. Rotterdam: Balkema, 19–26.

Ladd, C.C., and D.J. de Groot, 2003. Recommended Practice for Soft Ground Site Characterization. In: *Proc. 12th Pan American Conf. Soil Mech. and Geot. Eng*. Cambridge, MA.

Lam, S.Y., 2010. Ground Movements Due to Excavations in Clay: Physical and Analytical Models. PhD thesis, University of Cambridge, United Kingdom.

Lam, S.Y., and M.D. Bolton, 2009. Energy Conservation as a Principle Underlying Mobilizable Strength Design for Deep Excavations. *ASCE Journal of Geotechnical and Geo-environmental Engineering*, 137 (11), 1062–1074.

Lam, S.Y., M. Elshafie, S.K. Haigh, and M.D. Bolton, 2012. A New Apparatus for Modelling Excavations. *International Journal of Physical Modelling in Geotechnics,* 12 (1), 24–38.

Lanzano, G., E. Bilotta, G. Russo, F. Silvestri, and S.P.G. Madabhushi, 2012. Centrifuge Modelling of Seismic Loading on Tunnels in Sand. *ASTM Geotechnical Testing Journal*, 35 (6), 55–71.

Lau, B., S.Y. Lam, S.K. Haigh, and S.P.G. Madabhushi, 2014. Centrifuge Testing of Monopile in Clay under Monotonic Loads. In: *Proc. 8th International Conf. on Physical Modelling in Geotechnics*. Perth: Balkema.

Ledbetter, R., 1991. Large Centrifuge: A Critical Army Capability for the Future. Report No: GL-91-12. U.S. Army Corps of Engineers, Waterways Experiment Station, Vicksburg, Mississippi.

Lee, C.J., 2001. The Influence of Negative Skin Friction on Piles and in Pile Groups. PhD thesis, University of Cambridge, United Kingdom.

Lee, J.S., and J.C. Santamarina, 2005. Bender Elements: Performance and Signal Interpretation. *ASCE Journal of Geotechnical and Geo-environmental Engineering,* 131 (9), 1063–1070.

Li, Z., 2010. Piled Foundations Subjected to Cyclic Loads or Earthquakes. PhD thesis, University of Cambridge, United Kingdom.

Madabhushi, S.P.G., 1991. Response of Tower Structures to Earthquake Perturbations. PhD thesis, University of Cambridge, United Kingdom.

Madabhushi, S.P.G., 1994. Effect of Pore Fluid in Dynamic Centrifuge Modelling. In: *Proc. Centrifuge' 94*. Singapore: Balkema.

Madabhushi, S.P.G., 1995. Geotechnical Aspects of the Northridge Earthquake of 17 January. In: EEFIT Report. London: Institution of Structural Engineers.

Madabhushi, S.P.G., 2004. Modelling of Earthquake Damage Using Geotechnical Centrifuges. *Current Science*, 87 (10), 1405–1416.

Madabhushi, S.P.G., 2007. Geotechnical Aspects of the 1921 Ji-Ji Earthquake. In: EEFIT Report. London: Institution of Structural Engineers.

Madabhushi, S.P.G., and V.S. Chandrasekaran, 2005. Rotation of Cantilever Sheet Pile Walls. *ASCE Journal of Geotechnical and Geo-environmental Engineering*, 131 (2), 202–212.

Madabhushi, S.P.G., and V.S. Chandrasekaran, 2006. On Modelling the Behaviour of Flexible Sheet Pile Walls. *Indian Geotechnical Journal,* 36 (2), 160–180.

Madabhushi, S.P.G., C.H. Collison, and T. Wilmshurst, 1997. Development of a Mini Electro-Magnetic Earthquake Actuator. In: *Proc. I Intl. Symposium on Structures and Foundations in Civil Engineering*. Hong Kong.

Madabhushi, S.P.G., S.K. Haigh, and N.E. Houghton, 2006. A New CNC Sand Pourer for Model Preparation at the University of Cambridge. In: *Proc. International Conference on Physical Modelling in Geotechnics*. Hong Kong: Balkema.

Madabhushi, S.P.G., S.K. Haigh, N.E. Houghton, and E. Gould, 2012. Development of a Servo-Hydraulic Earthquake Actuator for the Cambridge Turner Beam Centrifuge. *International Journal of Physical Modelling in Geotechnics*, 12 (2), 77–88.

Madabhushi, S.P.G., J.A. Knappett, and S.K. Haigh, 2009. *Design of Pile Foundations in Liquefiable Soils.* London: Imperial College Press.

Madabhushi, S.P.G., D. Patel, and S.K. Haigh, 2005. Geotechnical Aspects of the Bhuj Earthquake. In: EEFIT Report. London: Institution of Structural Engineers.

Madabhushi, S.P.G., and A.N. Schofield, 1993. Centrifuge Modelling of Tower Structures Subjected to Earthquake Perturbations. *Géotechnique*, 43 (4), 555–565.

Madabhushi, S.P.G., A.N. Schofield, and S. Lesley, 1998. A New Stored Angular Momentum (SAM) Based Earthquake Actuator. In: *Proc. Centrifuge '98*. Tokyo, Japan: Balkema.

Madabhushi, S.P.G., M.E. Stringer, and U. Cilingir, 2010. Development of Teaching Resources for Physical Modelling Community. In: *Proc. 7th International Conference on Physical Modelling in Geotechnics.*

Madabhushi, S.P.G., W.A. Take, and A.J. Barefoot, 2002. Teaching of Geotechnical Engineering Using a Mini-Drum Centrifuge. In: R Phillips, P.J. Guo, and R. Popescu eds. *Proc. International Conference on Physical Modelling in Geotechnics.*

Madabhushi, S.P.G., and X. Zeng, 2006. Seismic Response of Flexible Cantilever Retaining Walls with Dry Backfill. *Geomechanics and Geoengineering: An International Journal,* 1 (4), 275–290.

Madabhushi, S.P.G., and X. Zeng, 2007. Simulating Seismic Response of Cantilever Retaining Walls with Saturated Backfill. *ASCE Journal of Geotechnical and Geo-environmental Engineering*, 133 (5), 539–549.

Mair, R.J., 1979. Centrifugal Modelling of Tunnel Construction in Soft Clay. PhD thesis, University of Cambridge, United Kingdom.

Mak, K.-W., 1983. Modelling the Effects of a Strip Load behind Rigid Retaining Walls. PhD thesis, University of Cambridge, United Kingdom.

Marshall, A.M., 2009. Tunnelling in Sand and Its Effect on Pipelines and Piles. PhD thesis, University of Cambridge, United Kingdom.

Matlock, H., and L.C. Reese, 1960. Generalised Solutions for Laterally Loaded Piles. *Journal of Soil Mechanics and Foundation Engineering*, 86 (SM5), 1220–1246.

McMahon, B., 2012. Deformation Mechanisms below Shallow Foundations. PhD thesis, University of Cambridge, United Kingdom.

Milligan, G.W.E., 1974. The Behaviour of Rigid and Flexible Retaining Walls in Sand. PhD thesis, University of Cambridge, United Kingdom.

Morris, D.V., 1979. Centrifuge Modelling of Dynamic Behaviour. PhD thesis, University of Cambridge, United Kingdom.

Newland, D.E., 2005. *An Introduction to Random Vibrations, Spectral and Wavelet Analysis*. 3rd edition. Dover Publications, New York.

Nunez, I.L., 1989. Centrifuge Model Tension Piles in Clay. PhD thesis, University of Cambridge, United Kingdom.

Osman, A.S., and M.D. Bolton, 2004. A New Design Method for Retaining Walls in Clay. *Canadian Geotechnical Journal*, 41 (3), 451–466.

Osman, A.S., and M.D. Bolton, 2006. Design of Braced Excavations to Limit Ground Movements. *Proceedings of Institution of Civil Engineers, Geotechnical Engineering*, 159 (3), 167–175.

Ovesen, N.K., 1975. Centrifugal Testing Applied to Bearing Capacity Problems of Footings on Sand. *Géotechnique*, 25 (2), 394–401.

Paolucci, R., and A. Pecker, 1997. Seismic Bearing Capacity of Shallow Strip Foundations on Dry Soils. *Soils & Foundations*, 37 (3), 95–105.

Peiris, L.M.N., S.P.G. Madabhushi, and A.N. Schofield, 1998. Behaviour of Gravel Embankments Founded on Loose Saturated Sand Deposits Subjected to Earthquakes and with Different Pore Fluids. In: *Proc. Centrifuge '98*. Tokyo, Japan: Balkema.

Piccoli, S., and F.P. Smits, 1991. Real Time Evaluation of Shear Wave Velocity During Seismic Cone Penetration Test. In: *Proceedings of the 3rd International Symposium on Field Measurements in Geomechanics*. Oslo, Norway, 159–166.

Potter, L.J., 1996. Contaminant Migration through Consolidating Soils. PhD thesis, University of Cambridge, United Kingdom.

Powrie, W., 1986. The Behaviour of Diaphragm Walls in Clays. PhD thesis, University of Cambridge, United Kingdom.

Rothman, M.A., 1989. *Discovering the Natural Laws: The Experimental Basis of Physics*. Courier Dover Publications, New York.

Rowe, P.W., 1972. Large Scale Laboratory Model Retaining Wall Apparatus. In: *Proc. Roscoe Memorial Symposium on Stress-Strain Behaviour of Soils*. Foulis, 441–449.

Schofield, A.N., 1980. Cambridge Geotechnical Centrifuge Operations. *Géotechnique*, 30 (3), 227–268.

Schofield, A.N., 1981. Dynamic and Earthquake Centrifuge Modelling. In: *Proc. Int. Conf. on Recent Advances in Geotech. Earthquake Eng. and Soil Dynamics*. St. Louis, Missouri, Vol. 3, 1081–1099.

Schofield, A.N., 2006. *Disturbed Soil Properties and Geotechnical Design*. London: Thomas Telford.

Scott, R.F., 1981. Pile Testing in a Centrifuge. In: *Proc. ICSMFE*. Stockholm, Sweden, Vol. 2, 839–842.

Scott, R.F., 1994. Review of Progress in Dynamic Geotechnical Centrifuge Research. In: R.J. Ebleher, V.P. Drnevich, and B.L. Kutter, eds. *Dynamic Geotechnical Testing* II, ASTM STP 1213. San Francisco: ASTM.

Shepley, P., 2014, Water injection to assist Pile Jacking, Ph.D. thesis, University of Cambridge, United Kingdom.

Shi, Q., 1988. Centrifuge Modelling of Surface Footings Subject to Combined Loading. PhD thesis, University of Cambridge, United Kingdom.

Soga, K., 1998. Soil Liquefaction Effects Observed in the Kobe Earthquake of 1995. *Proceedings of Institution of Civil Engineers, Geotechnical Engineering*, 146 (1), 34–51.

Springman, J. S.M., Laue, R. Boyle, J. White, and A. Zweidler, 1991. The ETH Zurich Geotechnical Drum Centrifuge. *International Journal of Physical Modelling in Geotechnics*, 1 (1), 59–70.

Springman, S.M., N. Philippe, R. Chikatmarla, and J. Laue, 2002. Use of Flexible Tactile Pressure Sensors in Geotechnical Centrifuges. In: R. Phillips, P.J. Guo, and R. Popescu, eds. *Proc. International Conference on Physical Modelling in Geotechnics*. St. Johns, Newfoundland: Balkema.

Steedman, R.S., and S.P.G. Madabhushi, 1991. Wave Propagation in Sand Medium. In: *Proc. 4th Intl. Conference on Seismic Zonation*. Palo Alto, CA: Stanford University.

Stewart, D.P., Y.R. Chen, and B.L. Kutter, 1998. Experience with the Use of Methylcellulose as a Viscous Pore Fluid in Centrifuge Models. *ASTM Geotechnical Testing Journal*, 21 (4), 365–369.

Stringer, M.E., C.M. Heron, and S.P.G. Madabhushi, 2010. Experience Using MEMS-Based Accelerometers in Dynamic Testing. In: S.M. Springman, J. Laue, and L. Seward, eds. *Proc. 7th International Conference on Physical Modelling in Geotechnics*. Zurich, Switzerland: Balkema, 389–394.

Stringer, M.E., and S.P.G. Madabhushi, 2010. Improving Model Quality in Dynamic Centrifuge Modelling through Computer Controlled Saturation. In: S.M. Springman, J. Laue, and L. Seaward, eds. *Proc. International Conference on Physical Modelling in Geotechnics*. Zurich, Switzerland: Balkema.

Stringer, M.E., and S.P.G. Madabhushi, 2009. Novel Computer Controlled Saturation of Dynamic Centrifuge Models Using High Viscosity Fluids. *ASTM Geotechnical Testing Journal*, 32 (6), 53–59.

Stringer, M.E., and S.P.G. Madabhushi, 2011. Effect of Driving Process on the Response of Piles in Liquefiable Soils. *International Journal of Physical Modelling in Geotechnics*, 11 (3), 87–99.

Stringer, M.E., and S.P.G. Madabhushi, 2012. Axial Load Transfer in Liquefiable Soils for Free-Standing Piles. *Géotechnique*, 63 (5), 400–409.

Stringer, M.E., and S.P.G. Madabhushi, 2013. Re-Mobilisation of Pile Shaft Friction after an Earthquake. *Canadian Geotechnical Journal*, 50 (9), 979–988.

Stroud, M.A., 1971. The Behaviour of Sand at Low Stress Levels in the Simple Shear Apparatus. PhD thesis, University of Cambridge, United Kingdom.

Swain, C.W., 1979. The Behaviour of Piled Foundations Supporting a Model Offshore Structure. PhD thesis, University of Cambridge, United Kingdom.

Take, W.A., and M.D. Bolton, 2003. Tensiometer Saturation and the Reliable Measurement of Matric Suction. *Géotechnique*, 53 (2), 159–172.

Take, W., and M.D. Bolton, 2011. Seasonal Ratcheting and Softening in Clay Slopes, Leading to First-Time Failure. *Géotechnique*, 61 (9), 757–769.

Tatsuoka, F., S. Nakamura, C.C. Huang, and K. Tani, 1990. Strength Anisotropy and Shear Band Direction in Plane Strain Tests of Sand. *Soils & Foundations*, 30 (1), 35–54.

Taylor, R.N., R.J. Grant, S. Robson, and J. Kuwano, 1998. An Image Analysis System for Determining Plane and 3-D Displacements in Soil Models. In: T. Kimura, O. Kusukabe, and J. Takemura, eds. *Proc. Centrifuge '98*. Tokyo, Japan: Balkema, 73–78.

Terazaghi, K., 1934a. Large Retaining Wall Tests, I: Pressure of Dry Sand. *Engineering News Record*, 136–140.

Terazaghi, K., 1934b. Large Retaining Wall Tests, II: Pressure of Dry Sand. *Engineering News Record*, 259–262.

Terazaghi, K., and R. Peck, 1967. *Soil Mechanics in Engineering Practice*. New York: John Wiley & Sons.

Terazaghi, K., R.B. Peck, and G. Mesri, 1963. *Soil Mechanics in Engineering Practice*. New York: John Wiley & Sons.

Teymur, B., and S.P.G. Madabhushi, 2003. Experimental Study of Boundary Effects in Dynamic Centrifuge Modelling. *Géotechnique*, 53 (7), 655–663.

Thusyanthan, N.I., S.P.G. Madabhushi, and S. Singh, 2006. Centrifuge Modelling of Solid Waste Landfill Systems—Part 2: Centrifuge Testing of Model Waste. *ASTM Geotechnical Testing Journal*, 29 (3), 223–229.

Tomlinson, M.J., 1986. *Foundation Design and Construction*. 5th edition. London: Longman Scientific & Technical.

Tricarico, M., S.P.G. Madabhushi, L. de Sanctis, and S. Aversa, 2013. Seismic Testing of Propped Retaining Walls: RELUIS-II Project. Unpublished Internal Report, Cambridge University Technical Services, Cambridge, United Kingdom.

Viggiani, G., and J.H. Atkinson, 1995. Interpretation of Bender Element Tests. *Géotechnique*, 45 (1), 149–154.

Viswanadham, B.V.S., S.P.G. Madabhushi, K.V. Babu, and V.S. Chandrasekaran, 2009. Modelling the Failure of a Cantilever Sheet Pile Wall. *International Journal of Geotechnical Engineering*, 2 (1), 215–231.

Vorster, T.E.B., 2005. The Effect of Tunnelling on Buried Pipes. PhD thesis, University of Cambridge, United Kingdom.

White, D.J., and W.A. Take, 2002. GeoPIV: Particle Image Velocimetry (PIV) Software for Use in Geotechnical Testing. Report No: CUED/D-Soils/TR322, University of Cambridge, United Kingdom.

White, D.J., W.J. Take, and M.D. Bolton, 2003. Soil Deformation Measurement Using Particle Image Velocimetry (PIV) and Photogrammetry. *Géotechnique*, 53 (7), 619–631.

White, D.J., K.L. Teh, C.F. Leung, Y.K. Chow, and D. White, 2008. A Comparison of the Bearing Capacity of Flat and Conical Circular Foundations on Sand. *Géotechnique*, 58 (10), 781–792.

Williams, A.B., and F.J. Taylors, 1988. *Electronic Filter Design Handbook*. New York: McGraw Hill.

Williams, D.J., 1979. The Behaviour of Model Piles in Dense Sand under Vertical and Horizontal Loading. PhD thesis, University of Cambridge, United Kingdom.

Wind, H.G., 1976. Interaction of Sand and L-Shaped Walls in Centrifuge Models. PhD thesis, University of Cambridge, United Kingdom.

Zeng, X., 1990. Modelling the Behaviour of Quay Walls in Earthquakes. PhD thesis, University of Cambridge, United Kingdom.

Zeng, X., and A.N. Schofield, 1996. Design and Performance of an Equivalent Shear Beam (ESB) Model Container for Earthquake Centrifuge Modelling. *Géotechnique*, 46 (1), 83–102.

Zhao, Y., K. Gafar, N.I. Thusyanthan, J.A. Knappett, H. Mohamed, A. Deeks, and S.P.G. Madabhushi, 2006. Calibration and Use of a New CNC Automatic Sand Pourer. In: *Proc. International Conference on Physical Modelling in Geotechnics*. Hong Kong: Balkema.

Subject Index